STATA USER'S GUIDE
RELEASE 9

A Stata Press Publication
StataCorp LP
College Station, Texas

Stata Press, 4905 Lakeway Drive, College Station, Texas 77845

For copyright information about the software, type `help copyright` within Stata.

The suggested citation for this software is

StataCorp. 2005. *Stata Statistical Software: Release 9*. College Station, TX: StataCorp LP.

Contents of User's Guide

ADVICE

x

Stata Basics

Chapters

1 Read this—it will help

Contents

(Continued on next page)

A complete Documentation Set contains over 6,000 pages of information in the following manuals:

[GS]   *Getting Started with Stata*

[G]   *User's Guide*

[R]   *Stata Base Reference Manual*
        Volume 1, A–J
        Volume 2, K–Q
        Volume 3, R–Z

[D]   *Stata Data Management Reference Manual*

[G]   *Stata Graphics Reference Manual*

[P]   *Stata Programming Reference Manual*

[XT]   *Stata Longitudinal/Panel Data Reference Manual*

[MV]   *Stata Multivariate Statistics Reference Manual*

[SVY]   *Stata Survey Data Reference Manual*

[ST]   *Stata Survival Analysis and Epidemiological Tables Reference Manual*

[TS]   *Stata Time-Series Reference Manual*

[I]   *Stata Quick Reference and Index*

[M]   *Mata Reference Manual*

Detailed information about each of these manuals and the other Stata manuals may be found online at

http://www.stata-press.com/manuals/

1.1   Getting Started with Stata

There are three *Getting Started* manuals:

Getting Started with Stata for Windows
Getting Started with Stata for Macintosh
Getting Started with Stata for Unix

1. Locate your *Getting Started* manual.

2. Install Stata. The instructions are found in the *Getting Started* manual.

3. Learn how to invoke Stata and use it—read the *Getting Started* manual.

4. Now turn to the other manuals; see [U] **1.2 The User's Guide and the Reference manuals**.

1.2   The User's Guide and the Reference manuals

The *User's Guide* is divided into three sections: *Basics*, *Elements*, and *Advice*. At the beginning of each section is a list of the chapters found in that section. In addition to helping you explore the fundamentals of Stata—information that all users should know—this manual will guide you to other sources for Stata information.

The other manuals are the *Reference* manuals. The Stata *Reference* manuals are each arranged like an encyclopedia—alphabetically. Look at the *Base Reference Manual*. Look under the name of a command. If you do not find the command, look in the index. A few commands are so closely related that they are documented together, such as `ranksum` and `median`, which are both documented in [R] **ranksum**.

Not all the entries in the *Base Reference Manual* are Stata commands; some contain technical information, such as [R] **maximize**, which details Stata's iterative maximization process, or [R] **error messages**, which provides information on error messages and return codes.

Like an encyclopedia, the *Reference* manuals are not designed to be read from cover to cover. When you want to know what a command does, complete with all the details, qualifications, and pitfalls, or when a command produces an unexpected result, read its description. Each entry is written at the level of the command. The descriptions assume that you have little knowledge of Stata's features when they are explaining simple commands, such as those for using and saving data. For more complicated commands, they assume that you have a firm grasp of Stata's other features.

If a Stata command is not in the *Base Reference Manual*, it can be found in one of the other *Reference* manuals. The titles of the manuals indicate the types of commands that they contain. The *Programming Reference Manual*, however, contains not only commands for programming Stata, but also the matrix-manipulation commands (not to be confused with the matrix programming language described in the *Mata Reference Manual*).

1.2.1 Sample datasets

Various examples in this manual use what is referred to as the automobile dataset, `auto.dta`. We previously created a dataset on the prices, mileages, weights, and other characteristics of 74 automobiles and saved it in a file called `auto.dta`. (These data originally came from the April 1979 issue of *Consumer Reports* and from the United States Government EPA statistics on fuel consumption; they were compiled and published by Chambers et al., 1983.)

In our examples, you will often see us type

        . use http://www.stata-press.com/data/r9/auto

We include the `auto.dta` file with Stata. If you want to use it from your own computer rather than via the Internet, you can type

        . sysuse auto

See [D] **sysuse**.

You can also access the `auto` dataset by selecting **File > Example Datasets...**, clicking on *Example datasets installed with Stata*, and clicking on `use` beside the `auto.dta` filename.

Here is a list of the example datasets included with Stata:

(Continued on next page)

auto.dta	1978 automobile data
autornd.dta	Subset of 1978 automobile data
bplong.dta	Blood pressure data, fictional
bpwide.dta	Blood pressure data, fictional
cancer.dta	Patient survival in drug trial
census.dta	1980 census data by state
citytemp.dta	U.S. city temperature data
citytemp4.dta	U.S. city temperature data
educ99gdp.dta	Education and GDP
gnp96.dta	U.S. GNP, 1967–2002
lifeexp.dta	Life expectancy
nlsw88.dta	U.S. National Longitudinal Study of Young Women
pop2000.dta	U.S. Census, 2000, extract
sp500.dta	S&P 500 historical data
uslifeexp.dta	U.S. life expectancy, 1900–1999
uslifeexp2.dta	U.S. life expectancy, 1900–1940
voter.dta	1992 presidential voter data

All of these datasets may be used or described from the **Example Datasets...** help file.

Most of the datasets that are used in the *Reference* manuals can be found at the Stata Press web site (*http://www.stata-press.com/data/*). You can download the datasets using your browser, or you can use them directly from the Stata command line; that is,

    . use http://www.stata-press.com/data/r9/nlswork

An alternative to the use command is webuse.

    . webuse auto

For additional information, see [D] **webuse**.

These sample datasets may also be used or described from the **Example Datasets...** help file.

1.2.2   Cross-referencing

The *Getting Started* manual, the *User's Guide*, and the *Reference* manuals cross-reference each other.

[R]   **regress**
[D]   **reshape**
[XT]   **xtmixed**

The first is a reference to the regress entry in the *Base Reference Manual*, the second is a reference to the reshape entry in the *Data Management Reference Manual*, and the third is a reference to the xtmixed entry in the *Longitudinal/Panel Data Reference Manual*.

[GSW] **A. More on starting and stopping Stata**
[GSM] **A. Starting and stopping Stata for Macintosh**
[GSU] **A. Starting and stopping Stata for Unix**

are instructions to see the appropriate section of the *Getting Started with Stata for Windows, Getting Started with Stata for Macintosh*, or *Getting Started with Stata for Unix* manuals.

1.2.3 The index

At the end of each manual is an index for that manual. The index for the *Stata Base Reference Manual* is found at the end of the third volume.

The *Stata Quick Reference and Index* contains a combined index for the *Getting Started* manuals, the *User's Guide*, and all the *Reference* manuals except the *Mata Reference Manual*. It also contains "quick reference" information on subjects, such as the estimation commands.

To find information and commands quickly, you can use Stata's `search` command; see [R] **search**. You can broaden your search to the Internet using `search, all` to find additional commands and extensions written by Stata users.

1.2.4 The subject table of contents

A subject table of contents for the *User's Guide* and all the *Reference* manuals except the *Mata Reference Manual* is located in the *Stata Quick Reference and Index*.

If you look under "Functions and expressions", you will see

[U]   User's Guide, Chapter 13 .... Functions and expressions
[D]   egen .... Extensions to generate
[D]   functions .... Functions

1.2.5 Typography

You will note that we mix the ordinary printing that you are reading now with a typewriter-style typeface `that looks like this`. When something is printed in the typewriter-style typeface, it means that something is a command or an option—it is something that Stata understands and something that you might actually type into your computer. Differences in typeface are important. If a sentence reads, "You could list the result ...", it is just an English sentence—you *could* list the result, but the sentence provides no clue as to how you might actually do that. On the other hand, if the sentence reads, "You could `list` the result ...", it is telling you much more—you could list the result, and you could do that using the `list` command.

We will occasionally lapse into periods of inordinate cuteness and write, "We `described` the data and then `listed` the data." You get the idea. `describe` and `list` are Stata commands. We purposely began the previous sentence with a lowercase letter. Because `describe` is a Stata command, it must be typed in lowercase letters. The ordinary rules of capitalization are temporarily suspended in favor of preciseness.

You will also notice that we mix in words printed in italic type, such as "To perform the rank-sum test, type `ranksum` *varname* `,` `by(`*groupvar*`)`". Italicized words are not supposed to be typed; instead, you are to substitute another word for them.

We would also like users to note our rule for punctuation of quotes. We follow a rule that is often used in mathematics books and British literature. The punctuation mark at the end of the quote is only included in the quote if it is a part of the quote. For instance, the pleased Stata user said she thought that Stata was a "very powerful program". Another user simply said, "I love Stata."

In this manual, however, there is little dialog, and we follow this rule to make precise what you are to type, as in, type "`cd c:`". The period is outside the quotation mark since you should not type the period. If we had wanted you to type the period, we would have included two periods at the end of the sentence: one inside the quotation and one outside, as in, type "`use myfile.`".

We have tried not to violate the other rules of English. If you find such violations, they were unintentional and resulted from our own ignorance or carelessness. We would appreciate hearing about them.

We have heard from Nicholas J. Cox of the Department of Geography at the University of Durham in the UK and wish to express our appreciation. Nicholas' efforts have gone far beyond dropping us a note and—without Nicholas' assistance—there is no way with words that we can express our gratitude.

1.2.6   Vignettes

If you look, for example, at the entry [R] **brier**, you will see a brief biographical vignette of Glenn Wilson Brier (1913–1998), who did pioneering work on the measures described in that entry. A few such vignettes were added without fanfare in the Stata 8 manuals, just for interest, and many more have been added in this release. In many cases, a vignette could appropriately go in several entries, so that G.E.P. Box deserves to be mentioned in entries other than [TS] **arima**, such as [R] **boxcox**. To save space, each is given once only, and an index of all vignettes is given in the *Quick Reference Manual and Index*.

All the vignettes were written by Nicholas J. Cox, University of Durham, and were compiled using a wide range of reference books, articles in the literature, Internet sources, and information from individuals. Especially useful were the dictionaries of Upton and Cook (2002) and Everitt (2002) and the compilations of statistical biographies edited by Heyde and Seneta (2001) and Johnson and Kotz (1997). Of these, only the first provides information on people living at the time of publication.

1.3   What's new

This section is intended for previous Stata users. If you are new to Stata, skip to *What's more* below.

As always, the new Stata is 100% compatible with the previous release of Stata, but we remind programmers that it is vitally important that you put `version 8` at the top of your old do-files and ado-files if they are to work. You were supposed to do that when you wrote them, but if you did not, go back and do it now. We have made a lot of changes (improvements) to Stata.

Some of the important new additions include

1. New matrix programming language Mata.

2. New survey features, including balanced repeated replications (BRR) and jackknife variance estimates, complete support for multistage designs, and poststratification.

3. Estimation of linear mixed models, including standard errors and confidence intervals for all variance components.

4. Estimation of multinomial probit models, including support for several correlation structures and for user-defined structures.

5. New multivariate analysis, including multidimensional scaling, correspondence analysis, and Procrustean analysis, along with the ability to analyze proximity matrices, as well as raw data.

6. Improved GUI, including multiple Do-file Editors, multiple Viewers, and multiple Graph windows; multiple windowing preferences; dockable windows; and much more.

There are other major features, and it will take us another 30 pages to mention everything.

What's new is presented under the headings

New matrix language
Survey statistics
Longitudinal/panel data
Time-series statistics
Multivariate statistics
    Analysis of proximity
    Factor and principal component analysis additions
Survival analysis
General-purpose statistics
New ML features
Functions and expressions
Data management
Graphics
User interface
Programming
Documentation

1.3.1   New matrix language

Stata has an all-new matrix language called Mata, which is the subject of its own manual, [M] **Mata Reference Manual**. Mata can be used by those who want to think in matrix terms and perform matrix calculations interactively, and it can be used by programmers who want to add features to Stata.

Mata has been used to implement many of the new features found in this release. Mata is compiled, optimized, and fast.

Stata's previously existing `matrix` command continues to be documented. There is an admittedly uneasy relationship between the two, but `matrix` continues to have its uses. For serious computation, however, you will definitely want to use the new language.

See [M-0] **intro**—or `help mata`—which provides an introduction and organized reading list. The first thing you will read is [M-1] **first**.

1.3.2   Survey statistics

Stata 9 substantially extends Stata's survey-analysis and correlated-data-analysis facilities by adding the remaining two methods of computing standard errors—balanced repeated replications (BRR) and survey jackknife.

Stata 9 also adds complete support for multistage sampling and poststratification.

A new, unified syntax is used for declaring the design of survey data and for fitting models. For an overview of all survey facilities, see [SVY] **survey**.

All the old syntax continues to work under version control, although the survey estimation commands do not require that, but if you use old syntax, the new features will not be available.

1. Existing command `svyset` for declaring the survey design has new syntax that supports a host of new features in Stata's survey-analysis facilities:

   a. BRR and jackknife variance estimators have been added to the previously available linearization variance estimator. Moreover, use of BRR or jackknife (or linearization) can now be specified when you `svyset` or at estimation time.

   b. Multistage designs can now be declared, and they may have primary, secondary, and lower-stage sampling units. The linearization variance estimator takes complete advantage of the information in multistage designs.

    c. Stratification is now allowed in all stages, making variance estimates more efficient wherever stratification can be exploited.

    d. Poststratification is now available and, like stratification, also makes variance estimates more efficient. Poststratification adjusts weights, improves variance estimates, and accounts for biases when demographic or other groupings are known.

    e. Finite-population corrections are now allowed in all stages.

    f. Sampling weights are handled under all three variance estimators.

    For details, see [SVY] **svyset**. The previous svyset syntax continues to work under version control.

2. New prefix command svy: is how you tell estimators that you have survey data. You no longer type svyregress; you type svy: regress. This is not just a matter of style; svy really is a prefix command, and in fact, you can even use it as a prefix on estimation commands you write. In addition, svy: provides a standard, unified syntax for accessing Stata's survey features. svy: is easy to use because it automatically applies everything you have previously svyset, including the design.

    The following estimators can be used with svy: prefix:

Descriptive statistics

svy: mean	Population and subpopulation means
svy: proportion	Population and subpopulation proportions
svy: ratio	Population and subpopulation ratios
svy: total	Population and subpopulation totals
svy: tabulate oneway	One-way tables for survey data
svy: tabulate twoway	Two-way tables for survey data

Regression models

svy: regress	Linear regression
svy: ivreg	Instrumental variables regression
svy: intreg	Interval regression
svy: logistic	Logistic regression, reporting odds ratios
svy: logit	Logistic regression, reporting coefficients
svy: probit	Probit regression
svy: mlogit	Multinomial logistic regression
svy: ologit	Ordered logistic regression
svy: oprobit	Ordered probit regression
svy: poisson	Poisson regression
svy: nbreg	Negative binomial regression
svy: gnbreg	Generalized negative binomial regression
svy: heckman	Heckman selection model
svy: heckprob	Probit model with selection

Previously existing survey-estimation commands, such as svyregress, svymean, and svypoisson, continue to work as they did before, but only if your survey design is declared using version 8: svyset or if you are working with an old Stata 8 dataset. For a mapping from old estimation commands to the new syntax, see help svy8. (The new prefix svy: works with datasets that were svyset under an earlier release of Stata.)

In addition to the three variance estimators and support for multistage sampling, the new `svy:` prefix provides other enhancements, including

   a. Option `subpop()` allows more flexible selection of subpopulations, meaning that more general `if` conditions are now allowed.

   b. Strata with only one sampling unit (sometimes called singleton PSUs) are now handled better—the coefficients are now reported, but with missing standard errors. `svydes` can now be used to find and describe these strata; see [SVY] **svydes**.

   c. With BRR variance estimation, a Hadamard matrix can be used in place of BRR weights, and Fay's adjustment may be specified; see [SVY] ***brr_options***.

3. New command `svy: proportion` replaces `svyprop`. (By the way, new command `proportion` can be used without the `svy:` prefix; see [R] **proportion**.) Unlike `svyprop`, `svy: proportion` is an estimation command and computes a full covariance matrix for all the estimated proportions, allowing postestimation features, such as tests of linear and nonlinear combinations of proportions (`test` and `testnl`) or creation of linear and nonlinear combinations with confidence intervals (`lincom` and `nlcom`).

4. New commands `ratio`, `total`, and `mean`, used with the `svy:` prefix, use casewise deletion and estimate full covariance matrices for the estimates.

5. New command `svy: tabulate oneway` addresses a missing feature. Previously, anyone wanting a one-way tabulation had to create a constant and perform two-way survey tabulation with that constant.

6. New command `estat` computes and reports additional statistics and information after estimation with `svy:` prefix:

   a. `estat svyset` reports complete information on the survey design.

   b. `estat effects` computes and reports the design effects—DEFF and DEFT—and the misspecification effects—MEFF, and MEFT—in any combination for each estimated parameter.

   c. `estat effects` can also compute DEFF and DEFT for subpopulations using simple random-sample estimates from either the overall population or from the subpopulation. `estat effects` replaces and extends the `deff`, `deft`, `meff`, and `meft` options previously available on survey estimators.

   d. `estat lceffects` computes and reports the survey design effects and misspecification effects for any linear combination of estimated parameters.

   e. `estat size` reports the sample and population sizes for each subpopulation after `svy: mean`, `svy: proportion`, `svy: ratio`, and `svy: total`.

For details on `estat` after survey estimation, see [SVY] **estat**.

7. Existing command `svydes` has several new features and options:

   a. New option `stage()` lets you select the sampling stage for which sample statistics are to be reported.

   b. New option `generate()` identifies strata with a single sampling unit.

   c. New option `finalstage` replaces `bypsu` and reports observation sample statistics by sampling unit in the final stage.

8. New options `stdize()` and `stdweight()` for commands `svy: mean`, `svy: ratio`, `svy: proportion`, `svy: tabulate oneway`, and `svy: tabulate twoway` allow direct standardization of means, ratios, proportions, and tabulations using any of the three survey variance estimators.

9. Programmers of estimation commands can get full support for estimation with survey and correlated data almost automatically. This support includes correct treatment of multistage designs, weighting, stratification, poststratification, and finite-population corrections, as well as access to all three variance estimators. See [P] **program properties**.

10. The [SVY] manual now has a glossary that defines commonly used terms in survey analysis and explains how these terms are used in the manual; see [SVY] **glossary**.

1.3.3   Longitudinal/panel data

1. The big news is new command `xtmixed`—Stata now fits linear mixed models, also known as hierarchical models or multilevel models.

   Mixed models include what social scientists call random-effects models, including one-way, two-way, multi-way, and hierarchical models, and it includes random-coefficient models.

   Estimates are obtained using maximum likelihood (ML), restricted maximum likelihood (REML), or expectation maximization (EM). Covariances among random effects are estimated and may be independent (no covariance), exchangeable (common covariance), or unstructured (unique covariance for each pair of effects).

   `xtmixed` estimates standard errors and confidence intervals for the fixed parameters, and it estimates the standard deviations (variances) and correlations (covariances) of the random effects and the full VCE matrix among them.

   For details, see [XT] **xtmixed**.

   After estimation with `xtmixed`,

   a. `estat recovariance` reports the estimated variance–covariance matrix of the random effects for each level.

   b. `estat group` summarizes the composition of the nested groups, providing minimum, average, and maximum group size for each level in the model.

   `predict` after `xtmixed` can compute best linear unbiased predictions (BLUPs) for each random effect. It can also compute the linear predictor, the standard error of the linear predictor, the fitted values (linear predictor plus contributions of random effects), the residuals, and the standardized residuals.

2. New features have been added to the maximum-likelihood estimators that do not have closed-form solutions and require numeric evaluation of the likelihood. These estimators include `xtlogit`, `xtprobit`, `xtpoisson`, `xtcloglog`, `xtintreg`, and `xttobit`.

   a. The likelihood may now be approximated using adaptive Gauss–Hermite quadrature (the new default) or nonadaptive quadrature (the previous default). Adaptive quadrature substantially increases the accuracy of the approximation, particularly on difficult problems such as data with large panel sizes or data with a large variance for the random effects.

   b. Linear constraints may now be imposed using the new option `constraints()`. Constraints are specified the standard way; see [R] **constraint**.

   c. New option `intpoints()` replaces old option `quad()`, although `quad()` continues to work. The new name is more meaningful, especially when used with estimators that integrate likelihoods using methods other than quadrature.

3. Existing command `xtreg` now allows options `robust` and `cluster()` when estimating fixed-effects (FE) and random-effects (RE) models; see [XT] **xtreg**.

4. Most [XT] commands that previously did not allow time-series operators now support them. These commands include `xtgls`, `xtreg`, `xtsum`, `xtcloglog`, `xtintreg`, `xtlogit`, `xtpoisson`, `xtprobit`, `xttobit`, and `xtgee`.

5. New command `xtrc` is old command `xtrchh`, renamed, and with new features. New option `beta` reports the best linear predictors (BLUPs) for the group-specific coefficients, along with their standard errors and confidence intervals. For details, see [XT] **xtrc**.

6. `predict` after `xtrc` has the new option `group()` to compute the BLUPs of the dependent variable using the BLUPs of the coefficients.

7. New command `xtline` plots panel data and allows either overlaid or separate graphs for each panel; see [XT] **xtline**

8. New section [XT] **glossary** defines commonly used terms and how they are used by us.

1.3.4   Time-series statistics

1. Existing command `arima` can now estimate multiplicative seasonal ARIMA (SARIMA) models; see new options `sarima()`, `mar()`, and `mma()` in [TS] **arima**.

2. New command `rolling` performs rolling-window or recursive estimations, including regressions, and collects statistics from the estimation on each window; see [TS] **rolling**.

3. The [TS] manual now has a glossary that defines commonly used terms in time-series analysis and explains how we use them in the manual; see the [TS] **glossary**.

4. Many existing commands that previously did not allow time-series operators now do. These commands include `areg`, `binreg`, `biprobit`, `boxcox`, `cloglog`, `cnsreg`, `glm`, `heckman`, `heckprob`, `hetprob`, `impute`, `intreg`, `logistic`, `logit`, `lowess`, `mvreg`, `nbreg`, `orthog`, `pcorr`, `poisson`, `probit`, `pwcorr`, `rreg`, `testparm`, `treatreg`, `truncreg`, `xtcloglog`, `xtgls`, `xtintreg`, `xtlogit`, `xtpoisson`, `xtprobit`, `xtgee`, `xtreg`, `xtsum`, and `xttobit`.

5. Many commands requiring time-series data now work on a single panel from a panel dataset when that panel is selected using an `if` expression or an `in` qualifier. Those commands include `ac`, `estat archlm`, `estat bgodfrey`, `corrgram`, `cumsp`, `dfgls`, `dfuller`, `estat durbinalt`, `estat dwatson`, `pac`, `pergram`, `pperron`, `wntestb`, `wntestq`, and `xcorr`.

6. The dialogs for analyzing IRF results are much improved. The dialogs now populate lists of models and variables from the current IRF results that may be chosen for producing tables and graphs. The improved dialogs include `irf cgraph`, `irf ctable`, `irf graph`, `irf ograph`, and `irf table`.

7. Existing command `dfuller` has new option `drift` for testing the null hypothesis of a random walk with drift. The algorithm for calculating MacKinnon's approximate p-values is also now more accurate in cases where the p-value is relatively large; see [TS] **dfuller**.

8. Existing commands `corrgram` and `pac` have new option `yw` that computes partial autocorrelations using the Yule–Walker equations instead of the default regression-based method; see [TS] **corrgram**.

9. Time-series operators are now better displayed in estimation and other result tables.

10. New command `estat`—used after `regress`—brings together what was previously done by commands `dwstat`, `durbina`, `bgodfrey`, and `archlm`. The new commands are `estat dwatson`, `estat durbina`, `estat bgodfrey`, and `estat archlm`. See [R] **regress postestimation time series**.

11. The ability of `arima` and `arch` to estimate standard errors using either the observed information matrix (OIM) or the outer product of gradients (OPG) has been consolidated under the new `vce()` option.

(What follows was first released in Stata 8.2.)

12. New command `vec` fits cointegrated vector error-correction models (VECMs) using Johansen's method; see [TS] **vec**.

13. New command `vecrank` produces statistics used to determine the number of cointegrating vectors in a VECM, including Johansen's trace and maximum-eigenvalue tests for cointegration; see [TS] **vecrank**.

14. New command `fcast`—which replaces old command `varfcast`—produces and graphs dynamic forecasts of the dependent variables after fitting a VAR, SVAR, or VECM; see [TS] **fcast compute** and [TS] **fcast graph**.

15. New command `irf`—which replaces the old command `varirf`—does everything the old command did and more. `irf` estimates the impulse–response functions, cumulative impulse–response functions, orthogonalized impulse–response functions, structural impulse–response functions, and forecast error-variance decompositions after fitting a VAR, SVAR, or VECM. `irf` can also make graphs and tables of the results. See [TS] **irf**.

    `varirf` continues to work but is no longer documented. `irf` accepts `.vrf` result files created by `varirf`.

16. Existing command `varsoc` can now be used to obtain lag-order selection statistics for VECMs, as well as VARs; see [TS] **varsoc**.

17. New command `veclmar` computes Lagrange-multiplier statistics for autocorrelation after fitting a VECM; see [TS] **veclmar**.

18. New command `vecnorm` tests whether the disturbances in a VECM are normally distributed. For each equation and for all equations jointly, three statistics are computed: a skewness statistic, a kurtosis statistic, and the Jarque–Bera statistic. See [TS] **vecnorm**.

19. New command `vecstable` checks the eigenvalue stability condition after fitting a VECM; see [TS] **vecstable**.

20. New command `vecstable` and the existing command `varstable` have a new graph option for presenting the stability results. See [TS] **vecstable** and [TS] **varstable**

21. The output of the following commands has been standardized to improve formatting: `var`, `svar`, `vargranger`, `varlmar`, `varnorm`, `varsoc`, `varstable`, and `varwle`.

22. New command `haver` makes it easy to load and analyze economic and financial databases available from Haver Analytics; see [TS] **haver**.

1.3.5 Multivariate statistics

Stata has four all-new methods for analyzing multivariate data and many more extensions to existing methods. In addition, most methods now support direct analysis of matrices, as well as raw data.

Be sure to check the postestimation documentation for the multivariate estimators you use; many important new features are documented there. In particular, all the multivariate commands make extensive use of new command `estat` for providing additional statistics and results after estimation.

1. New commands `mds`, `mdslong`, and `mdsmat` perform classic metric multidimensional scaling: `mds` performs the scaling with respect to the distances (dissimilarities) between observations, `mdslong` performs the scaling on a long dataset where each observation represents the distance between two points or objects, and `mdsmat` performs the scaling on a matrix of distances. See [MV] **mds**, [MV] **mdslong**, and [MV] **mdsmat**.

    `mds` supports all 33 similarity/dissimilarity measures available in Stata; see [MV] *measure_option*.

The following new `estat` commands work after `mds`, `mdslong`, or `mdsmat` and provide additional statistics and results:

a. `estat config` reports the coordinates of the approximating configuration.

b. `estat correlations` reports the Pearson and Spearman correlations between the dissimilarities and the approximating distances for each object.

c. `estat pairwise` reports a set of statistics for each pairwise comparison; it reports the dissimilarities, the approximating distances, and the raw residuals.

d. `estat quantiles` reports the quantiles of the residuals for each observation (after `mds`) or object (after `mdslong` or `mdsmat`).

e. `estat stress` reports the Kruskal stress (loss) measure between the transformed dissimilarities and fitted distances for each object.

See [MV] **mds postestimation** for more information.

In addition, there are two new commands for graphing results from a multidimensional scaling:

a. `mdsconfig` plots the approximating Euclidean configuration of the first two dimensions; see [MV] **mds postestimation**.

b. `mdsshepard` produces a Shepard diagram of the dissimilarities against the approximating Euclidean distances; see [MV] **mds postestimation**.

   `predict` after any multidimensional-scaling command produces

   a. variables containing the approximating configuration (`predict` *newvarlist*, `config`);

   b. variables containing the dissimilarity, distance, and raw residuals (`predict` *newvarlist*, `pairwise`)

See [MV] **mds postestimation** for more information.

2. New commands `ca` and `camat` perform two-way correspondence analysis using any of several available forms of normalization. `ca` performs the analysis on the cross-tabulation of two categorical variables; `camat` performs the analysis on a matrix of counts; see [MV] **ca** for more information on both commands.

The following new `estat` commands work after `ca` and `camat` and provide additional statistics and results

a. `estat coordinates` reports the coordinates in both the row space and the column space.

b. `estat distances` reports the chi-squared distances between the row profiles and between the column profiles, including the distances to the marginal distributions (commonly called centers). Both observed and fitted profiles are available.

c. `estat inertia` reports the inertia contributions of the individual cells.

d. `estat profiles` reports the row profiles and column profiles—the conditional distributions, given the other dimension.

e. `estat summarize` reports summary information of the row and column variables over the estimation sample.

f. `estat table` reports the fitted correspondence table, the observed "correspondence" table, or the expected table under the assumption of independence.

See [MV] **ca postestimation** for more information.

In addition, there are two new commands for graphing results from a correspondence analysis:

a. `cabiplot` produces a biplot of each row category and each column category; see [MV] **ca postestimation**.

b. `caprojection` produces a graph that shows the ordering of row categories and column categories on each principal dimension of the analysis. Each principal dimension is represented by a vertical line; markers are plotted on the lines where the row categories and column categories project onto the dimensions; see [MV] **ca postestimation**.

`predict` after `ca` and `camat` computes fitted values and row or column scores for any dimension; see [MV] **ca postestimation**.

3. The new command `procrustes` performs Procrustean analysis for comparing and measuring the similarity between two sets of variables: source and target. Two datasets can also be compared if the datasets are first merged by record.

The following new `estat` commands work after `procrustes` and provide additional statistics and results:

a. `estat compare` reports fit statistics of the three transformations available in Procrustean analysis: orthogonal, oblique, and unrestricted.

b. `estat mvreg` reports the multivariate regression that is related to the current Procrustean analysis.

c. `estat summarize` reports summary information of the two sets of variables over the estimation sample.

See [MV] **procrustes postestimation** for more information.

New command `procoverlay` after `procrustes` creates an overlay graph comparing the target variables with the fitted values derived from the source variables; see [MV] **procrustes postestimation**.

`predict` after `procrustes` produces fitted values for all variables, residuals for all variables, or residual sums of squares for a specified target variable; see [MV] **procrustes postestimation**.

4. New command `biplot` performs a biplot analysis of a dataset and produces a two-dimensional biplot of the results. A biplot simultaneously displays the observations (rows) and the relative positions of the variables (columns). Observations are projected to two dimensions such that the distance between the observations is approximately preserved. The variables are plotted as arrows, with the cosine of the angle between arrows approximating the correlation between the variables. See [MV] **biplot**.

5. New command `tetrachoric` computes a tetrachoric correlation matrix for a set of binary variables. `tetrachoric` is documented in [R] but is often used in multivariate analyses; see [R] **tetrachoric**.

`tetrachoric` results can be used in subsequent factor analyses or principal-component analyses using the new `factormat` and `pcamat` commands; see [MV] **factor** and [MV] **pca**.

6. Existing command `canon` now allows analysis and presentation of more than one linear combination and has new options for reporting the raw or standardized coefficients and for reporting significance tests of the canonical correlations; see [MV] **canon**.

The following new `estat` commands work after `canon` and provide additional statistics and results:

a. `estat correlations` reports the correlations among all variables.

b. `estat loadings` reports the matrices of canonical loadings.

See [MV] **canon postestimation** for more information.

7. Existing command `cluster dendrogram` has many new features, including horizontal dendrograms and the ability to label branch counts. The look of the graph can now be changed (titles, axes, colors, etc.); see [MV] **cluster dendrogram**.

8. The existing hierarchical cluster commands have new option `measure()` that specifies the proximity measure to use in computing dissimilarities between observations. Any of 33 measures may be specified; see [MV] *measure_option*. Previously most of the measures were available under other option names; those options continue to work but are undocumented. See [MV] **cluster**.

9. Existing command `cluster stop` has new option `varlist()` that specifies alternative variables to use when computing the stopping rules; see [MV] **cluster stop**.

Analysis of proximity matrices

All of Stata's multivariate analysis facilities that rely on pairwise comparisons of distance, similarity, dissimilarity, covariance, correlation, or other proximity measures can now work directly with proximity matrices that you compute or obtain from other sources.

Previously, all these facilities worked only with raw datasets. The new commands implement analyses on matrices. They share the common ability to accept either full matrices or vectors representing the lower or upper triangle of a symmetric proximity matrix.

10. New command `clustermat` extends all of Stata's hierarchical clustering facilities to the analysis of matrices of a dissimilarity measure (sometimes called a distance or proximity measure). This includes all seven linkage methods and the ability to create dendrograms of the results; see [MV] **clustermat**.

11. New command `factormat` performs factor analysis on a matrix of correlations, extending all the new and previously available capabilities of the existing command `factor` to precomputed matrices of correlations; see [MV] **factor**.

12. New command `pcamat` performs principal component analysis on an existing correlation or covariance matrix; see [MV] **pca**.

13. New `matrix` subcommand `dissimilarity` computes similarity, dissimilarity, or distance matrices using any of 19 proximity measures for continuous data and 14 measures for binary data; see [MV] *measure_option* and see [MV] **matrix dissimilarity**.

Factor and principal component analysis additions

In addition to allowing direct analysis of correlation and covariance matrices using `factormat` and `pcamat`, Stata's factor analysis and principal component analysis (PCA) methods have been expanded, particularly through the addition of postestimation commands for reporting and graphing results.

14. Command `factor` has new reporting option `altdivisor` that specifies the trace of the correlation matrix be used as the divisor for proportions, rather than the default (the sum of all eigenvalues).

15. New `estat` commands for use after `factor` and `factormat` provide additional statistics and results:

   a. `estat common` reports the correlation matrix of the common factors and is more of interest after oblique rotations.

   b. `estat factors` reports model-selection criteria (AIC and BIC) over all the factors retained in an analysis.

   c. `estat rotatecompare` reports the unrotated factor loadings next to the most-recent rotated loadings.

d. `estat structure` reports the factor structure—the correlations between the variables and the common factors.

See [MV] **factor postestimation** for more information.

16. Existing command `pca` allows several new options:

a. Option `vce(normal)` computes the VCE of the eigenvalues and eigenvectors, assuming multivariate normality.

   This gives you access to many of Stata's postestimation facilities for analyzing estimation results, including tests of eigenvalue and eigenvector significance, tests of linear and nonlinear combinations (`test` and `testnl`), linear and nonlinear combinations with confidence intervals (`lincom` and `nlcom`), and nonlinear predictions with confidence intervals (`predictnl`).

   `vce(normal)` also produces the ingredients for adding confidence intervals to screeplots; see [MV] **screeplot**.

b. Options `level()`, `blanks()`, `novce`, and `norotated` allow more flexible control of the displayed results.

c. Option `components(#)` specifies the number of components to retain and is a synonym for old option `factor()`.

d. Options `tol()` and `ignore` provide advanced control for computationally difficult problems.

See [MV] **pca** for more information.

17. New `estat` commands for use after `pca` and `pcamat` provide additional statistics and results:

a. `estat loadings` reports the component loading matrix in any of several available normalizations of the columns (eigenvectors).

b. `estat rotatecompare` reports the unrotated (principal) components next to the most recent rotated components.

See [MV] **pca postestimation** for more information.

18. New `estat` commands for use after any factor analysis or any principal component analysis (that is, after `factor` or `factormat` or after `pca` or `pcamat`) provide additional statistics and results:

a. `estat anti` reports the anti-image correlation and anti-image covariance matrices.

b. `estat kmo` reports the Kaiser–Meyer–Olkin measure of sampling adequacy.

c. `estat residuals` reports the difference between the observed correlation or covariance matrix and the fitted (reproduced) matrix using the retained factors.

d. `estat smc` reports the squared multiple correlations (SMC) between each variable and all other variables. SMC is a theoretical lower bound for communality, so it is an upper bound for the unexplained variance.

See [MV] **factor postestimation** and [MV] **pca postestimation** for more information.

19. Three new graphs are available after any factor analysis (`factor` and `factormat`) or after any principal component analysis (`pca` and `pcamat`):

a. `scoreplot` graphs scatterplots comparing each pair of factors or components; see [MV] **scoreplot**.

b. `loadingplot` graphs scatterplots comparing loadings for each pair of factors or components; see [MV] **scoreplot**.

c. `screeplot` plots the eigenvalues of a covariance or correlation matrix; see [MV] **screeplot**. (`screeplot` replaces `greigen` and has more features; `greigen` continues to work but is undocumented.)

20. New command `rotate` performs orthogonal and oblique rotations after `factor`, `factormat`, `pca`, and `pcamat`. Available rotations include varimax, quartimax, equamax, parsimax, minimum entropy, Comrey's tandem 1 and 2, promax power, biquartimax, biquartimin, covarimin, oblimin, factor parsimony, Crawford–Ferguson family, Bentler's invariant pattern, oblimax, quartimin, and target and partial-target matrices; see [MV] **rotate**.

New command `rotatemat` performs these same linear transformations (rotations) on any Stata matrix.

1.3.6   Survival analysis

1. The [ST] manual now has a glossary that defines commonly used terms in survival (or duration) analysis and often explains how these terms are used in the manual; see the [ST] **glossary**.

2. New command `estat` can be used after `stcox` and `streg`. In addition to the standard `estat` statistics—information criteria, estimation sample summary, and formatted variance–covariance matrix (VCE)—statistics specific to the proportional hazards estimator are available after `stcox`. These include

    a. `estat concordance` computes Harrell's C and Somer's D statistics measuring concordance—agreement of predictions with observed failure order.

    b. `estat phtest` replaces the existing `stphtest` for computing tests and graphs of the proportional hazards assumption. `stphtest` continues to work.

    See [ST] **stcox postestimation** and [ST] **streg postestimation**.

3. Existing command `sts graph` has new options `cihazard` and `per(#)`. `cihazard` draws pointwise confidence bands around the smoothed hazard function, and `per()` specifies the units used to report the survival or failure rate. See [ST] **sts graph**.

4. Existing command `stcurve` now plots over an evenly spaced grid, producing smooth curves, even in small samples; see [ST] **stcurve**.

5. Existing command `sts graph` has new options `atriskopts()` and `lostopts()` that let you control how the labels for at-risk and lost observations look (their color, font size, etc.); see [ST] **sts**.

6. Existing command `stci` has new options for controlling how the plotted survival line looks (color, thickness, etc.) and for adding titles, controlling legends, and all other characteristics of the graph; see [ST] **stci**.

1.3.7   General-purpose statistics

1. New estimation command `asmprobit` fits multinomial probit (MNP) models to categorical data and is frequently used in choice-based modeling. `asmprobit` allows several correlation structures for the alternatives, including completely unstructured, where all possible correlations are estimated. It also allows for either heteroskedastic or homoskedastic variances among the alternatives and allows arbitrary patterns within the alternative variances or correlations. `asmprobit`'s syntax makes specifying both case-specific and alternative-in-case-specific regressors easy.

    In addition to common postestimation commands, such as `mfx` for computing marginal effects, new command `estat` provides additional statistics and results:

    a. `estat alternatives` reports summary statistics about each of the alternatives and provides a mapping between the index numbers labeling the alternatives and their associated values and labels in the dataset.

b. `estat covariance` computes and reports the estimated covariance matrix for the alternatives.

c. `estat correlation` reports the correlations among the alternatives in matrix form.

Predicted statistics after `asmprobit` include the linear predictor, the probability that an alternative is selected, and the standard error of the linear predictor.

See [R] **asmprobit**, and [R] **asmprobit postestimation**.

2. New estimation command `mprobit` also fits multinomial probit models to categorical data but in the simplified situation of having only case-specific covariates (as with the multinomial logistic regression, `mlogit`). Maximizing the likelihood is much faster in such cases because the numeric approximation to the likelihood is simpler. See [R] **mprobit**.

3. New estimation command `slogit` fits the stereotype logistic regression model for categorical dependent variables. This model can be viewed as either a generalization of the multinomial logistic regression model (`mlogit`) or a generalization of the ordered logistic regression model (`ologit`) that relaxes the proportional-odds assumption. See [R] **slogit**.

Predicted statistics after `slogit` include the linear predictor, the probability of any or all outcomes, and the standard error of the linear predictor. See [R] **slogit postestimation**.

4. New estimation command `ivprobit` fits probit regression models of binary outcomes with endogenous regressors. Estimation can be performed by maximum likelihood estimation (MLE) or by Newey's minimum chi-squared two-step estimation, but note that some postestimation facilities, such as computing marginal effects with `mfx`, are available only after ML estimation—the two-step estimator imposes a transformation that invalidates many postestimation results. See [R] **ivprobit**.

5. New estimation command `ivtobit` fits linear regression models with censored dependent variables by maximum likelihood estimation or by Newey's minimum chi-squared two-step estimation (but see the note about the two-step estimator in 4 above). See [R] **ivtobit**.

6. New estimation command `ztp` fits a zero-truncated Poisson model of event counts with truncation at zero.

Predicted statistics after `ztp` include the linear predictor and its standard error, the predicted number of events, the incidence rate, the conditional mean, and the likelihood score See [R] **ztp** and [R] **ztp postestimation**.

7. New estimation command `ztnb` fits a zero-truncated negative binomial model of event counts with truncation at zero and over- or underdispersion.

Predicted statistics after `ztnb` include the linear predictor and its standard error, the predicted number of events, the incidence rate, the conditional mean, and the likelihood scores See [R] **ztnb** and [R] **ztnb postestimation**.

8. New estimation commands `mean`, `ratio`, `proportion`, and `total` estimate means, ratios, proportions, and totals over the entire sample or over groups within the sample. When estimating over groups, the entire covariance matrix (VCE) is estimated. These are full estimation commands that support a range of postestimation facilities, such as linear and nonlinear tests among the groups (`test` and `testnl`) and linear and nonlinear combinations of group-level statistics (`lincom` and `nlcom`). All four commands support several SE and VCE estimates: robust, cluster–robust, bootstrap, jackknife, and observed information matrix (the default).

`mean`, `ratio`, and `proportion` also support direct standardization across strata (groups) using the `stdize()` and `stdweight()` options.

See [R] **mean**, [R] **ratio**, [R] **proportion**, and [R] **total**.

9. To avoid conflict with the new `mean` command, existing command `means` has been renamed `ameans`, with synonyms `gmeans` and `hmeans`.

10. Existing command nl has a new syntax that makes estimating nonlinear least-squares regressions easier. For most models, estimation is now as easy as typing the nonlinear expression. Full programmability has been retained for complex models, and the old syntax continues to work.

    nl also now supports robust (Huber white/sandwich) and cluster–robust SE and VCE estimates, including two popular adjustments that can dramatically improve the small-sample performance of robust SE and VCE estimates.

    A number of new reporting and estimation options have also been added. See [R] **nl**.

11. New option vce() selects how standard errors (SEs) and covariance matrix of the estimated parameters are estimated by most estimation commands. Choices are vce(oim), vce(opg), vce(robust), vce(jackknife), and vce(bootstrap), although the choices can vary estimator by estimator. vce(robust) is a synonym for robust, and you can use either. What is new are vce(jackknife) and vce(bootstrap).

    vce(bootstrap) specifies that the standard errors, significance tests, and confidence intervals be normal-based bootstrap estimates, rather than the default analytic estimates based on the observed information matrix. You can also produce percentile-based or bias-corrected confidence intervals after estimation using estat bootstrap; see [R] **bootstrap postestimation**.

    vce(jackknife) specifies that the standard errors, significance tests, and confidence intervals be jackknife estimates.

    Both vce(bootstrap) and vce(jackknife) will automatically perform either observation or cluster sampling, whichever is appropriate for the estimator.

    Notably, both vce(bootstrap) and vce(jackknife) compute bootstrapped or jackknifed estimates of the complete VCE matrix. This means that many of Stata's postestimation commands are available. You can form linear and nonlinear combinations or functions of the parameters and obtain jackknife or normal-based bootstrap standard errors and confidence intervals for the combinations using [R] **lincom** and [R] **nlcom**. Similarly, you can perform linear and nonlinear tests using [R] **test** and [R] **testnl**.

12. New command estat centralizes the computing and reporting of additional statistics after estimation, just as predict does with predictions. estat allows subcommands; estat summarize, for instance, reports summary statistics for the estimation sample and can be used after any estimator. estat also allows subcommands that are specific to the estimation command. To find out what is available after a command, see the corresponding postestimation entry. For example, after [R] **regress**, see [R] **regress postestimation**; or after [XT] **xtmixed**, see [XT] **xtmixed postestimation**.

(*Continued on next page*)

Existing postestimation commands have been brought into the `estat` framework:

Estimation command	Old command	New estat command
`regress`	`ovtest`	`estat ovtest`
	`hettest`	`estat hettest`
	`szroeter`	`estat szroeter`
	`vif`	`estat vif`
	`imtest`	`estat imtest`
`regress`	`dwstat`	`estat dwatson`
(time series)	`durbina`	`estat durbinalt`
`bgodfrey`	`estat bgodfrey`	
`archlm`	`estat archlm`	
`anova`	`ovtest`	`estat ovtest`
`hettest`	`estat hettest`	
`logit` and	`lstat`	`estat classification`(*)
`logistic`	`lfit`	`estat gof`(*)
`poisson`	`poisgof`	`estat gof`
`stcox`	`stphtest`	`estat phtest`
`xtgee`	`xtcorr`	`estat wcorrelation`

(*) The new command works after `probit`, as well as after `logit` and `logistic`; the old command worked after `logit` and `logistic` only.

The original commands continue to work but are undocumented.

Three `estat` subcommands are available after almost all estimators:

a. `estat ic` reports Akaike's and Schwartz's Bayesian information criteria (AIC and BIC).

b. `estat summarize` reports summary statistics on the variables in the estimation model for the estimation sample.

c. `estat vce` reports the covariance (VCE) or correlation matrix estimates. (`estat vce` replaces the old `vce` command and has more features.)

13. Stata has many new prefix commands (commands that behave like `by:` and `xi:`). New prefix commands include `statsby:`, `bootstrap:`, `jackknife:`, `permute:`, `simulate:`, `stepwise:`, `svy:`, and `rolling:`. For instance, to obtain the standard error and confidence interval of the mean, you might type

    `. jackknife: mean earnings`

    or to obtain survey-adjusted estimates, you might type

    `. svy:  mean earnings`

    after svysetting your data.

    See [R] **bootstrap**, [R] **jackknife**, [R] **permute**, [TS] **rolling**, [R] **simulate**, [R] **stepwise**, [D] **statsby**, and [SVY] **svy**.

14. New prefix commands `bootstrap:` and `jackknife:` replace old commands `bs` and `jknife`, and in addition to having better syntax, they also provide new features:

a. They handle and report expressions better.

b. They post their results as estimation results with a complete VCE. Most postestimation facilities may now be used after them and will be based on the bootstrap or jackknife VCE. These include

adjust	adjusted predictions
estimates	cataloging estimation results
lincom	linear combinations with SEs, tests, and CIs
nlcom	nonlinear combinations with SEs, tests, and CIs
mfx	computing marginal effects and elasticities
predict	predictions, residuals, probabilities, etc.
predictnl	generalized nonlinear predictions with SEs and CIs
test	Wald tests of simple and composite linear hypotheses
testnl	Wald tests of nonlinear hypotheses

c. They produce a model test when applied to the coefficients of estimation commands.

d. They allow option seed(#) to set the random-number seed.

e. They allow option reject(*exp*) to reject replicates that explicitly match *exp*.

f. bootstrap: uses the normal distribution instead of the Student's *t* distribution to compute the normal-approximation confidence intervals.

g. jackknife: now allows fweights to be specified.

See [R] **bootstrap** and [R] **jackknife**.

15. New prefix command statsby: replaces old command statsby (not a prefix) and provides enhanced handling and reporting of expressions, allows weights, and allows string variables in the option by(). See [D] **statsby**.

16. New prefix command stepwise: replaces old command sw and, in addition to working with all the previous estimators, also works with [R] **intreg** and [R] **scobit**.

17. Existing prefix command xi: has new option noomit that prevents it from omitting a category when generating category indicators for group variables. See [R] **xi**.

18. New command tetrachoric computes a tetrachoric correlation matrix for set of binary variables. See [R] **tetrachoric**.

19. Existing command suest, which combines estimation results for subsequent testing, is easier to use and has new features:

a. Scores are now computed for the models you combine; you no longer need to save scores when estimating.

b. suest, used after svy: estimation, now accounts for your survey design.

c. suest now works more smoothly with certain estimation commands that previously required special treatment, including regress, ologit, and oprobit.

d. suest now works with all models fitted by clogit, rather than only those with a single positive outcome per group.

See [R] **suest**.

20. Existing command clogit has new features:

a. Robust and cluster–robust SE and VCE estimates are now supported through options robust and cluster().

b. Linear constraints on the parameters are now implemented via option constraints().

   c. New option `vce()` allows SE and VCE estimates to be computed using OIM (the default), OPG, bootstrap, and jackknife.

  See [R] **clogit**.

21. Option `level()` now allows noninteger confidence levels to be specified. See [R] **estimation options**.

22. Existing command `predict` now generates equation-level scores after most maximum-likelihood estimation commands; see the documentation of `predict` in the postestimation entry for each estimation command.

23. Existing command `cumul` has a new option `equal` to create equal cumulative values for ties. See [R] **cumul**.

24. Existing command `estimates table` now allows you to specify more models, and the command wraps the table if necessary. Also allowed are new options

   a. `equations()`, which matches equations by number rather than by name.

   b. `coded`, which displays the table in a compact, symbolic format.

   c. `modelwidth()`, which sets the number of characters for displaying model names.

  See [R] **estimates**.

25. `test` after `anova` and `manova` has two new options for performing Wald tests:

   a. `mtest()`, which implements three methods to adjust for multiple tests: Bonferroni, Holm, and Sidák.

   b. `test()`, which makes specifying contrasts easier by accepting a matrix containing the contrast.

  See [R] **anova postestimation**.

26. Commands `ci` and `cii` have new options `exact`, `wilson`, `agresti`, `jeffreys`, and `wald` for computing different types of binomial confidence intervals. See [R] **ci**.

27. Command `hausman` has new option `df()` for controlling the degrees of freedom. See [R] **hausman**.

28. After existing command `ivreg`, `predict` has a `score` option for generating equation-level scores; see [R] **ivreg postestimation**.

29. Command `mfx` is now faster and has new option `varlist()` for computing effects of specific variables. See [R] **mfx**.

30. Commands `tabulate` and `tabi` with the `exact` option are now significantly faster.

31. In existing command `mlogit`, option `basecat` has been renamed `baseoutcome()` for better consistency with the terminology of choice models. See [R] **mlogit**.

32. Existing commands `spearman` and `ktau` now allow more than two variables to be specified and have more flexible output. See [R] **spearman**.

33. Existing command `bsample` for sampling with replacement (bootstrap sampling) now supports weighted bootstrap resampling using the new `weight()` option. See [R] **bsample**.

34. Existing command `bstat` for reporting bootstrap results has a number of new reporting options. In addition, `bstat` previously computed percentile and other confidence intervals. This is now handled by `estat bootstrap`, which can be used after any bootstrap estimation, including `bstat`. See [R] **bstat** and [R] **bootstrap postestimation**.

35. Most maximum likelihood estimators now test for convergence using the Hessian-scaled gradient, $gH^{-1}g'$. This criterion ensures that the gradient is close to zero when scaled by the Hessian (the curvature of the likelihood or pseudolikelihood surface at the optimum) and provides greater

assurance of convergence for models whose likelihoods tend to be difficult to optimize, such as those for `arch`, `asmprobit`, and `scobit`. You can set the tolerance level for this test with new option `nrtolerance()`, show the Hessian-scaled gradient in the iteration log with option `shownrtolerance`, and turn the test off with option `nonrtolerance`. See [R] **maximize**.

36. Existing command `set` has new setting `maxiter`—default value 16000—that specifies the maximum number of iterations to be performed by all estimation commands. You change this setting by typing `set maxiter #`, and you may add option `permanently` to retain the setting in future Stata sessions.

1.3.8   New ML features

Command `ml`, for implementing user-written maximum-likelihood estimators, has many new features:

1. New option `technique()` sets the optimization technique. BHHH, DFP, and BFGS optimization techniques are now available; the default technique remains modified Newton–Raphson.

2. New option `vce()` sets the type of covariance-matrix calculations that will be made.

    `vce(oim)` specifies the observed information matrix (OIM), also called the Hessian-based estimator; this is (and always has been) the default.

    `vce(opg)` specifies the outer product of the gradients (OPG). This is new.

    `vce(robust)` specifies Taylor-series linearization, also known as the Huber/White sandwich estimator and, in Stata, as simply robust.

3. Most estimators written with `ml` now support estimation with survey data and correlated data with no additional programming. This support includes correct treatment of multistage designs, weighting, stratification, poststratification, and finite-population corrections, as well as access to linearization, jackknife, and bootstrap variance estimators. For a discussion, see [P] **program properties**.

4. `ml` has always allowed linear constraints to be applied using the option `constraints()` with no additional programming. It now handles irrelevant constraints more elegantly. Irrelevant constraints are those that have no impact on the model. Previously, irrelevant constraints caused an error message. Now they are flagged and ignored.

5. When linear constraints are imposed, `ml` now applies a Wald test for the overall fit of the model, rather than attempting a likelihood-ratio (LR) test, which is often inappropriate.

6. `ml` has new subcommand `score` for generating scores after fitting a model.

7. `ml` has new option `diparm_options()` that automatically performs transformations of ancillary parameters.

8. `ml` now saves the gradient vector in `e(gradient)`.

9. `ml` has new option `search(norescale)` that prevents rescaling when searching for starting values.

10. `ml` honors the new setting for maximum iterations, `set maxiter #`, and will iterate a maximum of # iterations, even if convergence has not been achieved.

11. `ml` now displays a prominent message in the footer of the estimation results when convergence is not achieved. This message continues to be shown on redisplay of estimation results.

12. `ml` has new option `nofootnote` to suppress printing the new message warning if convergence is not achieved.

13. `ml` tests for convergence using the Hessian-scaled gradient—$\mathbf{gH}^{-1}\mathbf{g}'$. This is a true convergence criterion that ensures that the gradient is close to zero when scaled by the Hessian (the curvature

of the likelihood or pseudolikelihood surface at the optimum). This new criterion is particularly important when maximizing difficult likelihoods to prevent stopping the maximization too soon.

14. New option `nrtolerance()` lets you change the tolerance for the Hessian-scaled gradient convergence criterion; the default is `nrtolerance(1e-5)`.

15. New option `shownrtolerance` displays the criterion value of the Hessian-scaled gradient at each iteration.

16. New undocumented command `mlmatbysum` helps you compute the Hessian of panel-data likelihoods and is of interest to those seeking the speed that comes with programming your own second-derivative calculations; see `help mlmatbysum`.

17. `ml` has two new undocumented subcommands—`ml hold` and `ml unhold`—to assist in solving nested optimization problems, see `help ml_hold`.

    See [R] **ml** for more information on these features. Anyone programming estimators using `ml` should read the book *Maximum Likelihood Estimation with Stata*, 2nd edition (Gould, Pitblado, and Sribney 2003). Many of the features mentioned above are discussed and applied to real problems in the book.

1.3.9   Functions and expressions

1. The limit for the number of dyadic operators has been increased from 200 to 500; see help limits.

2. The default matrix size (`matsize`) for Intercooled Stata is now 200, rather than 40. The default for Stata/SE remains 400, and for Small Stata, it is 40.

3. The following new functions have been added in the context of expressions, such as `generate` *newvar* = *exp* or if *exp*:

name	purpose
`binormal()`	bivariate normal cumulative
`atan2()`	two-argument arc tangent
`regexm()`	regular expression matching
`regexr()`	regular expression replacement
`regexs()`	regular subexpressions
`indexnot()`	first string *s1* not in *s2*

See [D] **functions** or type `help` followed by the function name, such as `help binormal()`.

In addition, a host of new functions are available through Mata; see [M-4] **intro**.

4. The following existing functions have been renamed:

old name	new name
index()	strpos()
binorm()	binormal()
match()	strmatch()
norm()	normal()
invnorm()	invnormal()
normd()	normalden()
lnfact()	lnfactorial()
issym()	issymmetric()
syminv()	invsym()

Old names continue to work. Functions were renamed because the new name is better and because Mata uses the new name, and you want to be able to use the same names in both environments.

5. The following existing functions now have two names, and you can use either:

Name 1	Name 2
lower()	strlower()
upper()	strupper()
proper()	strproper()
ltrim()	strltrim()
rtrim()	strrtrim()
trim()	strtrim()
reverse()	strreverse()
string()	strofreal()
int()	trunc()
length()	strlen()

In this case, throughout the Stata documentation, we use name 1, but you can use name 1 or name 2 in your Stata expressions. Name 2 matches the name of the Mata function that does the same thing, so you may want to standardize on name 2.

6. The following egen functions have been renamed:

old name	new name
any()	anyvalue()
eqany()	anymatch()
neqany()	anycount()
rfirst()	rowfirst()
rlast()	rowlast()
rmean()	rowmean()
rmin()	rowmin()
rmiss()	rowmiss()
robs()	rownonmiss()
rsd()	rowsd()
rsum()	rowtotal()
sum()	total()

The new names are more consistent. Old names continue to work but are not documented.

1.3.10   Data management

1. There is a new manual **[D] Data Management**, and the data-management commands have been moved from [R] to [D]. See [D] **intro** for an expanded what's new for data-management capabilities.

2. Existing command `set type` now has a `permanently` option. You can now permanently set the default help datatype to either `float` (the factory default) or `double`.

3. New commands `xmlsave` and `xmluse` save and restore datasets in Extended Markup Language (XML) format. Data may be saved or used in either Stata `dta` XML format or Microsoft Excel's SpreadsheetML format. See [D] **xmlsave**.

4. New commands `fdasave`, `fdause`, and `fdadescribe` save, use, and describe files in the format required by the U.S. Food and Drug Administration (FDA) for new drug and device applications (NDAs). These commands are designed to assist people making submissions to the FDA, but the commands are general enough for use in transferring data between SAS and Stata. The FDA format is identical to the SAS XPORT Transport format. See [D] **fdasave**.

5. Value labels may now be up to 32,000 characters long.

6. Existing command `label` has a new subcommand `language` that lets you create and use datasets containing different variable, value, and data labels, which might be in different languages. See [D] **label language**.

7. Datasets from the examples in the Stata manuals can now be browsed, described, and used. Type `help dta contents`, or select **File > Example datasets...** from the Stata menu.

8. `statsby` is now a prefix command; see [U] **11.1.10 Prefix commands**. For information on its new syntax, see [D] **statsby**. Enhancements to `statsby` include

    a. Rather than requiring a list of expressions for the statistics to collect, `statsby` now collects a default set.

    b. Expressions to be computed and saved can now be grouped together as equations; see `help exp_list`.

    c. String variables are now allowed.

    d. Weights are now allowed.

    e. New option `force` forces `statsby` to work with `survey` estimators. By default, this is prevented because the method `statsby` uses to select subsamples will generally not produce appropriate standard-error estimates with survey data (the `subpop` option must be used with survey data).

    f. Dots showing the progress of computations are now shown by default.

    g. New option `nolegend` suppresses the table reporting on what `statsby` is running.

9. New command `filefilter` copies an input file to an output file while converting a specified ASCII or binary pattern to another pattern; see [D] **filefilter**.

10. New command `expandcl` replicates clusters of unique observations, much like an `expand`, but for clustered data; see [D] **expandcl**.

11. New command `tostring` converts numeric variables to string; see [D] **destring**.

12. Existing command `codebook` now allows `if` and `in` qualifiers; see [D] **codebook**.

13. New command `rmdir` removes an existing directory (folder); see [D] **rmdir**.

14. New command `clonevar` makes an identical copy of an existing variable; see [D] **clonevar**.

15. Existing commands `icd9` and `icd9p` have been updated to use the V21 codes; see [D] **icd9**.

16. Existing command `encode` has new option `noextend` that prevents adding new value label mappings; see [D] **encode**.

17. Existing command `odbc` for accessing Open DataBase Connectivity (ODBC) data sources has the following enhancements:

   a. ODBC is now supported under Mac OS X and Linux systems that use the iODBC Driver Manager. For more information on configuring ODBC for Mac and Linux, see the FAQ at *http://www.stata.com/support/faqs/data/odbcmu.html*.

   b. `odbc` has new subcommands `odbc insert` and `odbc exec` for writing data to an ODBC data source. Positioned updates can be performed using the `odbc exec` command.

   c. `odbc` has a new subcommand `sqlfile` for batch processing SQL instructions.

   d. `odbc load` has a new option `sqlshow` for debugging SQL communication with ODBC drivers.

   e. `odbc load` has new options `allstring` and `datestring`, which import either all data or just dates as strings.

   See [D] **odbc**.

18. Existing command `merge` has the following new features:

   a. It now accepts multiple `using` files.

   b. New option `nosummary` suppresses creating variables that summarize how the records were merged.

   c. New option `sort` option sorts the master and using datasets if they are not already sorted.

   d. Existing options `unique`, `uniqmaster`, and `uniqusing` now require you to specify matching variables.

   e. Warning messages are now given when matching variables do not uniquely identify observations.

   See [D] **merge**.

19. Existing commands `merge` and `append` now incorporate all notes from the using dataset that do not already appear in the master dataset, unless new option `nonotes` is specified; see [D] **merge** and [D] **append**.

20. Existing command `contract` has new options `cfreq()`, `percent()`, `cpercent()`, `float`, and `format()` to create frequency and percentage variables; see [D] **contract**.

21. Existing commands `corr2data` and `drawnorm` now support triangular specification of the correlation or covariance matrix; see [D] **corr2data** and [D] **drawnorm**.

22. Existing command `separate` has new option `shortlabel` to specify that shorter variable labels be created; see [D] **separate**.

23. Existing command `outfile` has new option `missing` that preserves both standard and extended missing values when the `comma` option is also specified; see [D] **outfile**.

24. Existing command `clear` now performs `mata: mata clear` in addition to everything else; see [D] **drop**.

1.3.11 Graphics

1. Stata now allows multiple Graph windows. The existing `name()` option now creates a named graph and displays it in its own window. See *User interface* below.

2. New command `sunflower` draws sunflower density-distribution plots; see [R] **sunflower**.

3. `graph twoway` has two new *plottypes* for plotting time-series data, `tsline` and `tsrline`; see [TS] **tsline** and [G] **graph twoway tsline**.

4. Graphs have better axis labels when graphing dates.

5. `graph twoway` has seven new options that are useful when plotting time-formatted variables: `tscale()`, `tlabel()`, `tmlabel()`, `ttick()`, `tmtick()`, `tline()`, and `ttext()`; see [G] *axis_options*, [G] *added_line_options*, and [G] *added_text_options*.

6. `graph twoway` has seven new *plottypes* for plotting paired-coordinate data—data with 4 variables, where two variables form a starting x–y point and the other two variables form an ending x–y point. The new *plottypes* are

plottype	description
pcarrow	plots a directional arrow for each observation's paired coordinates
pcbarrow	plots a two-headed arrow for each observation's paired coordinates
pcspike	plots a line or spike for each observation
pccapsym	plots a line with symbols at each end for each observation
pcscatter	plots both pairs of x–y variables as a scatter, using a common style
pci	immediate form of paired coordinate plots; plots the specified coordinate pairs
pcarrowi	immediate form of pcarrow

See [G] **graph twoway pcarrow** (for pcarrow and pcbarrow), [G] **graph twoway pcspike** (for pcspike), [G] **graph twoway pccapsym** (for pccapsym), [G] **graph twoway pcscatter** (for pcscatter), [G] **graph twoway pci** (for pci), and [G] **graph twoway pcarrowi** (for pcarrowi).

7. `graph twoway`, `graph bar`, `graph box`, and `graph dot` have new option `aspectratio()` that controls the aspect ratio of a plot region; see [G] *aspect_option*.

8. `graph display` has new option `scale()` that allows all text, symbols, and line widths to be rescaled when a graph is redisplayed; see [G] **graph display**.

9. `graph export` now supports additional output formats:

   a. TIFF files
      These files are limited to the resolution of the display device and the size of the Graph window; see [G] **graph export**.

   b. PNG (portable network graphics)
      This format is especially useful for posting graphs on the Internet; see [G] **graph export**.

   c. TIFF previews for EPS files
      The new option `preview()` embeds a preview of the graph so that it can be viewed in publishing applications; see [G] **graph export** and [G] *eps_options*. The Graph window must be open so that it can be used to create the preview.

10. `graph` now supports CMYK output to Postscript and Encapsulated Postscript (EPS) files. CMYK stands for Cyan-Magenta-Yellow-blacK and is popular in the printing industry. See [G] **graph export** and [G] *ps_options*.

    `palette color` has the new option `cmyk`, specifying that color values be reported in CMYK; see [G] **palette**.

11. `graph box` can now label outside values using option `marker()`; see [G] **graph box** and [G] *marker_label_options*.

12. `graph bar` has new options `over(, reverse)` and `yvaroptions(reverse)` to specify that the categorical scale be reversed to run from maximum to minimum; see [G] **graph bar**.

13. graph twoway has new option pcycle() that specifies the maximum number of plots that may appear on a graph before the pstyles recycle to the first style; see [G] *advanced_options*.

14. graph combine has new option altshrink that provides alternate sizing of the text, markers, line thickness, and line patterns on the individual combined graphs; see [G] **graph combine**.

15. graph has improved control over whether the largest and smallest possible grid lines are drawn. This control is provided by improving the actions of the existing suboptions [no]gmin and [no]gmax; see [G] *axis_label_options*.

16. graph bar, graph dot, graph box, and graph pie have new option allcategories specifying that the legend include all over() groups, not just groups in the sample specified by if and in. See, for example, [G] **graph bar**.

17. graph and all other commands that draw graphs have new options for changing the color of objects and changing the appearance of lines:

    a. Options lstyle(), lcolor(), lwidth(), and lpattern() are now accepted anywhere cl⟨*attribute*⟩ and the bl⟨*attribute*⟩ were allowed. Specifically, the new options replace the following original options:

new options	original options
lstyle()	clstyle(), blstyle()
lcolor()	clcolor(), blcolor()
lwidth()	clwidth(), blwidth()
lpattern()	clpattern(), blpattern()

    The new options can be applied to all lines—lines connecting points, lines outlining bars, lines around text boxes, etc. The original option names continue to work but are undocumented.

    b. New option fcolor() changes area fill colors and can be used anywhere bfcolor() or afcolor() were allowed. bfcolor() and afcolor() continue to work but are undocumented.

    c. New option color(*arg*) sets all of a plot's colors; it is the equivalent of specifying mcolor(*arg*), lcolor(*arg*), and fcolor(*arg*).

18. The syntax of the ROC curve commands is now consistent across all the ROC commands—roctab, roccomp, rocgold, and rocplot—with some new options added and some old options changing names. The original options continue to work but are undocumented. See [R] **roc** and [R] **rocfit postestimation**

19. Existing commands fracplot and lowess have new option lineopts() that replaces the confusingly named rlopts(); see [R] **fracpoly postestimation** and [R] **lowess**.

20. Option plot(), available on many graph commands, has been renamed addplot(). addplot() allows twoway plots, such as scatters, lines, or function plots to be added to most statistical graph commands.

21. Command kdensity has new option epan2 providing an alternate Epanechnikov kernel; see [R] **kdensity**. Accordingly, sts graph and stcurve now allow kernel(epan2) for specifying this new kernel.

22. The base margin for histogram graphs is now zero.

1.3.12   User interface

Stata 9 has a number of new features in the graphical user interface (GUI) that are shared across all platforms, such as multiple Viewer and Graph windows. There are also some significant improvements that affect only Windows, such as dockable windows. Most GUI features are documented in the *Getting Started Manual.*

1. New versions of Stata are available:

    a. Stata for Intel Itanium-based PCs running 64-bit Windows.

    b. Stata for x86-64 standard systems, including those based on AMD Opteron chips, Athlon-64 chips, and Intel Xeon emt64 chips running 64-bit Windows.

    c. Stata for Intel Itanium-based PCs running 64-bit Linux.

    d. Stata for x86-64 standard systems running 64-bit Linux.

2. Stata for Windows and Stata for Macintosh now have automatic update checking (nothing is ever downloaded without your confirmation). The first time you start Stata and every 7th day afterward, you will be prompted whether to check for updates.

    To control how often you are prompted, or to turn the feature off, select **Prefs > General Preferences**, and select **Internet**; or you can type `set update_interval #` or `set update_query off` at the Stata prompt; see [R] **update**.

3. Stata now allows multiple Viewer windows so that you can, for example, simultaneously view the help for several commands and the results from several logs or search queries.

    There are several ways to open another Viewer window.

    a. While viewing something in a Viewer, hold down the shift key, and click on any link. A new Viewer will appear displaying the contents of the link.

    b. Right-click on the link, and choose **Open Link in New Viewer**. That does the same thing.

    c. Click with the middle mouse button on the link. That also does the same thing.

    d. Right-click anywhere in an open Viewer, and choose **Open New Viewer**. This will open a new Viewer displaying `help contents`.

    See [GS] **5 Using the Viewer**.

4. The Viewer also has the following new features:

    a. It supports links within documents, including help files. You will see this feature used extensively in Stata's online help.

    b. It has the ability to search for text within the window. Click on the find icon that looks like a pair of binoculars at the top right of the Viewer.

    c. It now remembers its position in the document when you click **Refresh**.

    In addition, both the Viewer and Results windows no longer underline links when they are displayed on a white background. You can change this by selecting **Prefs > General Preferences**.

5. Stata now allows multiple Graph windows. The existing `name()` option of [G] **graph** now creates a named graph and displays it in its own window of the same name.

    Graph-management commands do what you would expect with the named windows; `graph drop` drops the graph and closes its window; `graph rename` renames both the graph and its window; and so on. Note that closing a Graph window does not delete the underlying graph and the graph can be redisplayed with `graph display`.

6. The **Window** menu now supports multiple Viewer and Graph windows:

   a. You can switch to specific Viewers or Graphs from this menu.

   b. Menu item **Window > Viewer > Close All Viewers** closes the Viewers.

   c. Menu item **Window Graph > Close All Graphs** closes the Graphs.

7. There are a number of enhancements to the toolbar:

   a. The **Open** button now has a menu that shows recently opened datasets and allows you to reopen those datasets with a click. This even includes datasets loaded over the web from **File > Example Datasets...** or with webuse.

   b. The **Print** button has a new menu that lets you select the window to print.

   c. The **Viewer** button lets you switch to any Viewer or close all Viewers.

   d. The **Graph** button lets you switch to any graph or close all Graphs.

   e. The **Do-file Editor** button lets you switch to any Do-file Editor (Windows and Macintosh).

8. A number of new features and improvements are available under the **File** menu:

   a. Recently opened datasets can now be reopened by selecting **File > Open Recent**, and recently opened do-files or ado-files can likewise be reopened from within the Do-file Editor by selecting **File > Open Recent**.

   b. **File > Print** lets you select the window to print.

   c. All the datasets shipped with Stata and all the datasets used in the examples in the manuals can be browsed and loaded by selecting **File > Example Datasets...**

9. Stata now allows multiple Do-file Editors under Windows and Macintosh. See [GS] **14 Using the Do-file Editor**.

10. Contextual menus for common tasks, such as setting preferences, copying to the clipboard, and printing, are now available in all windows; right-click in the window to access them.

11. You can now define multiple windowing preferences and switch easily among those preferences. For example, you might use small fonts and large Review and Variables windows for your normal work but use large fonts with hidden Review and Variables windows for presentation. Access this new feature by selecting **Prefs > Manage Preferences**.

12. The **Data Editor** has several enhancements:

   a. The contents of string variables and variables with value labels are now shown in different colors so that they can be easily distinguished.

   b. Variables with value labels can now be displayed as either the values of the variables or the labels.

   c. For variables with value labels, you now may change the value of a variable by right-clicking on the cell and selecting **Select Value from Value Label**. You may then select a value and label from a list.

   d. You may now associate an existing value label with a variable by right-clicking on the variable's column and selecting a value label from **Assign Value Label to Variable**.

   e. You may now define or modify value labels from within the Data Editor by right-clicking and selecting **Define/Modify Value Labels...**.

   f. You can now access and modify the preferences for the Data Editor by right-clicking in the editor and selecting **Preferences...**.

13. Dialogs have new features:

   a. Keyboard shortcuts for **Copy**, **Paste**, and **Cut** now work.

   b. Anywhere that you need to select a variable or variables for a *varlist*, you may now select those variables from a drop-down list (Windows and Macintosh).

   c. The new copy button will copy the command built by the dialog to the clipboard. The button appears just right of the refresh button at the bottom left of each dialog. It works just like **Submit**, but rather than executing the command, it pastes the command.

   d. Pressing the **Return** key now works the same as clicking **OK**; pressing **Shift+Return** works the same as clicking **Submit**. Pressing the **Escape** key works the same as clicking **Cancel**.

   e. Pressing the space bar when the keyboard focus is on a radio button works the same as clicking on the radio button.

   f. Keyboard arrow keys now work with dialog spinner controls.

   g. Estimation-command dialogs are laid out better, with the model specification always appearing on the **Model** tab. You can also now select standard error (SE) types with a single click in the **SE/Robust** tab (which includes bootstrap and jackknife SEs as options for most estimators).

   h. The twoway graph dialog boxes are laid out better, with easier selection of the plottype (scatter, line, range bar, etc.) and the addition of the new paired-coordinate time-series plottypes.

   In addition, the printed manuals and online documentation do a better job of describing the options and controls available on a dialog. The option entries in the manual and online are grouped into categories that match the tabs on the dialog.

14. Stata for Windows has vastly improved flexibility for managing your work environment:

   a. Most windows—the dockable ones—can now be docked with the main Stata window or with each other. By dragging a dockable window over another dockable window, you may create either a single-paned window, containing both the original windows with a separator in between or a single window with tabs for each of the original windows. The Viewer, Command, Review, and Variables windows are all dockable.

   In addition, any of these windows can either be attached (docked) to the main Stata window or detached and made free floating. Each also has a **pin** icon in the title bar that makes the window always shown or makes it roll up into its title bar when undocked, or makes it shown only as a tab when docked. For an overview of these features, see [GS] **4. The Stata user interface**.

   b. Most windows can be moved outside the main Stata window. These include the Graph, Viewer, Browse, and Edit windows and include all dialogs.

   c. The toolbar can be detached and repositioned.

   d. Double-clicking the Results window, when it is docked, merges it with the main Stata window as the primary document. This saves some screen real estate, and we suggest that you try it. Double-click again to undo it.

   e. A number of new window preferences available from the **Windowing** tab under **Prefs > General Preferences...** let you control how windows behave and how they dock. You can lock paned windows so that they cannot be resized, turn on or off docking, turn on or off the docking guides, make all windows floating, make the contents of Viewers persistent so that they maintain their contents between Stata sessions, and even turn off all the advanced windowing features to lock your current settings.

f. As with Stata on all other platforms, you can now save multiple windowing preferences and choose the one most appropriate for what you are doing, e.g., working at home, giving a presentation, etc.

If you are fond of the way Stata for Windows worked prior to Stata 9, or if you like to maximize your Stata window, we suggest that you select from the menu **Prefs > Manage Preferences > Load Preferences > Maximized**. Even so, we recommend that you try using the new layout without maximizing the Stata window.

15. You may now copy the **Review** window to the Clipboard. Right-click in the window to access the contextual menu.

16. `help` now displays in the Viewer window; new command `chelp` displays in the Results window. `help` also has two new options:

   a. `nonew` displays help in the topmost Viewer rather than in a new one.

   b. `name(`*viewername*`)` displays the help in the specified Viewer. If that name does not exist, a new Viewer will be created with that name.

17. You may now define and access `notes` for a variable by right-clicking on the variable name in the Variables window. Right-clicking on an empty space allows you to define and access notes for the dataset.

18. The Do-file Editor has a new SMCL preview button on its toolbar that displays the current file in the Viewer as rendered SMCL.

19. (Windows and Macintosh) You can now copy selected text as an HTML table using **Edit > Copy Table as HTML**.

20. (Unix) The minimize keyboard shortcut **<Ctrl>-m** has been added to all windows.

21. (Unix) You can now use the Window menu's keyboard shortcuts from any window.

22. (Macintosh) You can now increase or decrease the font size in a window by pressing **Apple +** and **Apple -**.

23. (Macintosh) The ability to undo or redo multiple actions has been added to the Do-file Editor.

24. (Macintosh) You can now have Stata automatically bring all windows to the front when it is active by selecting **Prefs > General Preferences...**.

25. (Macintosh) You can now have Stata automatically snap windows to the edge of the main Stata window or to the edge of the screen when you move or resize them. The setting is under **Prefs > General Preferences...**.

26. (Macintosh) You can move all of Stata's currently open windows simultaneously by holding down the Control key while dragging one of the windows. This will also bring all of Stata's open windows to the foreground.

27. (Macintosh) The toolbar may be a floating window or may be anchored to the menubar. The advantage of making the toolbar float is that it takes up less room on the screen and can be moved. You can access this feature using **Window > Toolbar**.

1.3.13   Programming

1. Mata, Stata's new matrix-programming language, can be used to code ado-file subroutines; see [M-1] **ado**.

2. New command viewsource displays official and user-written source code. viewsource searches for the specified file along the adopath and displays the file in the Viewer. This works not only for ado programs, but also for Mata functions that are programmed themselves in Mata. See [P] **viewsource**.

3. Programmers of estimation commands or commands that work with estimation results can tie postestimation analysis facilities into estat, making their postestimation facilities behave just like those shipped with Stata; see help estat programming.

4. New command matlist provides extensive format control for displaying a matrix; see [P] **matlist**.

5. Macro-extended functions that work on matrices now work on the matrices stored in r() and e(), including e(b) and e(V). These extended functions include rownames, colnames, roweq, coleq, rowfullnames, and colfullnames. See [P] **matrix**.

6. c() (c-class returned values) has the following new items:

item	description
c(Wdays)	Sun Mon ... Sat
c(Weekdays)	Sunday Monday Tuesday ... Saturday
c(alpha)	a b c d e f h j ... x y z
c(ALPHA)	A B C D E F H J ... X Y Z
c(Mons)	Jan Feb ... Dec
c(Months)	January February March ... December
c(tracehilite)	pattern to be highlighted in trace log
c(maxiter)	maximum iterations for maximum likelihood estimators
c(varabbrev)	whether variable abbreviation is on

7. A program can now be assigned properties when the program is declared, and those properties can be checked using macro extended functions. Specifically,

   a. program has the new option properties(), which attaches properties to programs; see [P] **program**.

   b. A new properties macro extended function allows programmers to obtain the list of properties attached to a program; see [P] **macro**.

   To learn more, see [P] **program properties**.

8. Estimation results can now be assigned properties using new option properties() of ereturn post and ereturn repost. These property settings can be checked with the new function has_eprop(). See [P] **ereturn** and [D] **functions**.

9. ereturn post now allows posting results without a beta vector, e(b), or a covariance matrix, e(V).

10. version has new option born() to prevent the program from running if the date of the Stata executable is earlier than the specified date. version also issues more-descriptive error messages. See [P] **version**.

11. On Microsoft Windows and Unix platforms, the new command window manage maintitle allows you to reset the main title of the Stata window; see manage maintitle under [P] **window programming**.

12. New command `levelsof` displays a sorted list of the distinct values of a variable. This is especially useful for looping over the values of a variable with, say, `foreach`. See [P] **levelsof**.

13. Plugins (also known as DLLs or shared objects) written in C can now be incorporated into Stata to create new Stata commands; see [P] **plugin**.

14. The maximum number of description lines in a `stata.toc` file has been increased from 10 to 50; see [R] **net**

15. New undocumented command `_coef_table` is a programmer's tool for displaying coefficient tables; see `help _coef_table`.

16. `trace` has new setting `set tracehilite` to highlight a specified pattern in the trace output; see [P] **trace**.

17. The functionality of `macval()` has been extended to macro dereferencing of values in a class. For example, `'macval(.a.b.c)'` causes the class reference `.a.b.c` to be macro expanded only once, rather than being recursively re-expanded when the result itself contained a macro reference.

18. Variable abbreviation can now be turned on and off using the new `set varabbrev`; see [R] **set** or type `help set varabbrev`.

19. Command `syntax` has new specifier `syntax anything(everything)` that specifies that `anything` include `if`, `in`, and `using`; see [P] **syntax**.

20. Command `syntax` has new option descriptor `cilevel` that restricts valid arguments to a standard confidence level and issues appropriate error messages for invalid entries; see [P] **syntax**.

21. A number of new directives and extensions to existing directives have been added to SMCL. See [P] **smcl** for complete documentation.

    See *What's new* in [P] for a table of new tags.

22. Existing command `window manage` has the following changes and additions:

    `window manage close graph` [*graphname* | `_all`]
        closes the graph window named *graphname*, if it exists. Specifying `_all` closes all Graph windows.

    `window manage forward graph` [*graphname*]
        now brings the Graph window named *graphname* to the top of the other windows and otherwise works as before.

    `window manage close viewer` [*viewername* | `_all`]
        closes the Viewer window named *viewername*. Specifying `_all` closes all Viewer windows.

    `window manage forward viewer` [*viewername*]
        now brings the Viewer window named *viewername* to the top of other windows and continues to work as before when no *viewername* is specified.

    `window manage minimize`
        minimizes the main Stata window.

    `window manage restore`
        restores the main Stata window, if it is minimized.

23. Existing command `window menu` now has new subcommand `append_recentfiles` to add `.dta` or `.gph` files to the **Open Recent** menu.

24. Existing command `confirm variable` has new option `exact` that disallows variable abbreviations.

25. New command `svymarkout` resets the value of a supplied 0/1 variable to 0 when any of the survey-characteristic variables set by `svyset` contain missing values; see [SVY] **svymarkout**.

26. Help files now allow include files. Syntax is INCLUDE help *helptopic* to include file *helptopic*.ihlp.

27. String scalars are now supported, meaning that a scalar can contain either a numeric or string value. The maximum length of a string scalar is the same as the maximum length of a string—80 characters in Small or Intercooled Stata and 512 characters in Stata/SE. See [P] **scalar**.

28. In addition to coding "local x : all scalars" to obtain a list of all defined scalars, you can now code "local x : all numeric scalars" and "local x : all string scalars" to obtain the list restricted to numeric or string scalars. See [P] **macro**.

29. In macro expansion, double backslash (\\) used to become single backslash (\). Now (but under version control) it becomes single backslash only if the second backslash precedes macro-expansion punctuation (` or $).

1.3.14   Documentation

1. There are new manuals: **[D] Data Management**, **[MV] Multivariate Statistics**, and **[M] Mata**.

2. Documentation (printed and online) groups related options into categories. In addition, the categories match the tabs on dialog boxes.

3. For all estimation commands, there is now an entry called postestimation following the estimation command. For instance, following [R] **regress** is [R] **regress postestimation**. The postestimation entry documents command-specific postestimation facilities to further analyze the results and also directs you to other relevant postestimation features.

   In the online help system, go to help for the estimation command, and click on postestimation in the upper-right corner.

4. There are now glossaries in the [M], [SVY], [TS], and [XT] manuals. The glossaries define commonly used terms and explain how these terms are used in the documentation.

5. Stata's help command and online help facility have new features:

   a. Spaces and colons are now allowed in help topics, for example, help graph intro, help regress postestimation, or help svy: logistic (with or without the colon).

   b. Typing help sqrt() now gives you help for Stata's sqrt() function. Typing help mata sqrt() gives you help for Mata's sqrt() function.

   c. Many command abbreviations are now recognized; for example, help reg post is understood to mean help regress postestimation, and help tw con is understood to mean help graph twoway connected.

1.3.15   What's more

We have not listed all the changes, but we have listed the important ones. The remaining changes—a list of about equal length as the one above—are all implications of what has been listed.

What is important to know is that Stata is continually being updated and those updates are available for free over the Internet. All you have to do is type

    . update query

and follow the instructions.

To learn what has been added since this manual was printed, select **Help > What's new?** or type

    . help whatsnew

We hope that you enjoy Stata 9.

1.4   References

Heyde, C. C. and E. Seneta, ed. 2001. *Statisticians of the Centuries*. New York: Springer.

Johnson, N. L. and S. Kotz, ed. 1997. *Leading Personalities in Statistical Sciences*. New York: Wiley.

Upton, G. and I. Cook. 2002. *A Dictionary of Statistics*. Oxford: Oxford University Press.

Gould, W., J. Pitblado, and W. Sribney. 2003. *Maximum Likelihood Estimation with Stata*. 2nd ed. College Station, TX: Stata Press.

2 A brief description of Stata

Stata is a statistical package for managing, analyzing, and graphing data.

Stata is available for a variety of platforms. Stata may be used either as a point-and-click application or as a command-driven package.

Stata's GUI provides an easy interface for those new to Stata, and for experienced Stata users who wish to execute a command that they seldom use.

The command language provides a fast way to communicate with Stata, and a way to communicate more complex ideas.

Here is an extract of a Stata session using the GUI:

(Throughout the Stata manuals, we will refer to various datasets. These datasets are all available from *http://www.stata-press.com/data/r9/*. For easy access to them from within Stata, you may type `webuse` *dataset_name*, or select **File > Example Datasets...** and click on Stata 9 manual datasets.)

```
. webuse lbw2
(Hosmer & Lemeshow data)
```

We select **Data > Describe data > Summary statistics**, and choose to summarize variables `low`, `age`, and `smoke`, which names we obtained from the Variables window. We press **OK**.

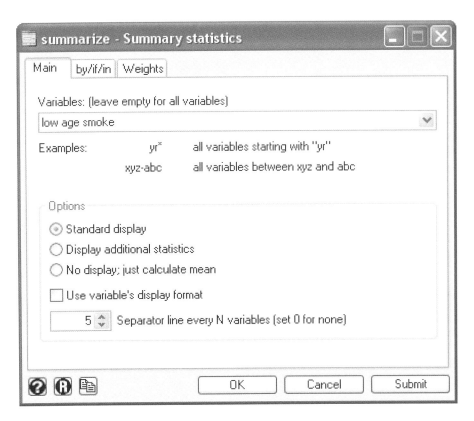

```
. summarize low age smoke
    Variable |       Obs        Mean    Std. Dev.       Min        Max
-------------+--------------------------------------------------------
         low |       189    .3121693    .4646093          0          1
         age |       189     23.2381    5.298678         14         45
       smoke |       189    .3915344    .4893898          0          1
```

Stata shows us the command that we could have typed in command mode—summarize low age smoke—before displaying the results of our request.

Next, we fit a logistic regression model of low on age and smoke. We select **Statistics > Binary outcomes > Logistic regression**, fill in the fields, and press **OK**.

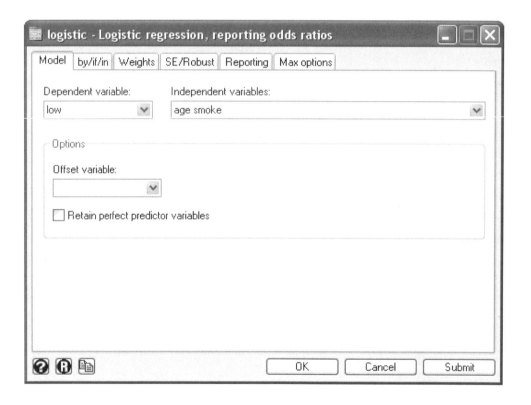

```
. logistic low age smoke
Logistic regression                             Number of obs   =        189
                                                LR chi2(2)      =       7.40
                                                Prob > chi2     =     0.0248
Log likelihood = -113.63815                     Pseudo R2       =     0.0315

-------------------------------------------------------------------------------
         low | Odds Ratio   Std. Err.      z    P>|z|     [95% Conf. Interval]
-------------+-----------------------------------------------------------------
         age |  .9514394   .0304194    -1.56   0.119     .8936481    1.012968
       smoke |  1.997405    .642777     2.15   0.032     1.063027    3.753081
-------------------------------------------------------------------------------
```

Here's an extract of a Stata session using the command language:

```
. use http://www.stata-press.com/data/r9/auto
(1978 Automobile Data)

. summarize mpg weight
```

Variable	Obs	Mean	Std. Dev.	Min	Max
mpg	74	21.2973	5.785503	12	41
weight	74	3019.459	777.1936	1760	4840

The user typed `summarize mpg weight` and Stata responded with a table of summary statistics. Other commands would produce different results:

```
. correlate mpg weight
(obs=74)
```

	mpg	weight
mpg	1.0000	
weight	-0.8072	1.0000

```
. gen w_sq = weight^2

. regress mpg weight w_sq
```

Source	SS	df	MS		
Model	1642.52197	2	821.260986	Number of obs = 74
Residual	800.937487	71	11.2808097	F(2, 71) = 72.80
				Prob > F = 0.0000
				R-squared = 0.6722
				Adj R-squared = 0.6630
Total	2443.45946	73	33.4720474	Root MSE = 3.3587

mpg	Coef.	Std. Err.	t	P>|t|	[95% Conf. Interval]
weight	-.0141581	.0038835	-3.65	0.001	-.0219016 -.0064145
w_sq	1.32e-06	6.26e-07	2.12	0.038	7.67e-08 2.57e-06
_cons	51.18308	5.767884	8.87	0.000	39.68225 62.68392

```
. scatter mpg weight, by(foreign, total row(1))
```

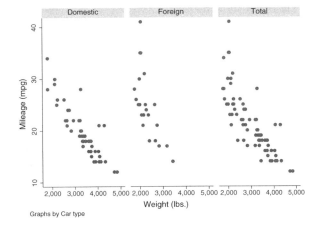

The user-interface model is type a little, get a little, etc., so that the user is always in control.

Stata's model for a dataset is that of a table—the rows are the observations and the columns are the variables:

```
. list mpg weight in 1/10
```

	mpg	weight
1.	22	2,930
2.	17	3,350
3.	22	2,640
4.	20	3,250
5.	15	4,080
6.	18	3,670
7.	26	2,230
8.	20	3,280
9.	16	3,880
10.	19	3,400

Observations are numbered; variables are named.

Stata is very fast. Partly, that speed is due to clever programming, and, partly, it is because Stata keeps the data in memory. Stata's data model is that of a word processor: a dataset may exist on disk, but that is just a copy. The dataset is loaded into memory, where it is worked on, analyzed, changed, and then perhaps stored back on disk.

Working on a copy of the data in memory makes Stata safe for interactive use. The only way to harm the permanent copy of your data on disk is if you explicitly save over it.

Having the data in memory means that the dataset size is limited by the amount of memory. Stata stores the data in memory in a very compressed format—you will be surprised how much data can fit into a given region of memory. Nevertheless, if you work with large datasets, you may run into memory constraints. There are two solutions to this problem:

1. By default, Stata/SE allocates 10 megabytes to Stata's data areas, and you can change it; see [U] **6 Setting the size of memory**.

   By default, Intercooled Stata allocates 1 megabyte to Stata's data areas, and you can change it; see [U] **6 Setting the size of memory**.

   By default, Small Stata allocates about 300K to Stata's data areas, and you cannot change it.

2. You will want to learn how to compress your data as much as possible; see [D] **compress**.

3 Resources for learning and using Stata

Contents

3.1   Overview

The *Getting Started* manual, *User's Guide*, and *Reference* manuals are the primary tools for learning about Stata; however, there are many other sources of information. A few are

1. Stata itself. Stata has a subject table-of-contents online with links to the help system and dialog boxes that make it easy to find and to execute a Stata command. See [U] **4 Stata's online help and search facilities**.

2. The Stata web site. Visit *http://www.stata.com*. Much of the site is dedicated to user support; see [U] **3.2 The http://www.stata.com web site**.

3. The Stata Press web site. Visit *http://www.stata-press.com*. This site contains the datasets used throughout the Stata manuals; see [U] **3.3 The http://www.stata-press.com web site**.

4. The Stata listserver. An active group of Stata users communicate over an Internet listserver, which you can join for free; see [U] **3.4 The Stata listserver**.

5. The Stata software distribution site and other user-provided software distribution sites. Stata itself can download and install updates and additions. We provide official updates to Stata—type `update` or select **Help > Official Updates**. We also provide user-written additions to Stata and links to other user-provided sites—type `net` or select **Help > SJ and User-written Programs**; see [U] **3.6 Updating and adding features from the web**.

6. The *Stata Journal* and the *Stata Technical Bulletin*. The *Stata Journal* contains reviewed papers, regular columns, book reviews, and other material of interest to researchers applying statistics in a variety of disciplines. The *Stata Technical Bulletin*, the predecessor to the *Stata Journal*, contains articles and user-written commands. See [U] **3.5 The Stata Journal and the Stata Technical Bulletin**.

7. NetCourses. We offer training via the Internet. Details are in [U] **3.7 NetCourses** below.

8. Books and support materials. Supplementary Stata materials are available; see [U] **3.8 Books and other support materials**.

9. Technical support. We provide technical support by email, telephone, and fax; see [U] **3.9 Technical support**.

3.2   The http://www.stata.com web site

Point your browser to *http://www.stata.com* and click on **User Support**. Over half of our web site is dedicated to providing support to users.

1. The web site provides FAQs (Frequently Asked Questions) on Windows, Macintosh, Unix, statistics, programming, Internet capabilities, graphics, and data management. These FAQs run the gamut from "I cannot save/open files" to "What does 'completely determined' mean in my logistic-regression output?" Everyone will find something of interest.

2. Visiting the web site is one way that you can subscribe to the Stata listserver; see [U] **3.4 The Stata listserver**.

3. The web site provides detailed information about NetCourses, along with the current schedule; see [U] **3.7 NetCourses**.

4. The web site provides information about Stata courses and meetings, both in the United States and elsewhere.

5. The web site provides an online bookstore for Stata-related books and other supplementary materials; see [U] **3.8 Books and other support materials**.

6. The web site provides links to information about statistics: other statistical software providers, book publishers, statistical journals, statistical organizations, and statistical listservers.

7. The web site provides links to Stata resources for learning Stata at *http://www.stata.com/links/resources.html*. Be sure to look at these materials, as many valuable resources about Stata are listed here, including the UCLA Stata portal, which includes a set of links about Stata, and the SSC archive, which has become the premier Stata download site for user-written software on the web.

In short, the web site provides up-to-date information on all support materials and, where possible, provides the materials themselves. Visit *http://www.stata.com* if you can.

3.3   The http://www.stata-press.com web site

Point your browser to *http://www.stata-press.com*. This site is devoted to the publications and activities of Stata Press.

1. Datasets that are used in the Stata *Reference* manuals and other books published by Stata Press may be downloaded. Visit *http://www.stata-press.com/data*. These datasets can be used in Stata by simply typing use `http://www.stata-press.com/data/r9/auto`. Alternatively, you could type `webuse auto`; see [D] **webuse**.

2. An online catalog of all of our books and multimedia products is at *http://www.stata-press.com/catalog.html*. We have tried to include enough information, such as table of contents, preface material, etc., so that you may tell if the book is appropriate for you.

3. Information about forthcoming publications is posted at *http://www.stata-press.com/forthcoming.html*.

3.4   The Stata listserver

The Stata listserver (Statalist) is an independently operated, real-time list of Stata users on the Internet. Anyone may join. Instructions for doing so can be found at *http://www.stata.com* by clicking on User Support and then Statalist or by emailing *stata@stata.com*.

Many knowledgeable users are active on the list, as are the StataCorp technical staff. We recommend that new users subscribe, observe the exchanges, and, if it turns out not to be useful, unsubscribe.

3.5   The Stata Journal and the Stata Technical Bulletin

The *Stata Journal* (SJ) is a printed and electronic journal, published quarterly, containing articles about statistics, data analysis, teaching methods, and effective use of Stata's language. The *Journal* publishes reviewed papers together with shorter notes and comments, regular columns, tips, book reviews, and other material of interest to researchers applying statistics in a variety of disciplines. The *Journal* is a publication for all Stata users, both novice and experienced, with different levels of expertise in statistics, research design, data management, graphics, reporting of results, and of Stata, in particular.

Tables of contents for past issues and abstracts of the articles are available at *http://www.stata-journal.com/archives.html*.

We recommend that all users subscribe to the SJ. Visit *http://www.stata-journal.com* to learn more about the *Stata Journal* and to order your subscription.

To obtain any programs associated with articles in the SJ, type

```
. net from http://www.stata-journal.com/software
```

or

1. Select **Help > SJ and User-written Programs**
2. Click on *Stata Journal*

The Stata Technical Bulletin

For ten years, the *Stata Technical Bulletin* (STB) served as the means of distributing new commands and Stata upgrades, both user-written and "official". After ten years of continuous publication, the STB evolved into the *Stata Journal*. The Internet provided an alternative delivery mechanism for user-written programs, so the emphasis shifted from user-written programs to more expository articles. Although the STB is no longer published, many of the programs and articles that appeared in it are still valuable today. Reprints of past issues are available from *http://www.stata.com/bookstore/stbr.html*. To obtain the programs that were published in the STB, type

```
. net from http://www.stata.com
. net cd stb
```

(Continued on next page)

3.6   Updating and adding features from the web

Stata itself is web-aware.

First, try this:

```
. use http://www.stata.com/manual/oddeven.dta, clear
```

That will load an uninteresting dataset into your computer from our web site. If you have a homepage, you can use this feature to share datasets with coworkers. Save a dataset on your homepage, and researchers worldwide can use it. See [R] **net**.

3.6.1   Official updates

Although we follow no formal schedule for the release of updates, we typically provide updates to Stata approximately every two weeks. Installing the updates is easy. Type

```
. update
```

or select **Help > Official Updates**. Do not be concerned; nothing will be installed unless and until you say so. Once you have installed the update, you can type

```
. help whatsnew
```

or select **Help > What's New** to find out what has changed. We distribute official updates to fix bugs and to add new features.

3.6.2   Unofficial updates

There are also "unofficial" updates—additions to Stata written by Stata users, which includes members of the StataCorp technical staff. Stata is programmable, and even if you never write a Stata program, you may find these additions useful, and some of them spectacularly so. You start by typing

```
. net from http://www.stata.com
```

or select **Help > SJ and User-written Programs**.

Be sure to try visit the SSC-Archive. The `ssc` command makes it easy for you to install and uninstall packages from SSC. Type

```
. ssc whatsnew
```

to find out what's new at the site. If you find something that interests you, type

```
. ssc describe pkgname
```

for more information.

Periodically, you can type

```
. news
```

or select **Help > News** to display a short message from our web site telling you what is newly available.

See [U] **28 Using the Internet to keep up to date**.

3.7   NetCourses

We offer courses on Stata at the introductory and advanced levels. Courses on software are typically expensive and time-consuming. They are expensive because, in addition to the direct costs of the course, participants must travel to the course site. We have found it is better to organize courses over the Internet—saving everyone time and money.

We offer courses over the Internet and call them Stata NetCourses^{TM}.

1. **What is a NetCourse?**

    A NetCourse is a course offered through the Stata web site that varies in length from seven to eight weeks. You must have an email address and web browser to participate.

2. **How does it work?**

    Every Friday a "lecture" is posted on a password-protected web site. After reading the lecture over the weekend or perhaps on Monday, participants then post questions and comments on a message board. Course leaders typically respond to the questions and comments on the same day they are posted. The other participants are encouraged to amplify or otherwise respond to the questions or comments as well. The next lecture is then posted on Friday, and the process repeats.

3. **How much of my time does it take?**

    It depends on the course, but the introductory courses are designed to take roughly 3 hours per week.

4. **There are three of us here—can just one of us enroll and we redistribute the NetCourse materials ourselves?**

    We ask that you not. NetCourses are priced to cover the substantial time input of the Course Leaders. Moreover, enrollment is typically limited to prevent the discussion from becoming unmanageable. The value of a NetCourse, just like a real course, is the interaction of the participants, both with each other and with the Course Leaders.

5. **I've never taken a course by Internet before. I can see that it might work, but then again, it might not. How do I know I'll benefit?**

    All Stata NetCourses come with a 30-day satisfaction guarantee. The 30 days begin after the conclusion of the final lecture.

You can learn more about the current NetCourse offerings by visiting *http://www.stata.com*. Our offerings include

> NC-101 An introduction to Stata
> NC-151 An introduction to Stata programming
> NC-152 Advanced Stata programming

NetCourseNow

A NetCourseNow offers the same material as NetCourses, but it allows you to choose the time and pace of the course, and you have a personal NetCourse instructor.

1. **What is a NetCourseNow?**

    A NetCourseNow offers the same material as a Netcourse, but allows you to move at your own pace and to specify a starting data. With a NetCourseNow, you also have the added benefit of a personal NetCourse instructor whom you can email directly with questions about lectures and exercises. You must have an email address and a web browser to participate.

2. **How does it work?**

    All course lectures and exercises are posted at once, and you are free to study at your own pace. You will be provided with the email address of your personal NetCourse instructor to contact when you have questions.

3. **How much of my time does it take?**
   A NetCourseNow allosws you to set your own pace. How long the course takes and how much time you spend per week is up to you.

3.8   Books and other support materials

There are books published on Stata, both by us and by others. Visit the bookstore at *http://www.stata.com* for an up-to-date list and for the table of contents of each. For books that we carry in the Stata bookstore, we post a comment written by a member of our technical staff, explaining why we think this book might interest you.

3.9   Technical support

We are committed to providing superior technical support for Stata software. In order to assist you as efficiently as possible, please follow the procedures listed below.

3.9.1   Register your software

You must register your software in order to be eligible for technical support, updates, special offers, and other benefits. By registering, you will receive the *Stata News*, and you may access our support staff for free with any question that you encounter. You may register your software either electronically or by mail.

Electronic registration: After installing Stata and successfully entering your License and Authorization Key, your default web browser will open to the online registration form at the Stata web site. You may also manually point your web browser to *http://www.stata.com/register/* if you wish to register your copy of Stata at a later time.

Mail-in registration: Fill in the registration card that came with Stata and mail it to StataCorp.

3.9.2   Before contacting technical support

Before you spend the time gathering the information our technical support department needs, make sure that the answer does not already exist in the help files. You can use the `help` and `search` commands to find all the entries in Stata that address a given subject. Be sure to try selecting **Help > Contents**. Check the manual for a particular command. There are often examples that address questions and concerns. Another good source of information is our web site. You should keep a bookmark of our frequently asked questions page (*http://www.stata.com/support/faqs/*) and check it occasionally for new information.

Our technical department will need some information from you in order to provide detailed assistance. Most important is your serial number, but they will also need the following information:

1. The system information on the computer that you are using is especially important if you are having hardware problems. This includes the make and model of various hardware components such as the computer manufacturer, the video driver, the operating system and its version number, relevant peripherals, and the version number of any other software with which you experience a conflict.

2. The version of Stata that you are running. Type `about` at the Stata prompt, and Stata will display this information.

3. The types of variables in your dataset and the number of observations.

4. The command that is causing the error along with the exact error message and return code (error number).

3.9.3 Technical support by email

This is the preferred method of asking a technical support question. It has the following advantages:

- You will receive a prompt response from us saying that we have received your question and that it has been forwarded to *Technical Services* to answer.

- We are able to route your question to a specialist for your particular question.

- Questions submitted via email may be answered after normal business hours, or even on weekends or holidays. Although we cannot promise that this will happen, it may, and your email inquiry is bound to receive a faster response than leaving a message on Stata's voice mail.

- If you are receiving an error message or an unexpected result, it is easy to include a log file that demonstrates the problem.

Please see the FAQ at *http://www.stata.com/support/faqs/techsup/* for some suggestions to follow that will aid *Technical Services* in promptly answering your question.

3.9.4 Technical support by phone or fax

Our technical support telephone number is 979-696-4600. Please have your serial number handy. It is also best if you are at your computer when you call. If your question involves an error message from a command, please note the error message and number, as this will greatly help us in assisting you. Telephone support is reserved for nonstatistical questions. If your question requires the attention of a statistician, the question should be submitted via email or fax.

Send fax requests to 979-696-4601. If possible, collect the relevant information in a log file and include the file in your fax.

Please see the FAQ at *http://www.stata.com/support/faqs/techsup/* for some suggestions to follow that will aid *Technical Services* in promptly answering your question.

3.9.5 Comments and suggestions for our technical staff

By all means, send in your comments and suggestions. Your input is what determines the changes that occur in Stata between releases, so if we don't hear from you, we may not include your most desired new estimation command! Email is preferred, as this provides us with a permanent copy of your request. When requesting new commands, please include any references that you would like us to review should we develop those new commands. Email your suggestions to *service@stata.com*.

4 Stata's online help and search facilities

Contents

4.1   Introduction

Stata has online help and a lot of it. You have two ways to access this:

1. Select **Help > Stata Command...**.

2. Type the `help` and `search` commands.

Stata for Unix(console) users only have the second approach available to them.

Both methods access the same underlying information, which is displayed in a Viewer window or in the Results window. Blue text indicates a hypertext link, so you can click on it to go directly to the help. Stata for Unix(console) users must type `help` followed by the command name or topic, and information will be displayed on the console.

4.2   help: Stata's online manual pages

The `help` command provides access to Stata's interactive help files. These files are a shortened version of what is in the printed manuals. Try it. Ask for help on the `help` command:

(Continued on next page)

53

```
. help help
```

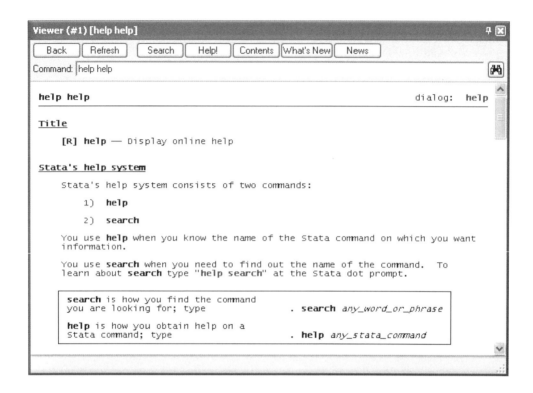

The help system consists of two commands: you use `help` when you know the name of the command and `search` when you need to find out the name of the command.

> **Rule:** If you know the name of the command and want to learn more about it, type `help` followed by the command name.

Try it. Two other Stata's commands are `use` and `regress`. Type `help use`. Type `help regress`.

Some commands in Stata consist of more than one word or contain nonalphanumeric characters. For instance, Stata's estimation commands for survey data contain two words and a colon, such as `svy: regress`. Stata is understanding: you can type `help svy: regress` or `help svy regress` (with or without the colon), or even `help svy reg`. Stata understands many command abbreviations, such as `reg` for `regress`.

You can also obtain help on a topic. For instance, you can obtain help on the postestimation commands that are allowed after, say, `logistic`, by typing `help logistic postestimation`, or just `help logistic post`. Stata understands that `post` is an abbreviation for `postestimation`.

4.3   search: Stata's online index

Our rule works fine when you know the name of the command, but what if you do not? In that case, you use `search`. `search`'s syntax is

`search` *anything you want*

or you may select **Help > Search**, check **Search documentation and FAQs**, and then type *anything you want* in the Keywords input field. Either way, `search` is very understanding. Here is what happens when you `search logistic regression`:

```
. search logistic regression
```

Keyword search

```
        Keywords:  logistic regression
          Search:  (1) Official help files, FAQs, Examples, SJs, and STBs
```

Search of official help files, FAQs, Examples, SJs, and STBs

```
[U]     Chapter 26 . . . . . . . . . . Overview of Stata estimation commands
        (help estcom)

[R]     clogit . . . . . . Conditional (fixed-effects) logistic regression
        (help clogit)

 (output omitted )

[R]     logistic . . . . . . . . Logistic regression, reporting odds ratios
        (help logistic)

[R]     logit . . . . . . . . . Logistic regression, reporting coefficients
        (help logit)

 (output omitted )

FAQ     . Standard error of the predicted probability with logistic regression
        . . . . . . . . . . . . . . . . . . . . . . . . . . . . R. Gutierrez
        3/01    How do I obtain the standard error of the predicted
                probability with logistic regression analysis?
                http://www.stata.com/support/faqs/stat/delta.html

FAQ     . . . . . . . . . . . . . . . . Logistic regression with grouped data
        . . . . . . . . . . . . . . . . . . . . . . . . . . . . . W. Sribney
        1/00    How can I do logistic regression or multinomial
                logistic regression with grouped data?
                http://www.stata.com/support/faqs/stat/grouped.html

 (output omitted )

SJ-5-1  st0081   Visualizing main effects & interactions for binary logit mod.
        . . . . . . . . . . . . . . . . . . . . M. N. Mitchell and X. Chen
        (help vibl, viblicc, viblidb, vibligraph, viblmcc, viblmdb,
        viblmgraph if installed)
        Q1/05   SJ 5(1):64--82
        presents new package vibl as visualization tool for
        interpreting main effects and interactions in logit models
        when using predicted probabilities

SJ-4-4  st0075   . Controlling time-dep. confounding using marg. struct. models
        . Z. Fewell, M.A. Hernan, F. Wolfe, K. Tilling, H. Choi, J.A.C. Sterne
        Q4/04   SJ 4(4):402--420                         (no commands)
        describes the use of marginal structural models to
        estimate exposure or treatment effects in the presence
        of time-dependent confounders affected by prior treatment

 (output omitted )

STB-53  sg124 . . . . . . Interpreting logistic regression in all its forms
        . . . . . . . . . . . . . . . . . . . . . . . . . . . . . W. Gould
        1/00    pp.19--29; STB Reprints Vol 9, pp.257--270    (no commands)
        the interpretation of ordinary, conditional, ordered, and
        multinomial logistic regression is explained; guidance is
        provided in interpreting odds ratios and risk ratios

 (output omitted )
```

search responds by providing a list of references—references to the online help, references to the printed documentation, references to FAQs at the *http://www.stata.com* web site, references to articles that have appeared in the *Stata Journal* (SJ), and references to articles that have appeared in the *Stata Technical Bulletin* (STB). Moreover, if you install the official updates—see [U] **28 Using the Internet to keep up to date**—the references to the FAQs, SJs, and STBs will even be up to date.

Anyway, you are supposed to look over the list and find what looks relevant to you. If you are really interested in logistic regression, `logistic` and `logit` seem particularly appropriate, so you might next type `help logistic` and `help logit` or click on the blue hypertext links for the `logistic` and `logit` help files.

4.4   Accessing help and search from the Help menu

You will get exactly the same output if you select **Help > Search**, click on **Search documentation and FAQs**, and type `logistic regression` in the Keyword(s) Search box.

There is an advantage because a separate window, the Viewer, is invoked instead of presenting the resulting output in the Results window. The Viewer can remain open as a reference while you continue to use Stata.

Some of the resulting output will be highlighted in blue to indicate hyperlinks. *help logit* and *help logistic*, for instance, will be displayed in blue. Click on either one, and you will go to the help file for the command.

There will be other places you can click, too. When `search` mentions a *Stata Journal* or STB article, the insert number (such as *st0001* or *pr0006* for the *Stata Journal* or *sbe14* or *sg63* for the STB) will be in blue. Click on one, and Stata will go to *http://www.stata-journal.com* (or *http://www.stata.com*) and show you a detailed description of the addition, leaving you just one click away from installing the addition if it interests you.

When `search` mentions a FAQ, the URL will be in blue. Click on that, and Stata will launch your browser, and you will be looking right at the answer to the frequently asked question.

You can pull down **Help** at any time, not just when Stata is idle. You can leave the Viewer up while you use Stata, which is especially convenient when viewing syntax diagrams and options.

4.5   More on search

However you access `search`—command or menu—it does the same thing. You tell `search` what you want information about, and it searches for relevant entries. If you want `search` to look for the topic across all sources, including the online help, the FAQs at the Stata web site, the *Stata Journal*, and all Stata-related Internet sources including user-written additions, then use `findit`, which is a synonym for `search, all`.

`search` can be used broadly or narrowly. For instance, if you want to perform the Kolmogorov–Smirnov test for equality of distributions, you could type

```
. search Kolmogorov-Smirnov test of equality of distributions

[R]     ksmirnov . . . . . . Kolmogorov-Smirnov equality of distributions test
        (help ksmirnov)
```

In fact, we did not have to be nearly so complete—typing `search Kolmogorov-Smirnov` would have been adequate. Had we specified our request more broadly—looking up `equality of distributions`—we would have obtained a longer list that included `ksmirnov`.

Here are guidelines for using `search`.

1. Capitalization does not matter. Look up Kolmogorov-Smirnov or kolmogorov-smirnov.

2. Punctuation does not matter. Look up kolmogorov smirnov.

3. Order of words does not matter. Look up smirnov kolmogorov.

4. You may abbreviate, but how much depends. Break at syllables. Look up kol smir. search tends to tolerate a lot of abbreviation; it is better to abbreviate than to misspell.

5. The prepositions for, into, of, on, to, and with are ignored. Use them—look up equality of distributions—or omit them—look up equality distributions—it makes no difference.

6. search is tolerant of plurals, especially when they can be formed by adding an s. Even so, it is better to look up the singular. Look up normal distribution, not normal distributions.

7. Specify the search criterion in English, not in computer jargon.

8. Use American spellings. Look up color, not colour.

9. Use nouns. Do not use -ing words. Look up median tests, not testing medians.

10. Use few words. Every word specified further restricts the search. Look up distribution, and you get one list; look up normal distribution, and the list is a sublist of that.

11. Sometimes words have more than one context. The following words can be used to restrict the context:

    a. data, meaning in the context of data management. Order could refer to the order of data or to order statistics. Look up order data to restrict order to its data-management sense.

    b. statistics (abbreviation stat), meaning in the context of statistics. Look up order statistics to restrict order to the statistical sense.

    c. graph or graphs, meaning in the context of statistical graphics. Look up median graphs to restrict the list to commands for graphing medians.

    d. utility (abbreviation util), meaning in the context of utility commands. The search command itself is not data management, not statistics, and not graphics; it is a utility.

    e. programs or programming (abbreviation prog), to mean in the context of programming. Look up programming scalar to obtain a sublist of scalars in programming.

search has other features, as well; see [U] **4.8 search: All the details**.

4.6   More on help

Both help and search are understanding of some mistakes. For instance, you may abbreviate a command name. If you type either help regres or help regress, you will bring up the help file for regress.

When help cannot find the command you are looking for, try the search feature. In this case, typing search regres will also find the command (because 'regres' is an abbreviation of the word regression), but, in general, that will not be the case.

Stata can run into some problems with abbreviations. For instance, Stata has a command with the inelegant name ksmirnov. You forget and think the command is called ksmir:

```
. help ksmir
help for ksmir not found
try help contents or search ksmir
```

This is a case where help gives bad advice because typing search ksmir will do you no good. You should type search followed by what you are really looking for: search kolmogorov smirnov.

4.7   help contents: Stata's online table of contents

Typing help contents or selecting **Help > Contents** provides another way of locating entries in the documentation and online help. Either way, you will be presented with a long table of contents, organized topically.

```
. help contents
```

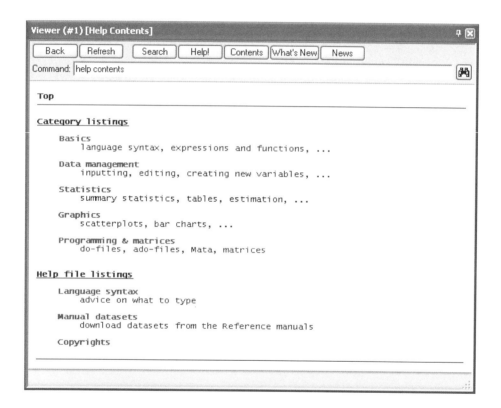

4.8   search: All the details

The search command actually provides a few features that are not available from the **Help** menu. The full syntax of the search command is

search *word* [*word* ...] [, [local | net | all] <u>au</u>thor <u>en</u>try <u>ex</u>act faq

<u>h</u>istorical or <u>man</u>ual sj]

where <u>under</u>lining indicates the minimum allowable abbreviation and [brackets] indicate optional.

local, the default (unless changed by set searchdefault), specifies that the search be performed using only Stata's keyword database.

net specifies that the search be performed across the materials available via Stata's net command. Using search *word* [*word* ...], net is equivalent to typing net search *word* [*word* ...] (without options); see [R] **net**.

all specifies that the search be performed across both the local keyword database and the net materials.

author specifies that the search be performed on the basis of author's name rather than keywords.

entry specifies that the search be performed on the basis of entry IDs rather than keywords.

exact prevents matching on abbreviations.

faq limits the search to entries found in the FAQs at *http://www.stata.com*.

historical adds to the search entries that are of historical interest only. By default, such entries are not listed. Past entries are classified as historical if they discuss a feature that later became an official part of Stata. Updates to historical entries will always be found, even if historical is not specified.

or specifies that an entry be listed if any of the words typed after search are associated with the entry. The default is to list the entry only if all the words specified are associated with the entry.

manual limits the search to entries in the *User's Guide* and all the *Reference* manuals.

sj limits the search to entries in the *Stata Journal* and the *Stata Technical Bulletin*.

4.8.1   How search works

search has a database—a file—containing the titles, etc. of every entry in the *User's Guide*, *Reference* manuals, articles in the *Stata Journal* and in the *Stata Technical Bulletin*, and FAQs at *http://www.stata.com*. In this file is a list of words associated with each entry, called keywords.

When you type search *xyz*, search reads this file and compares the list of keywords with *xyz*. If it finds *xyz* in the list or a keyword that allows an abbreviation of *xyz*, it displays the entry.

When you type search *xyz abc*, search does the same thing but displays an entry only if it contains both keywords. The order does not matter, so you can search linear regression or search regression linear.

Obviously, how many entries search finds depends on how the search database was constructed. We have included a plethora of keywords under the theory that, for a given request, it is better to list too much rather than risk listing nothing at all. Still, you are in the position of guessing the keywords. Do you look up normality test, normality tests, or tests of normality? Normality test would be best, but all would work. In general, use the singular, and strike the unnecessary words. We provide guidelines for specifying keywords in [U] **4.5 More on search** above.

4.8.2 Author searches

search ordinarily compares the words following search with the keywords for the entry. If you specify the author option, however, it compares the words with the author's name. In the search database, we have filled in author names for *Stata Journal* and STB articles and for FAQs.

For instance, in [R] **kdensity**, you will discover that Isaías H. Salgado-Ugarte wrote the first version of Stata's kdensity command and published it in the STB. Assume that you have read his original and find the discussion useful. You might now wonder what else he has written in the STB. To find out, type

        . search Salgado-Ugarte, author
        (*output omitted*)

Names like Salgado-Ugarte are confusing to some people. search does not require you specify the entire name; what you type is compared with each "word" of the name, and, if any part matches, the entry is listed. The dash is a special character, and you can omit it. Thus, you can obtain the same list by looking up Salgado, Ugarte, or Salgado Ugarte without the dash.

Actually, to find all entries written by Salgado-Ugarte, you need to type

        . search Salgado-Ugarte, author historical
        (*output omitted*)

Prior inserts in the STB that provide a feature that later was superseded by a built-in feature of Stata are marked as historical in the search database and, by default, are not listed. The historical option ensures that all entries are listed.

4.8.3 Entry ID searches

If you specify the entry option, search compares what you have typed with the entry ID. The entry ID is not the title—it is the reference listed to the left of the title that tells you where to look. For instance, in

        [R]   regress . . . . Linear regression
                (help **regress**)

"[R] regress" is the entry ID. In

        GS   . . . . Getting Started manual

"GS" is the entry ID. In

        SJ-4-4   **st0078** . . . . Boolean logit and probit in Stata
                 (help **mlboolean** if installed) . . . . B. F. Braumoeller
                 Q4/04   SJ 4(4): 436--441
                 discusses Boolean logit and probit models that allow
                 researchers to model binary outcomes as the results of
                 Boolean interactions among interdependent causal processes

"SJ-4-4 st0078" is the entry ID.

search with the entry option searches these entry IDs.

Thus, you could generate a table of contents for the *Reference* manuals by typing

        . search [R], entry
        (*output omitted*)

You could generate a table of contents for the 16th issue of the STB by typing

        . search STB-16, entry historical
        (*output omitted*)

The `historical` option in this case is possibly important. STB-16 was published in November 1993, and perhaps some of its inserts have been marked as historical.

You could obtain a complete list of all inserts associated with *dm36* by typing

```
. search dm36, entry historical
(output omitted )
```

Again, we include the `historical` option in case any of the relevant inserts have been marked historical.

4.8.4   FAQ searches

To search across the FAQs, specify the `faq` option:

```
. search logistic regression, faq
(output omitted )
```

4.8.5   Return codes

In addition to indexing the entries in the *User's Guide* and all of the *Stata Reference* manuals, `search` also can be used to look up return codes.

To see information about return code 131, type

```
. search rc 131

[R]     error messages  . . . . . . . . . . . . . . . . . . .  Return code 131
        not possible with test;
        You requested a test of a hypothesis that is nonlinear in the
        variables.  test tests only linear hypotheses.  Use testnl.
```

To get a list of all Stata return codes, type

```
. search rc
(output omitted )
```

4.9   net search: Searching net resources

When you select **Help > Search**, you will notice that there are two types of searches to choose. The first, which has been discussed in the previous sections, is to **Search documentation and FAQs**. The second is to **Search net resources**. This feature of Stata searches resources over the Internet.

When you choose **Search net resources** in the search dialog box and enter *keywords* in the field, Stata searches all user-written programs on the Internet, including user-written additions published in the *Stata Journal* and the STB. The results are displayed in the Viewer, and you can click to go to any of the matches found.

Equivalently, you can type `net search` *keywords* on the Stata command line to display the results in the Results window. For the full syntax for using the `net search` command, see [R] **net search**.

5 Flavors of Stata

5.1   Platforms

Stata is available for a variety of different computers, including

> Stata for Windows (32-bit Intel)
> Stata for Windows (64-bit x86-64)
> Stata for Windows (64-bit Itanium)
>
> Stata for Macintosh
>
> Stata for AIX
> Stata for HP-UX
> Stata for Linux (32-bit Intel)
> Stata for Linux (64-bit x86-64)
> Stata for Linux (64-bit Itanium)
> Stata for SGI Irix
> Stata for Solaris (Sparc)
> Stata for Tru64 Unix

At this time, eight 64-bit systems are supported by Stata: Windows on both x86-64 and Itanium, Linux on both x86-64 and Itanium, AIX, SGI Irix, Solaris, and Tru64 Unix.

Which version of Stata you run does not matter—Stata is Stata. You instruct Stata in the same way and Stata produces the same results, right down to the random-number generator.

Even files can be shared. For instance, a dataset created with Stata for Macintosh can be used on any other computer, and the same goes for graphs, programs, and any other file Stata uses or produces.

Moving a dataset or any other file across platforms requires no translation. If you do share datasets or graphs with other users across platforms, be sure that you make exact binary copies.

5.2   Stata/SE, Intercooled Stata, and Small Stata

Stata for Windows and Stata for Macintosh are available in three "flavors": Stata/SE (Special Edition), Intercooled Stata, and Small Stata. Stata for Unix is available only in the Stata/SE and Intercooled Stata flavors. All three flavors of Stata have the same features, but Stata/SE and Intercooled Stata are able to work with larger datasets and are faster. How much faster depends on the platform, but the advantage ranges from 50 to 600 percent. Stata/SE has much larger limits for matrix sizes and string lengths, and will accommodate up to 16 times more variables than Intercooled Stata.

Stata/SE is the version that we recommend to users who frequently analyze large datasets. Intercooled Stata is the version that we recommend for a typical user doing serious data analysis and statistics with small to moderate size datasets.

Small Stata would perhaps be better named Stata for Small Computers.

5.2.1 Determining which version you own

Included with every copy of Stata is a paper license that contains important codes that you will input during installation. This license also determines which flavor of Stata you have—SE, Intercooled, or Small. Look at the license to see if it says Stata/SE, Intercooled Stata, or Small Stata.

If you purchased Intercooled Stata and you want Stata/SE, your Intercooled Stata version can be upgraded to Stata/SE. In fact, we put all three flavors on the same installation media, so you won't have to wait to receive another CD. All you need is an upgraded paper license with the appropriate codes.

By the way, even if you purchased Stata/SE or Intercooled Stata, you may use Small Stata with your Stata/SE or Intercooled Stata license. This might be useful if you had a large computer at work and a smaller computer at home. Please remember, however, that you have only one license (or however many licenses you purchased). You may, both legally and morally, use one, the other, or both, but you should not subject the pair to simultaneous use.

5.2.2 Determining which version is installed

If Stata is already installed, you can find out which Stata you are using by entering Stata as you normally do and typing about:

```
. about
Intercooled Stata 9 for Windows
Born 22 Apr 2005
Copyright (C) 1985-2005
10-user Windows (network) perpetual license:
        Serial number:  199040000
        Licensed to:  Marsha Martinez
                      StataCorp
```

You are running Intercooled Stata 9 for Windows.

5.3   Size limits comparison of Stata/SE, Intercooled Stata, and Small Stata

Here are some of the different size limits for Stata/SE, Intercooled Stata, and Small Stata. Type help limits for a longer list.

Maximum size limits for Stata/SE, Intercooled Stata, and Small Stata

	Stata/SE	Intercooled	Small
Number of observations	limited only by memory	limited only by memory	fixed at approx. 1,000
Number of variables	32,766	2,047	fixed at 99
Width of a dataset	393,192	24,564	200
Maximum matrix size (matsize)	11,000	800	fixed at 40
Number of characters in a macro	1,081,511	67,784	8,681
Number of characters in a command	1,081,527	67,800	8,697

That is, Stata/SE allows more variables, larger matrices, longer macros (and longer strings), and a longer command line than Intercooled Stata. Intercooled Stata allows larger datasets, fits models with more independent variables, has longer macros, and allows a longer command line (required because of the increased number of variables allowed) than Small Stata.

5.4   Speed comparison of Stata/SE, Intercooled Stata, and Small Stata

Why the difference in speed for Stata/SE, Intercooled Stata, and Small Stata? In part, it is due to different options we specified when we compiled Stata. However, it is also due to differences in the code.

For instance, in Stata's `test` command, there comes a place where it must compute the matrix calculation $\mathbf{RZR}'$ (where $\mathbf{Z} = (\mathbf{X}'\mathbf{X})^{-1}$). Stata/SE and Intercooled Stata make the calculation in a straightforward way, which is to form $\mathbf{T} = \mathbf{RZ}$ and then calculate $\mathbf{TR}'$. This requires temporarily storing the matrix $\mathbf{T}$. Small Stata, on the other hand, goes into more complicated code to form the result directly—code that requires temporary storage of only one scalar! This code, in effect, recalculates intermediate results over and over again, and so it is slower.

Another difference is that Small Stata, since it is designed to work with smaller datasets, uses different memory-management routines. These memory-management routines use 2-byte rather than 4-byte offsets, and therefore require only half the memory to track locations.

In any case, the differences are all technical and internal. From the user's point of view, Stata/SE, Intercooled Stata, and Small Stata work the same way.

5.5   Features of Stata/SE

For those familiar with Stata, a table will say it all:

Parameter	Intercooled Stata			Stata/SE		
	Default	min	max	Default	min	max
maxvar	2,047	2,047	2,047	**5,000**	2,047	**32,766**
matsize	200	10	800	**400**	10	**11,000**
memory	1M	500K	...	**10M**	500K	...
str#	.	1	80	.	1	**244**

In other words, this means that Stata/SE

1. allows datasets with more variables—up to 32,766 variables;

2. allows datasets to contain longer string variables—variables up to 244 characters long;

3. allows larger matrices—matrices up to 11,000 x 11,000. (An implication of this is that Stata/SE can fit models with more independent variables and can fit certain panel-data models with longer time series within panel.)

To learn how to exploit Stata/SE to its fullest, type `help SpecialEdition` after invoking Stata.

6 Setting the size of memory

Contents

6.1 Memory size considerations

Stata works with a copy of the data that it loads into memory.

By default, Stata/SE allocates 10 megabytes and Intercooled Stata allocates 1 megabyte to Stata's data areas, and you can change it.

By default, Small Stata allocates about 300K to Stata's data areas, and you cannot change it.

You can even change the allocation to be larger than the physical amount of memory on your computer because Windows, Macintosh, and Unix systems provide virtual memory.

Virtual memory is slow but adequate in rare cases when you have a dataset that is too large to load into real memory. If you use large datasets frequently, we recommend that you add more memory to your computer.

One way to change the allocation is when you start Stata. Instructions for doing this are provided in

Windows	[GSW]	**A.6 Specifying the amount of memory allocated**
Macintosh	[GSM]	**A.6 Specifying the amount of memory allocated**
Unix	[GSU]	**A.6 Specifying the amount of memory allocated**

In addition, if you use Stata/SE or Intercooled Stata for Windows, Unix, or Macintosh, you can change the total amount of memory allocated while Stata is running. That is the topic of this chapter.

Understand that it does not matter which method you use. Being able to change the total on the fly is convenient, but even if you cannot do this, it just means that you specify it ahead of time. If later you need more, simply exit Stata and reinvoke it with the larger total.

6.2 Setting memory on the fly: Stata/SE

There are three limits in Stata/SE that affect memory allocation and usage. The three limits are

1. `maxvar`, the maximum number of variables allowed in a dataset. This limit is initially set to 5,000; you can increase it up to 32,767.

2. `matsize`, the largest dimension of a matrix: this limit is initially set to 400, and you can increase it up to 11,000. In most cases, this relates to the maximum number of independent variables allowed in the models that you fit, and thus, the dimension of the estimated variance–covariance matrix. However, in some panel-data models, covariance or correlation matrices must be fitted, and their dimensions depend on either the number of panels (groups) in your data or the number of observations in your dataset. Note that this limit does not apply to Mata matrices; see the *Mata Reference Manual*.

3. `memory`, the amount of memory Stata requests from the operating system to store your data. This limit is initially set to 10 megabytes in Stata/SE. You may set it to as large a number as your operating system will allow.

You set the limits using the

$$\text{set maxvar} \ \# \qquad \big[\, , \ \underline{\text{permanently}}\big]$$

$$\text{set matsize} \ \# \qquad \big[\, , \ \underline{\text{permanently}}\big]$$

$$\text{set memory} \ \#\big[\text{b}|\text{k}|\text{m}|\text{g}\big] \ \big[\, , \ \underline{\text{permanently}}\big]$$

commands. For instance, you might type

```
. set maxvar  5000
. set matsize 900
. set memory  50m
```

The order in which you set the limits does not matter. If you specify the `permanently` option when you set a limit, in addition to making the change right now, Stata will remember the new limit and use it in the future when you invoke Stata:

```
. set maxvar  5000, permanently
. set matsize 900, permanently
. set memory  50m, permanently
```

You can reset the current or permanent limits whenever and as often as you wish.

6.2.1   Advice on setting maxvar

$$\text{set maxvar} \ \# \ \big[\, , \ \underline{\text{permanently}}\big]$$

where $2{,}048 \le \# \le 32{,}767$

Why is there a limit on `maxvar`? Why not just set `maxvar` to 32,767 and be done with it? Because simply allowing room for variables, even if they do not exist, causes Stata to consume memory and, if you will only be using datasets with a lot fewer variables, you will be wasting memory.

The formula for the amount of memory consumed by `set maxvar` is approximately

$$\text{megs} = .3147 * (\text{maxvar}/1000) + .002$$

For instance, if you set `maxvar` to 20,000, the memory would be approximately

$$\text{megs} = .3147 * 20 + .002 = 6.296 \ \text{megs}$$

and if you left it at the default, the memory use would be roughly

$$\text{megs} = .3147 * 5 + .002 = 1.575 \text{ megs}$$

Thus, how big you set `maxvar` does not dramatically affect memory usage. Still, at `maxvar=32,000`, memory use is 10.072M.

Recommendation: Think about datasets with the most variables that you typically use. `set maxvar` to a few hundred or even a 1,000 above that. (Note that the memory cost of an extra 1,000 variables is only .315 megs.)

Remember, you can always reset `maxvar` temporarily by typing `set maxvar`.

❏ Technical Note

The formula above is only approximate, and the formula given is the formula appropriate for 32-bit computers. When you `set maxvar`, Stata/SE will give you a memory report showing the exact amount of memory used:

```
. set maxvar 10000
```

Current memory allocation

settable	current value	description	memory usage (1M = 1024k)
set maxvar	10000	max. variables allowed	3.149M
set memory	10M	max. data space	10.000M
set matsize	400	max. RHS vars in models	1.254M
			14.403M

❏

6.2.2 Advice on setting matsize

set matsize # $\left[\,, \underline{\text{perman}}\text{ently}\right]$

where $10 \leq \# \leq 11{,}000$

Although `matsize` can theoretically be set up to 11,000, on all but 64-bit computers, you will be unable to do that, and, even if you succeeded, Stata/SE would probably run out of memory subsequently. The value of `matsize` has a dramatic effect on memory usage, the formula being

$$\text{megs} = (8 * \text{matsize}^2 + 88 * \text{matsize})/(1024^2)$$

This formula is valid across all computers, 32-bit and 64-bit. For instance, the above formula states

matsize	memory use
400	1.254M
800	4.950M
1,600	19.666M
3,200	78.394M
6,400	313.037M
11,000	924.080M

The formula, in fact, is an understatement of the amount of memory certain Stata commands use, and is an understatement of what you will certainly use yourself if you use matrices directly. The formula gives the amount of memory required for one matrix and 11 vectors. If two matrices are required, the numbers above are nearly doubled. When you `set matsize`, if you specify too large a value, Stata will refuse, but remember that just because Stata does not complain, you still may run into problems later. What might happen is that Stata could be running a command and then complain, "op. sys. refuses to provide memory"; r(909).

For `matsize` = 11,000, nearly 1 gigabyte of memory is required, and doubling that would require nearly 2 gigabytes of memory. On most 32-bit computers, 2 gigabytes is the maximum amount of memory the operating system will allocate to a single task, so nearly nothing would be left for all the rest of Stata.

Why, then, is `matsize` allowed to be set so large? Because on 64-bit computers, such large amounts cause no difficulty.

For "reasonable" values of matsize (say up to 3,200), memory consumption is not too great. Choose a reasonable value given the kinds of models you fit, and remember that you can always reset the value.

6.2.3 Advice on setting memory

set memory #[b|k|m|g] [, permanently]

where # ≥ 500k and (k, m, and g may be typed in uppercase.)

The advice for setting memory is the same as for Intercooled Stata: set enough so that your datasets fit easily, and do not set so much that you exceed physical memory present on your computer, except in emergencies.

You may set memory in bytes (b), kilobytes (k), megabytes (m), or gigabytes (g), but the number specified must be an integer, so if you want to set 1.5g, you set 1500m. Actually, 1.5g is 1536m because the formulas for a kilobyte, megabyte, and gigabyte are

```
1024 bytes     = 1 kilobyte
1024 kilobytes = 1 megabyte
1024 megabytes = 1 gigabyte
```

This detail does not matter, but this is the rule that Stata uses when presenting numbers, so do not be surprised when 2000k is not displayed as 2M, or 2M is displayed as 2048k.

If you have a very large 32-bit computer, the maximum amount of memory you can set may surprise you. Many people think that 32-bit computers can allow up to 4 gigabytes of memory and that, in a sense, is true. Some 32-bit computers will even allow you to install 4 gigabytes of physical memory. Nevertheless, most modern operating systems will allocate a maximum of one-half the theoretical maximum to individual tasks, which is to say, most operating systems will allow Stata only 2 gigabytes of memory (even if they have 4 gigabytes of memory)!

The same one-half rule applies to 64-bit computers, but half the theoretical limit is still 536,870,912 gigabytes, so no one much cares.

This one-half limit is imposed by the operating system, not by Stata, and the operating system developers have good technical reasons for imposing the rule.

6.3   Setting memory size on the fly: Intercooled Stata

You can reallocate memory on-the-fly. Assume you have changed nothing about how Stata starts, so you get the default 1 megabyte of memory allocated to Stata's data areas. You are working with a large dataset and now wish to increase it to 32 megabytes. You can type

```
. set memory 32m
(32768k)
```

and, if your operating system can provide the memory to Stata, Stata will work with the new total. Later in the session, if you want to release that memory and work with only 2 megabytes, you could type

```
. set memory 2m
(2048k)
```

There is only one restriction on the set memory command: whenever you change the total, there cannot be any data already in memory. If you have a dataset in memory, save it, clear memory, reset the total, and then use it again. We are getting ahead of ourselves, but you might type

```
. save mydata, replace
file mydata.dta saved
. drop _all
. set memory 32m
(32768k)
. use mydata
```

When you request the new allocation, your operating system might refuse to provide it:

```
. set memory 128m
op. sys. refuses to provide memory
r(909);
```

If that happens, you are going to have to take the matter up with your operating system. In the above example, Stata asked for 128 megabytes and the operating system said no.

For most 32-bit computers, the absolute maximum amount of memory that can theoretically be allocated will be approximately $2^{(32-1)}$ bytes (2 gigabytes), regardless of the operating system. In practice, the amount of memory available to any one application on a system is affected by many factors, and may be somewhat less than the theoretical maximum.

64-bit computers can theoretically have up to $2^{(64-1)}$ bytes (over 2 billion gigabytes). In practice, 64-bit computers will have as much memory as is affordable. We are aware of sites using Stata with datasets consuming over 10 gigabytes of memory.

6.4   The memory command

memory helps you figure out whether you have sufficient memory to do something.

(Continued on next page)

```
. use http://www.stata-press.com/data/r9/regsmpl
(NLS Women 14-26 in 1968)

. memory
```

	bytes	
Details of set memory usage		
overhead (pointers)	114,136	10.88%
data	913,088	87.08%
data + overhead	1,027,224	97.96%
free	21,344	2.04%
Total allocated	1,048,568	100.00%
Other memory usage		
system overhead	677,289	
set matsize usage	16,320	
programs, saved results, etc.	505	
Total	694,114	
Grand total	1,742,682	

21,344 bytes free is not much. You might increase the amount of memory allocated to Stata's data areas by specifying set memory 2m.

```
. save regsmpl
file regsmpl.dta saved

. clear

. set memory 2m
(2048k)

. use regsmpl
(NLS Women 14-26 in 1968)

. memory
```

	bytes	
Details of set memory usage		
overhead (pointers)	114,136	5.44%
data	913,088	43.54%
data + overhead	1,027,224	48.98%
free	1,069,920	51.02%
Total allocated	2,097,144	100.00%
Other memory usage		
system overhead	677,289	
set matsize usage	16,320	
programs, saved results, etc.	667	
Total	694,276	
Grand total	2,791,420	

Over 1 megabyte free; that's better. See [D] **memory** for more information.

6.5   Virtual memory and speed considerations

When you use more memory than is physically available on your computer, Stata slows down. If you are only using a little more memory than on your computer, performance is probably not too bad. On the other hand, when you are using a lot more memory than is on your computer, performance will be noticeably affected. In these cases, we recommend that you

```
. set virtual on
```

Virtual memory systems exploit locality of reference, which means that keeping objects closer together allows virtual memory systems to run faster. `set virtual` controls whether Stata should perform extra work to arrange its memory to keep objects close together. By default, `virtual` is set `off`. `set virtual` can be used with Stata/SE and Intercooled Stata on all supported operating systems.

In general, you want to leave `set virtual` set to the default of `off` so that Stata will run faster.

When you `set virtual on`, you are asking Stata to arrange its memory so that objects are kept closer together. This requires Stata to do a substantial amount of work. We recommend setting `virtual on` only when the amount of memory in use drastically exceeds what is physically available. In these cases, setting `virtual on` will help, but keep in mind that performance will still be slow. If you are using virtual memory frequently, you should consider adding memory to your computer.

6.6   An issue when returning memory to Unix

There is a surprising issue of returning memory that Unix users need to understand. Let's say that you set memory to 128 megabytes, went along for a while, and then, being a good citizen, returned most of it:

```
. set memory 2m
(2048k)
```

Theoretically, 126 megabytes just got returned to the operating system for use by other processes. If you use Windows, that is exactly what happens and, with some Unixes, that is what happens, too.

Other Unixes, however, are strange about returned memory in a misguided effort to be efficient: they do not really take the memory back. Instead, they leave it allocated to you in case you ask for it back later. Still other Unixes sort of take the memory back: they put it in a queue for your use, but, if you do not ask for it back in 5 or 10 minutes, then they return it to the real system pool!

The unfortunate situation is that we at Stata cannot force the operating system to take the memory back. Stata returns the memory to Unix, and then Unix does whatever it wants with it.

So, let's review: You make your Stata smaller in an effort to be a good citizen. You return the memory so that other users can use it, or perhaps so you can use it with some other software.

If you use Windows, the memory really is returned and all works exactly as you anticipated.

If you use Unix, it might go back immediately, it might go back in 5 or 10 minutes, or it might never go back. In the last case, the only way to really return the memory is to exit Stata. All Unixes agree on that: when a process ends, the memory really does go back into the pool.

To find out how your Unix works, you need to experiment. We would publish a table and just tell you, but we have found that within manufacturer the way their Unix works will vary by subrelease! The experiment is tedious but not difficult:

1. Bring up a Stata and make it really big; use a lot of memory, so much that you are virtually hogging the computer.

2. Go to another window or session and bring up another Stata. Verify that you cannot make it big—that you get the "system limit exceeded" message.

3. Go back to the first Stata, leaving the second running, and make it smaller.

4. Go to the second Stata and try again to make it big. If you succeed, then your Unix returns memory instantly.

5. If you still get the "system limit exceeded" message, wait 5 minutes and try again. If it now works, your system delays accepting returned memory for about 5 minutes.

6. If you still get the "system limit exceeded" message, wait another 5 minutes and try again. If it now works, your system delays accepting returned memory for about 10 minutes.

7. Go to the first Stata and exit from it.

8. Go to the second Stata and try to make it big again. If it now works, your system never really accepts returned memory. If it still does not work, start all over again. Some other process took memory and corrupted your experiment.

If you are one of the unfortunates who have a Unix that never accepts returned memory, you will just have to remember that you must exit and reenter Stata to really give memory back.

7   –more– conditions

Contents

7.1   Description

When you see —more— at the bottom of the screen,

Press ...	and Stata...
letter *l* or *Enter*	displays the next line
letter *q*	acts as if you pressed *Break*
space bar or any other key	displays the next screen

In addition, you can press the *clear –more– condition* button, the button labeled **Go** with a circle around it.

—more— is Stata's way of telling you that it has something more to show you, but showing you that something more will cause the information on the screen to scroll off.

7.2   set more off

If you type set more off, —more— conditions will never arise and Stata's output will scroll by at full speed.

If you type set more on, —more— conditions will be restored at the appropriate places.

Programmers: Do-file writers sometimes include set more off in their do-files because they do not care to interactively watch the output. They want Stata to proceed at full speed because they plan on making a log of the output that they will review later. Do-filers need not bother to set more on at the conclusion of their do-file. Stata automatically restores the previous set more when the do-file (or program) concludes.

7.3   The more programming command

Ado-file programmers need take no special action to have —more— conditions arise when the screen is full. Stata handles that automatically.

If, however, you wish to force a —more— condition early, you can include the more command in your program. The syntax of more is

```
more
```

more takes no arguments.

For additional information, see [P] **more**.

8 Error messages and return codes

Contents

8.1 Making mistakes

When an error occurs, Stata produces an error message and a *return code*. For instance,

```
. list myvar
no variables defined
r(111);
```

We ask Stata to list the variable named `myvar`. Since we have no data in memory, Stata responds with the message "no variables defined" and a line that reads "`r(111)`".

The "no variables defined" is called the error message.

The 111 is called the return code. You can click on blue return codes to get a detailed explanation of the error.

8.1.1 Mistakes are forgiven

Having said "no variables defined" and `r(111)`, all is forgiven; it is as if the error never occurred.

Typically, the message will be enough to guide you to a solution, but, if it is not, the numeric return codes are documented in [P] **error**.

8.1.2 Mistakes stop user-written programs and do-files

Whenever an error occurs in a user-written program or do-file, the program or do-file immediately stops execution and the error message and return code are displayed.

For instance, consider the following do-file:

── top of myfile.do ────────

```
use http://www.stata-press.com/data/r9/auto
decribe
list
```

── end of myfile.do ────────

Note the second line—you meant to type `describe` but typed `decribe`. Here is what happens when you execute this do-file by typing `do myfile`:

```
. do myfile

. use http://www.stata-press.com/data/r9/auto
(1978 Automobile Data)
```

77

```
. decribe
unrecognized command:  decribe
r(199);

end of do-file
r(199);

.
```

The first error message and return code were caused by the illegal `decribe`. This then caused the do-file itself to be aborted; the valid `list` command was never executed.

8.1.3   Advanced programming to tolerate errors

Errors are not only of the typographical kind; some are substantive. A command that is valid in one dataset might not be valid in another. Moreover, in advanced programming, errors are sometimes anticipated: use one dataset if it is there, but use another if you must.

Programmers can access the return code to determine whether an error occurred, which they can then ignore, or, by examining the return code, code their programs to take the appropriate action. This is discussed in [P] **capture**.

You can also prevent do-files from stopping when errors occur by using the `do` command's `nostop` option.

```
. do myfile, nostop
```

8.2   The return message for obtaining command timings

In addition to error messages and return codes, there is something called a return message, which you normally do not see. Normally, if you typed `summarize tempjan`, you would see

```
. use http://www.stata-press.com/data/r9/citytemp
(City Temperature Data)
. summarize tempjan
```

Variable	Obs	Mean	Std. Dev.	Min	Max
tempjan	954	35.74895	14.18813	2.2	72.6

If you were to type

```
. set rmsg on
r; t=0.00 10:21:22
```

sometime during your session, Stata would display return messages:

```
. summarize tempjan
```

Variable	Obs	Mean	Std. Dev.	Min	Max
tempjan	954	35.74895	14.18813	2.2	72.6

```
r; t=0.01 10:21:26
```

The line that reads `r; t=0.01 10:21:26` is called the return message.

The `r;` indicates that Stata successfully completed the command.

The `t=0.01` shows the amount of time, in seconds, it took Stata to perform the command (timed from the point you pressed *Return* to the time Stata typed the message). This command took a hundredth of a second. In addition, Stata shows the time of day using a 24-hour clock. This command completed at 10:21 a.m.

You will learn that Stata has the ability to run commands stored in files (called do-files) and the ability to log output. Some users find the detailed return message helpful with do-files. They construct a lengthy program and let it run overnight, logging the output. They come back the next morning, look at the output, and discover a mistake in some portion of the job. They can look at the return messages to determine how long it will take to rerun that portion of the program.

You may `set rmsg` on whenever you wish.

When you want Stata to stop displaying the detailed return message, type `set rmsg off`.

9 The Break key

9.1 Making Stata stop what it is doing

When you want to make Stata stop what it is doing and return to the Stata dot prompt, you press *Break*:

Stata for Windows:	click the **Break** button (it is the button with the big red X), or hold down *Ctrl* and press the *Pause/Break* key
Stata for Macintosh:	click the **Break** button or hold down *Command* and press period
Stata for Unix(GUI):	click the **Break** button or hold down *Ctrl* and press k
Stata for Unix(console):	hold down *Ctrl* and press c or press q

Elsewhere in this manual, we describe this action as simply pressing *Break*. Break tells Stata to cancel what it is doing and return control to you as soon as possible.

If you press *Break* in response to the input prompt or while you are typing a line, Stata ignores it, since you are already in control.

If you press *Break* while Stata is doing something—creating a new variable, sorting a dataset, making a graph, etc.—Stata stops what it is doing, undoes it, and issues an input prompt. The state of the system is the same as if you had never issued the command.

▷ Example 1

You are fitting a logit model, type the command, and, as Stata is working on the problem, realize that you omitted an important variable:

```
. logit foreign mpg weight
Iteration 0:   log likelihood =-1801.3284
Iteration 1:   log likelihood =-1197.7089
—Break—
r(1);

. _
```

When you pressed *Break*, Stata responded by typing —Break— and then typing r(1);. Pressing *Break* always results in a return code of 1—that is why return codes are called return codes and not error codes. The 1 does not indicate an error, but it does indicate that the command did not complete its task.

◁

9.2   Side-effects of pressing Break

In general, there are none. We said above that Stata undoes what it is doing so that the state of the system is the same as if you had never issued the command. There are two exceptions to this.

If you are reading data from disk using `insheet`, `infile`, or `infix`, whatever data have already been read will be left behind in memory, the theory being that perhaps you stopped the process so you could verify that you were reading the right data correctly before sitting through the whole process. If not, you can always `drop _all`.

```
. infile v1-v9 using workdata
(eof not at end of obs)
(4 observations read)
—Break—
r(1);
```

The other exception is `sort`. You have a large dataset in memory, decide to sort it, and then change your mind.

```
. sort price
—Break—
r(1);
```

If the dataset was previously sorted by, say, the variable `prodid`, it is no longer. When you press *Break* in the middle of a `sort`, Stata marks the data as unsorted.

9.3   Programming considerations

There are basically no programming considerations for handling Break because Stata handles it all automatically. If you write a program or do-file, execute it, and then press *Break*, Stata stops execution just as it would with an internal command.

Advanced programmers may be concerned about cleaning up after themselves; perhaps, they have generated a temporary variable they intended to drop later or a temporary file they intended to erase later. If a Stata user presses *Break*, how can you ensure that these temporary variables and files will be erased?

If you obtain names for such temporary items from Stata's `tempname`, `tempvar`, and `tempfile` commands, Stata will automatically erase the temporary items; see [U] **18.7 Temporary objects**.

There are instances, however, when a program must commit to executing a group of commands without interruption, or the user's data would be left in an intermediate or undefined state. In these instances, Stata provides a

```
nobreak {
        ...
}
```

construct; see [P] **break**.

10 Keyboard use

Contents

10.1 Description

The keyboard should operate very much the way you would expect, with a few additions:

1. There are some unexpected keys you can press to obtain previous commands you have typed. In addition, you can click once on a command in the Review window to reload it, or click it twice to reload and execute; this feature is discussed in the *Getting Started* manuals.

2. There are a host of command-editing features for Stata for Unix(console) users since their user interface does not offer such features.

3. Regardless of operating system or user interface, if there are *F*-keys on your keyboard, they have special meaning and you can change the definitions of the keys.

10.2 F-keys

Note to Windows users: *F10* is reserved internally by Windows; you cannot program this key.

By default, Stata defines the *F*-keys to mean

F-key	Definition
F1	help
F2	#review;
F3	describe;
F7	save
F8	use

The semicolons at the end of some of the entries indicate the presence of an implied *Return*.

help shows a Stata help file—you use it by typing help followed by the name of a Stata command; see [U] **4 Stata's online help and search facilities**. You can type out help or you can press *F1*, type the Stata command, and press *Return*.

#review is the command to show the last few commands you issued. It is described in [U] **10.6 Editing previous lines**. Rather than typing out #review and pressing *Return*, you can simply press *F2*. You do not press *Return* following *F2* because the definition of F2 ends in a semicolon—Stata presses the *Return* key for you.

describe is the Stata command to report the contents of data loaded into memory. It is explained in [D] **describe**. Normally, you type describe and press *Return*. Alternatively, you can press *F3*.

save is the command to save the data in memory into a file, and use is the command to load data; see [D] **use** and [D] **save**. The syntax of each is the same: save or use followed by a filename. You can type out the commands or you can press *F7* or *F8* followed by the filename.

You can change the definitions of the *F*-keys. For instance, the command to list data is list; you can read about it in [D] **list**. The syntax is list to list all the data, or list followed by the names of some variables to list just those variables (there are other possibilities).

If you wanted *F3* to mean list, you could type

    . global F3 "list "

In the above, F3 refers to the letters *F* followed by *3*, not the *F3* key. Note the capitalization and spacing of the command.

You type global in lowercase, type out the letters F3, and then type "list ". The blank at the end of list is important. In the future, rather than typing out list mpg weight, you want to be able to press the *F3* key and then type only mpg weight. You put a blank in the definition of F3 so that you would not have to type a blank in front of the first variable name following pressing *F3*.

Now say you wanted *F5* to mean list all the data—list followed by *Return*. You could define

    . global F5 "list;"

Now you would have two ways of listing all the data: (1) press *F3*, then press *Return*, or (2) press *F5*. The semicolon at the end of the definition of *F5* will press *Return* for you.

If you really want to change the definitions of *F3* and *F5*, you will probably want to change the definition every time you invoke Stata. One way would be to type out the two global commands every time you invoked Stata. Another way would be to type the two commands into an ASCII text file called profile.do. Stata executes the commands in profile.do every time it is launched if profile.do is placed in the appropriate directory:

Windows:	put profile.do in the "start-in" directory;
	see [GSW] **A.7 Executing commands every time Stata is started**
Macintosh:	put profile.do in your home directory;
	see [GSM] **A.7 Executing commands every time Stata is started**
Unix:	put profile.do someplace along your shell's PATH;
	see [GSU] **A.7 Executing commands every time Stata is started**

You can use the *F*-keys any way you desire: They contain a string of characters, and pressing the *F*-key is equivalent to typing out those characters.

❑ Technical Note

(*Stata for Unix users.*) Sometimes Unix assigns a special meaning to the *F*-keys, and, if it does, those meanings will take precedence over our meanings. Stata provides a second way to get to the *F*-keys. Hold down *Ctrl*, press F, release the keys, and then press a number from 0 through 9. Stata interprets *Ctrl-F* plus 1 as equivalent to the *F1* key, *Ctrl-F* plus 2 as *F2*, and so on. *Ctrl-F* plus 0 means *F10*.

❑

❏ Technical Note

On some international keyboards, the left single quote is used as an accent character. In this case, we recommend mapping this character to one of your function keys. In fact, you might find it convenient to map both the left single quote (`) and right single quote (') characters so that they are next to each other.

Within Stata, open the Do-file Editor. Type the following two lines in the Do-file Editor:

```
global F4 `
global F5 '
```

Save the file as `profile.do` into your Stata directory. If you already have a `profile.do` file, append the two lines to your existing `profile.do` file.

Exit Stata, and restart it. You should see the startup message

```
running C:\Program Files\Stata9\profile.do ...
```

or some variant of it depending on where your Stata is installed. Press the function keys *F4* and *F5* to verify that they work.

If you did not see the startup message, you did not save the `profile.do` in your home folder.

You can of course map to any other function keys, but *F1*, *F2*, *F3*, *F7*, and *F8* are already used.

❏

10.3   Editing keys in Stata for Windows, Macintosh, and Unix(GUI)

Users have available to them the standard editing keys for their operating system. So, Stata should just edit what you type in the natural way—the Stata Command window is a standard edit window.

In addition, you can fetch commands from the Review window into the Command window. Click on a command in the Review window, and it is loaded into the Command window, where you can edit it. Alternatively, if you double-click a line in the Review window, it is loaded and executed.

Another way to get lines from the Review window into the Command window is with the *PgUp* and *PgDn* keys. Tap *PgUp* and Stata loads the last command you typed into the Command window. Tap it again and Stata loads the line before that, and so on. *PgDn* goes the opposite direction.

Another editing key that may interest users is *Esc*. This key clears the Command window.

In summary,

Press	Result
PgUp	Steps back through commands and moves command from Review window to Command window
PgDn	Steps forward through commands and moves command from Command window to Review window
Esc	Clears Command window

10.4   Editing keys in Stata for Unix(console)

Certain keys allow you to edit the line you are typing. Since Stata supports a variety of computers and keyboards, the location and the names of the editing keys are not the same for all Stata users.

Every keyboard has the standard alphabet keys (*QWERTY* and so on), and every keyboard has a *Ctrl* key. Some keyboards go further and have extra keys located to the right, above, or left, with names like *PgUp* and *PgDn*.

Throughout this manual we will refer to Stata's editing keys using names that appear on nobody's keyboard. For instance, PrevLine is one of the Stata editing keys—it retrieves a previous line. Hunt all you want, but you will not find it on your keyboard. So, where is PrevLine? We have tried to put it where you would naturally expect it. On keyboards with a key labeled *PgUp*, *PgUp* is the PrevLine key, but on everybody's keyboard, no matter which version of Unix, brand of keyboard, or anything else, *Ctrl-R* also means PrevLine.

When we say press PrevLine, now you know what we mean: press *PgUp* or *Ctrl-R*. With that introduction, the editing keys are

Name for Editing Key	Editing Key	Function
Kill	*Esc* on PCs and *Ctrl-U*	Deletes the line and lets you start over.
Dbs	*Backspace* on PCs and *Backspace, Rubout,* or *Delete* on other computers	Backs up and deletes one character.
Lft	←, *4* on the numeric keypad for PCs, and *Ctrl-H*	Moves the cursor left one character without deleting any characters.
Rgt	→, *6* on the numeric keypad for PCs, and *Ctrl-L*	Moves the cursor forward one character.
Up	↑, *8* on the numeric keypad for PCs, and *Ctrl-O*	Moves the cursor up one physical line on a line that takes more than one physical line. Also see PrevLine
Dn	↓, *2* on the numeric keypad for PCs, and *Ctrl-N*	Moves the cursor down one physical line on a line that takes more than one physical line. Also see NextLine.
PrevLine	*PgUp* and *Ctrl-R*	Retrieves a previously typed line. You may press PrevLine multiple times to step back through previous commands.
NextLine	*PgDn* and *Ctrl-B*	The inverse of PrevLine
Seek	*Ctrl-Home* on PCs and *Ctrl-W*	Goes to the line number specified. Before pressing Seek, type the line number. For instance, typing *3* and then pressing Seek is the same as pressing PrevLine three times.
Ins	*Ins* and *Ctrl-E*	Toggles insert mode. In insert mode, characters typed are inserted into the line at the position of the cursor.
Del	*Del* and *Ctrl-D*	Deletes the character at the position of the cursor.
Home	*Home* and *Ctrl-K*	Moves the cursor to the start of the line.
End	*End* and *Ctrl-P*	Moves the cursor to the end of the line.
Hack	*Ctrl-End* on PCs, and *Ctrl-X*	Hacks off the line at the cursor.
Tab	⊣ on PCs, *Tab,* and *Ctrl-I*	Moves the cursor forward eight spaces.
Btab	⊢ on PCs, and *Ctrl-G*	The inverse of Tab.

▷ Example 1

It is difficult to demonstrate the use of editing keys on paper. You should try each of them. Nevertheless, here is an example:

. summarize price w<u>a</u>ht

You typed `summarize price waht` and then tapped the *Lft* key ($\leftarrow$ key or *Ctrl-H*) three times to maneuver the cursor back to the `a` of `waht`. If you were to press *Return* right now, Stata would see the command `summarize price waht`, so where the cursor is does not matter when you press *Return*. If you wanted to execute the command `summarize price`, you could back up one more character and then press the Hack key. We will assume, however, that you meant to type `weight`.

If you were now to press the letter *e* on the keyboard, an e would appear on the screen to replace the a, and the cursor would move under the character h. We now have w<u>e</u>ht. You press *Ins*, putting Stata into insert mode, and press *i* and *g*. The line now says `summarize price weight`, which is correct, so you press *Return*. Notice that we did not have to press *Ins* before every character we wanted to insert. The *Ins* key is a toggle: If we press it again, Stata turns off insert mode, and what we type replaces what was there. When we press *Return*, Stata forgets all about insert mode, so we do not have to remember from one command to the next whether we are in insert mode.

◁

❏ Technical Note

Stata performs its editing magic based on the information about your terminal recorded in /etc/termcap(5) or, under System V, /usr/lib/terminfo(4). If some feature does not appear to work, it is probable that the entry for your terminal in the `termcap` file or `terminfo` directory is incorrect. Contact your system administrator.

❏

10.5   Editing previous lines in Stata for all operating systems

In addition to what is said below, remember that the Review window also shows the contents of the review buffer.

You may edit previously typed lines, or at least any of the last 25 or so lines. Stata records every line you type in a wraparound buffer. A wraparound buffer is a buffer of finite length in which the most recent thing you type replaces the oldest thing stored in the buffer. Stata's buffer is 8,697 characters long for Small Stata, 67,800 characters long for Intercooled Stata, and can range from 67,800 to 1,081,527 for Stata/SE.

One way to retrieve lines is with the PrevLine and NextLine keys. Remember, PrevLine and NextLine are the names we attach to these keys—there are no such keys on your keyboard. You have to look back at the previous section to find out which keys correspond to PrevLine and NextLine on your computer. To save you the effort this time, PrevLine probably corresponds to *PgUp* and NextLine probably corresponds to *PgDn*.

Suppose you wanted to reissue the third line back. You could press PrevLine three times and then press *Return*. If you made a mistake and pressed PrevLine four times, you could press NextLine to go forward in the buffer. You do not have to count lines because, each time you press PrevLine or NextLine, the current line is displayed on your monitor. Simply tap the key until you find the line you want.

Another method for reviewing previous lines, #review, is convenient when you want to see the lines in context.

▷ Example 2

Typing #review by itself causes Stata to list the last five commands you typed. (You need not type out #review—pressing *F2* has the same effect.) For instance,

```
. #review
5 list make mpg weight if abs(res)>6
4 list make mpg weight if abs(res)>5
3 tabulate foreign if abs(res)>5
2 regress mpg weight weight2
1 test weight2=0
. _
```

We can see from the listing that the last command typed by the user was test weight2=0.

◁

▷ Example 3

Perhaps the command you are looking for is not among the last five commands you typed. You can tell Stata to go back any number of lines. For instance, typing #review 15 tells Stata to show you the last 15 lines you typed:

```
. #review 15
15 replace resmpg=mpg-pred
14 summarize resmpg, detail
13 drop predmpg
12 describe
11 sort foreign
10 by foreign: summarize mpg weight
9 * lines that start with a * are comments.
8 * they go into the review buffer too.
7 summarize resmpg, detail
6 list make mpg weight
5 list make mpg weight if abs(res)>6
4 list make mpg weight if abs(res)>5
3 tabulate foreign if abs(res)>5
2 regress mpg weight weight2
1 test weight2=0
. _
```

If you wanted to resubmit the tenth previous line, you could type 10 and press Seek, or you could press PrevLine ten times. No matter which of the above methods you prefer for retrieving lines, you may edit previous lines using the editing keys.

◁

10.6   Tab expansion of variable names

Another way to quickly enter a variable name is to take advantage of Stata's variable name completion feature. Simply type the first few letters of the variable name in the Command window and press the *Tab* key. Stata will automatically type the rest of the variable name for you. If more than one variable name matches the letters you have typed, Stata will complete as much as it can and beep at you to let you know that you have typed a nonunique variable abbreviation.

Elements of Stata

Chapters

11 Language syntax

Contents

11.1   Overview

With few exceptions, the basic Stata language syntax is

$$\left[\text{by } \textit{varlist}:\right] \; \textit{command} \; \left[\textit{varlist}\right] \; \left[\text{=}\textit{exp}\right] \; \left[\text{if } \textit{exp}\right] \; \left[\text{in } \textit{range}\right] \; \left[\textit{weight}\right] \; \left[, \textit{options}\right]$$

where square brackets denote optional qualifiers. In this diagram, *varlist* denotes a list of variable names, *command* denotes a Stata command, *exp* denotes an algebraic expression, *range* denotes an observation range, *weight* denotes a weighting expression, and *options* denotes a list of options.

11.1.1   varlist

Most commands that take a subsequent *varlist* do not require that you explicitly type one. If no *varlist* appears, these commands assume a *varlist* of _all, the Stata shorthand for indicating all the variables in the dataset. In commands that alter or destroy data, Stata requires that the *varlist* be specified explicitly. See [U] **11.4 varlists** for a complete description.

▷ Example 1

The summarize command lists the mean, standard deviation, and range of the specified variables. In [R] **summarize**, we see that the syntax diagram for summarize is

s̲ummarize $\left[\,varlist\,\right]$ $\left[\,if\,\right]$ $\left[\,in\,\right]$ $\left[\,weight\,\right]$ $\left[\,,\;options\,\right]$

Further down on the manual page is a table summarizing options, but let's focus on the syntax diagram itself, first. Since everything except the word summarize is enclosed in square brackets, the simplest form of the command is 'summarize'. Typing summarize without arguments is equivalent to typing summarize _all; all the variables in the dataset are summarized. Underlining denotes the shortest allowed abbreviation, so we could have typed just su; see [U] **11.2 Abbreviation rules**.

The table that defines options looks like this:

options	description
Main	
d̲etail	display additional statistics
meanonly	suppress the display; only calculate the mean; programmer's option
f̲ormat	use variable's display format
separator(#)	draw separator line after every # variables; default is separator(5)

Thus we learn we could also type, for instance, summarize, detail or summarize, detail format.

As another example, the drop command eliminates variables or observations from a dataset. When dropping variables, its syntax is

drop *varlist*

drop has no option table because it has no options.

In fact, nothing is optional. Typing drop by itself would result in the error message "varlist or in range required". To drop all the variables in the dataset, we must type drop _all.

Even before looking at the syntax diagram, we could have predicted that the *varlist* would be required—drop is destructive, so we are required to spell out our intent. The syntax diagram informs us that the *varlist* is required since *varlist* is not enclosed in square brackets. Since drop is not underlined, it cannot be abbreviated.

◁

11.1.2   by varlist:

The by *varlist*: prefix causes Stata to repeat a command for each subset of the data for which the values of the variables in the *varlist* are equal. When prefixed with by *varlist*:, the result of the command will be the same as if you had formed separate datasets for each group of observations, saved them, and then given the command on each dataset separately. The data must already be sorted by *varlist*, although by has a sort option; see [U] **11.5 by varlist: construct** for more information.

▷ Example 2

Typing `summarize marriage_rate divorce_rate` produces a table of the mean, standard deviation, and range of `marriage_rate` and `divorce_rate`, using all the observations in the data:

```
. use http://www.stata-press.com/data/r9/census12
(1980 Census data by state)
. summarize marriage_rate divorce_rate
    Variable |      Obs        Mean    Std. Dev.        Min        Max
-------------+--------------------------------------------------------
  marriage_r~e |       50    .0133221    .0188122    .0074654    .1428282
  divorce_rate |       50    .0056641    .0022473    .0029436    .0172918
```

Typing `by region: summarize marriage_rate divorce_rate` produces one table for each region of the country:

```
. sort region
. by region: summarize marriage_rate divorce_rate

-> region = N Cntrl
    Variable |      Obs        Mean    Std. Dev.        Min        Max
-------------+--------------------------------------------------------
  marriage_r~e |       12    .0099121    .0011326    .0087363    .0127394
  divorce_rate |       12    .0046974    .0011315    .0032817    .0072868

-> region = NE
    Variable |      Obs        Mean    Std. Dev.        Min        Max
-------------+--------------------------------------------------------
  marriage_r~e |        9    .0087811     .001191    .0075757    .0107055
  divorce_rate |        9     .004207    .0010264    .0029436    .0057071

-> region = South
    Variable |      Obs        Mean    Std. Dev.        Min        Max
-------------+--------------------------------------------------------
  marriage_r~e |       16    .0114654    .0025721    .0074654    .0172704
  divorce_rate |       16     .005633    .0013355    .0038917    .0080078

-> region = West
    Variable |      Obs        Mean    Std. Dev.        Min        Max
-------------+--------------------------------------------------------
  marriage_r~e |       13    .0218987    .0363775    .0087365    .1428282
  divorce_rate |       13    .0076037    .0031486    .0046004    .0172918
```

As mentioned, the dataset must be sorted on the by variables:

```
. use http://www.stata-press.com/data/r9/census12
(1980 Census data by state)
. by region: summarize marriage_rate divorce_rate
not sorted
r(5);
. sort region
. by region: summarize marriage_rate divorce_rate
( output appears)
```

Alternatively, we could have asked that by sort the data:

```
. by region, sort: summarize marriage_rate divorce_rate
( output appears)
```

by *varlist*: can be used with most Stata commands; we can tell which ones by looking at their syntax diagrams. For instance, we could obtain the correlations by `region`, between `marriage_rate` and `divorce_rate`, by typing `by region: correlate marriage_rate divorce_rate`.

◁

❏ Technical Note

The *varlist* in by *varlist*: may contain up to 32,766 variables with Stata/SE or 2,047 variables with Intercooled Stata; these are the maximum allowed in the dataset. For instance, if we had data on automobiles and wished to obtain means according to market category (`market`) broken down by manufacturer (`origin`), we could type `by market origin: summarize`. That *varlist* contains two variables: `market` and `origin`. If the data were not already sorted on `market` and `origin`, we would first type `sort market origin`.

❏

❏ Technical Note

The *varlist* in by *varlist*: may contain string variables, numeric variables, or both. In the example above, `region` is a string variable, in particular, a `str7`. The example would have worked, however, if `region` were a numeric variable with values 1, 2, 3, and 4, or even 12.2, 16.78, 32.417, and 152.13.

❏

11.1.3   if exp

The `if` *exp* qualifier restricts the scope of a command to those observations for which the value of the expression is *true* (which is equivalent to the expression being nonzero; see [U] **13 Functions and expressions**).

▷ Example 3

Typing `summarize marriage_rate divorce_rate if region=="West"` produces a table for the western region of the country:

```
. summarize marriage_rate divorce_rate if region == "West"

    Variable |     Obs        Mean    Std. Dev.        Min         Max
-------------+--------------------------------------------------------
 marriage_r~e |      13    .0218987    .0363775    .0087365    .1428282
 divorce_rate |      13    .0076037    .0031486    .0046004    .0172918
```

The double equal sign in `region=="West"` is not an error. Stata uses *double* equal signs to denote equality testing and a *single* equal sign to denote assignment; see [U] **13 Functions and expressions**.

A command may have at most one `if` qualifier. If you want the summary for the West restricted to observations with values of `marriage_rate` in excess of 0.015, do *not* type `summarize marriage_rate divorce_rate if region=="West" if marriage_rate>.015`. Type instead

```
. summarize marriage_rate divorce_rate if region == "West" & marriage_rate >.015

    Variable |     Obs        Mean    Std. Dev.        Min         Max
-------------+--------------------------------------------------------
 marriage_r~e |       1    .1428282           .    .1428282    .1428282
 divorce_rate |       1    .0172918           .    .0172918    .0172918
```

You may not use the word *and* in place of the symbol '&' to join conditions. To select observations that meet one condition *or* another, use the '|' symbol. For instance, summarize marriage_rate divorce_rate if region=="West" | marriage_rate>.015 summarizes all observations for which region is West *or* marriage_rate is greater than 0.015.

◁

▷ Example 4

if may be combined with by. Typing by region: summarize marriage_rate divorce_rate if marriage_rate>.015 produces a set of tables, one for each region, reflecting summary statistics on marriage_rate and divorce_rate among observations for which marriage_rate exceeds 0.015:

```
. by region: summarize marriage_rate divorce_rate if marriage_rate >.015
```

-> region = N Cntrl

Variable	Obs	Mean	Std. Dev.	Min	Max
marriage_r~e	0				
divorce_rate	0				

-> region = NE

Variable	Obs	Mean	Std. Dev.	Min	Max
marriage_r~e	0				
divorce_rate	0				

-> region = South

Variable	Obs	Mean	Std. Dev.	Min	Max
marriage_r~e	2	.0163219	.0013414	.0153734	.0172704
divorce_rate	2	.0061813	.0025831	.0043548	.0080078

-> region = West

Variable	Obs	Mean	Std. Dev.	Min	Max
marriage_r~e	1	.1428282	.	.1428282	.1428282
divorce_rate	1	.0172918	.	.0172918	.0172918

The results indicate that there are no states in the Northeast and North Central regions for which marriage_rate exceeds 0.015, while there are two such states in the South and one state in the West.

◁

11.1.4 in range

The in *range* qualifier restricts the scope of the command to a specific observation range. A range specification takes the form $\#_1[/\#_2]$, where $\#_1$ and $\#_2$ are positive or negative integers. Negative integers are understood to mean "from the end of the data", with -1 referring to the last observation. The implied first observation must be less than or equal to the implied last observation.

The first and last observations in the dataset may be denoted by f and l (lowercase letter), respectively. A range specifies absolute observation numbers within a dataset. As a result, the in qualifier may not be used when the command is preceded by the by *varlist*: prefix; see [U] **11.5 by varlist: construct**.

▷ Example 5

Typing summarize marriage_rate divorce_rate in 5/25 produces a table based on the values of marriage_rate and divorce_rate in observations 5 through 25:

```
. summarize marriage_rate divorce_rate in 5/25
    Variable |      Obs        Mean    Std. Dev.        Min        Max
-------------+--------------------------------------------------------
 marriage_r~e |       21    .0096001    .0013263    .0075757    .0125884
 divorce_rate |       21     .004726    .0012025    .0029436    .0072868
```

This is, admittedly, a rather odd thing to want to do. It would not be odd, however, if we substituted list for summarize. If we wanted to see the states with the 10 lowest values of marriage_rate, we could type sort marriage_rate followed by list marriage_rate in 1/10.

Typing summarize marriage_rate divorce_rate in f/l is equivalent to typing summarize marriage_rate divorce_rate—all observations are summarized.

◁

▷ Example 6

Typing summarize marriage_rate divorce_rate in 5/25 if region=="South" produces a table based on the values of the two variables in observations 5 through 25 for which the value of region is South:

```
. summarize marriage_rate divorce_rate in 5/25 if region=="South"
    Variable |      Obs        Mean    Std. Dev.        Min        Max
-------------+--------------------------------------------------------
 marriage_r~e |        4    .0108886    .0015061    .0089201    .0125884
 divorce_rate |        4    .0054448    .0011166    .0041485    .0068685
```

The ordering of the in and if qualifiers is not significant. The command could also have been specified as summarize marriage_rate divorce_rate if region=="South" in 5/25.

◁

▷ Example 7

Negative in ranges can be usefully employed with sort. For instance, we have data on automobiles and wish to list the five with the highest mileage ratings:

```
. use http://www.stata-press.com/data/r9/auto
(1978 Automobile Data)
. sort mpg
. list make mpg in -5/l
```

	make	mpg
70.	Toyota Corolla	31
71.	Plym. Champ	34
72.	Subaru	35
73.	Datsun 210	35
74.	VW Diesel	41

◁

11.1.5   =exp

=exp specifies the value to be assigned to a variable and is most often used with `generate` and `replace`. See [U] **13 Functions and expressions** for details on expressions and [D] **generate** for details on the `generate` and `replace` commands.

▷ Example 8

Expression	Meaning
generate newvar=oldvar+2	creates a new variable named `newvar` equal to `oldvar`+2
replace oldvar=oldvar+2	changes the contents of the existing variable `oldvar`
egen newvar=rank(oldvar)	creates `newvar` containing the ranks of `oldvar` (see [D] **egen**)

◁

11.1.6   weight

weight indicates the weight to be attached to each observation. The syntax of *weight* is

$$[weightword=exp]$$

where you actually type the square brackets, and where *weightword* is one of

weightword	Meaning
<u>we</u>ight	default treatment of weights
<u>fw</u>eight *or* <u>frequency</u>	frequency weights
<u>pw</u>eight	sampling weights
<u>aw</u>eight *or* <u>cellsize</u>	analytic weights
<u>iw</u>eight	importance weights

The underlining indicates the minimum acceptable abbreviation. Thus, `weight` may be abbreviated `w` or `we`, etc.

▷ Example 9

Before explaining what the different types of weights mean, let's obtain the population-weighted mean of a variable called `median_age` from data containing observations on all 50 states of the United States. The dataset also contains a variable named `pop`, which is the total population of each state.

```
. use http://www.stata-press.com/data/r9/census12
(1980 Census data by state)
. summarize median_age [weight=pop]
(analytic weights assumed)
```

Variable	Obs	Weight	Mean	Std. Dev.	Min	Max
median_age	50	225907472	30.11047	1.66933	24.2	34.7

In addition to telling us that our dataset contains 50 observations, Stata informs us that the sum of the weight is 225,907,472, which was the number of people living in the U.S. as of the 1980 census. The weighted mean is 30.11. We were also informed that Stata assumed that we wanted "analytic" weights.

◁

`weight` is each command's idea of what the "natural" weights are and is one of `fweight`, `pweight`, `aweight`, or `iweight`. When you specify the vague `weight`, the command informs you which kind it assumes. Not every command supports every kind of weight. A note below the syntax diagram for a command will tell you which weights the command supports.

Stata understands four kinds of weights:

1. `fweight`s, or frequency weights, indicate duplicated observations. `fweight`s are always integers. If the `fweight` associated with an observation is 5, that means there are really 5 such observations, each identical.

2. `pweight`s, or sampling weights, denote the inverse of the probability that this observation is included in the sample due to the sampling design. A `pweight` of 100, for instance, indicates that this observation is representative of 100 subjects in the underlying population. The scale of these weights does not matter in terms of estimated parameters and standard errors, except when estimating totals and computing finite-population corrections with the `svy` commands; see [SVY] **survey**.

3. `aweight`s, or analytic weights, are inversely proportional to the variance of an observation; i.e., the variance of the jth observation is assumed to be σ^2/w_j, where w_j are the weights. Typically, the observations represent averages, and the weights are the number of elements that gave rise to the average. For most Stata commands, the recorded scale of `aweight`s is irrelevant; Stata internally rescales them to sum to N, the number of observations in your data, when it uses them.

4. `iweight`s, or importance weights, indicate the relative "importance" of the observation. They have no formal statistical definition; this is a catchall category. Any command that supports `iweight`s will define how they are treated. In most cases, they are intended for use by programmers who want to produce a certain computation.

See [U] **20.16 Weighted estimation** for a thorough discussion of weights and their meaning.

❑ Technical Note

When you do not specify a weight, the result is equivalent to specifying [fweight=1].

❑

11.1.7 options

Many commands take command-specific options. These are described along with each command in the *Reference* manuals. Options are indicated by typing a comma at the end of the command, followed by the options you want to use.

▷ Example 10

Typing `summarize marriage_rate` produces a table of the mean, standard deviation, minimum, and maximum of the variable `marriage_rate`:

```
. summarize marriage_rate

    Variable |     Obs        Mean    Std. Dev.       Min        Max
-------------+--------------------------------------------------------
 marriage_r~e |      50    .0133221    .0188122    .0074654    .1428282
```

The syntax diagram for `summarize` is

$\underline{\text{su}}\text{mmarize}\ \left[\textit{varlist}\right]\ \left[\textit{if}\right]\ \left[\textit{in}\right]\ \left[\textit{weight}\right]\ \left[\,,\ \textit{options}\right]$

followed by the option table

options	description
Main	
<u>d</u>etail	display additional statistics
<u>mean</u>only	suppress the display; only calculate the mean; programmer's option
<u>f</u>ormat	use variable's display format
<u>sep</u>arator(*#*)	draw separator line after every *#* variables; default is `separator(5)`

Thus, the options allowed by `summarize` are `detail` or `meanonly`, `format`, and `separator()`. The shortest allowed abbreviations for these options are `d` for `detail`, `mean` for `meanonly`, `f` for `format`, and `sep` for `separator()`; see [U] **11.2 Abbreviation rules**.

Typing `summarize marriage_rate, detail` produces a table that also includes selected percentiles, the four largest and four smallest values, the skewness, and the kurtosis.

```
. summarize marriage_rate, detail

                        marriage_rate
-------------------------------------------------------------
      Percentiles      Smallest
 1%    .0074654        .0074654
 5%    .0078956        .0075757
10%    .0080043        .0078956       Obs                  50
25%    .0089399        .0079079       Sum of Wgt.          50

50%    .0105669                       Mean           .0133221
                       Largest        Std. Dev.      .0188122
75%    .0122899        .0146266
90%    .0137832        .0153734       Variance       .0003539
95%    .0153734        .0172704       Skewness       6.718494
99%    .1428282        .1428282       Kurtosis       46.77306
```

◁

❏ Technical Note

Once you have typed the *varlist* for the command, you can place options anywhere in the command. You can type summarize marriage_rate divorce_rate if region=="West", detail, or you can type summarize marriage_rate divorce_rate, detail, if region=="West". Note that you use a second comma to indicate a return to the command line as opposed to the option list. Leaving out the comma after the word detail would cause an error because Stata would attempt to interpret the phrase if region=="West" as an option rather than as part of the command.

You may not type an option in the middle of a *varlist*. Typing summarize marriage_rate, detail, divorce_rate will result in an error.

Options need not be specified contiguously. You may type summarize marriage_rate divorce_rate, detail, if region=="South", noformat. Both detail and noformat are options.

❏

❏ Technical Note

Most options are toggles—they indicate that something either is or is not to be done. Sometimes it is difficult to remember which is the default. The following rule applies to all options: If *option* is an option, then no*option* is an option as well, and vice versa. Thus, if we could not remember whether detail or nodetail were the default for summarize but we knew that we did not want the detail, we could type summarize, nodetail. Typing the nodetail option is unnecessary, but Stata will not complain.

Some options take *arguments*. The Stata kdensity command has an n(#) option that indicates the number of points at which the density estimate is to be evaluated. When an option takes an argument, the argument is enclosed in parentheses.

Some options take more than one argument. In such cases, arguments should be separated from one another by commas. For instance, you might see in a syntax diagram

saving(*filename*[, replace])

In this case, replace is the (optional) second argument. *Lists*, such as lists of variables (varlists) and lists of numbers (numlists), are considered to be one argument. If a syntax diagram reported

powers(*numlist*)

the list of numbers would be one argument, so the elements would not be separated by commas. You would type, for instance, powers(1 2 3 4). In fact, Stata will tolerate commas in this case, so you could type powers(1,2,3,4).

Some options take string arguments. regress has an eform() option that works this way—for instance, eform("Exp Beta"). To play it safe, you should type the quotes surrounding the string, although it is not required. If you do not type the quotes, any sequence of two or more consecutive blanks will be interpreted as a single blank. Thus, eform(Exp   beta) would be interpreted the same as eform(Exp beta).

❏

11.1.8   numlist

A *numlist* is a list of numbers. Stata allows certain shorthands to indicate ranges:

numlist	meaning
2	just one number
1 2 3	three numbers
3 2 1	three numbers in reversed order
.5 1 1.5	three different numbers
1 3 -2.17 5.12	four numbers in jumbled order
1/3	three numbers: 1, 2, 3
3/1	the same three numbers in reverse order
5/8	four numbers: 5, 6, 7, 8
-8/-5	four numbers: -8, -7, -6, -5
-5/-8	four numbers: -5, -6, -7, -8
-1/2	four numbers: -1, 0, 1, 2
1 2 to 4	four numbers: 1, 2, 3, 4
4 3 to 1	four numbers: 4, 3, 2, 1
10 15 to 30	five numbers: 10, 15, 20, 25, 30
1 2:4	same as 1 2 to 4
4 3:1	same as 4 3 to 1
10 15:30	same as 10 15 to 30
1(1)3	three numbers: 1, 2, 3
1(2)9	five numbers: 1, 3, 5, 7, 9
1(2)10	the same five numbers, 1, 3, 5, 7, 9
9(-2)1	five numbers: 9, 7, 5, 3, and 1
-1(.5)2.5	the numbers -1, $-.5$, 0, .5, 1, 1.5, 2, 2.5
1[1]3	same as 1(1)3
1[2]9	same as 1(2)9
1[2]10	same as 1(2)10
9[-2]1	same as 9(-2)1
-1[.5]2.5	same as -1(.5)2.5
1 2 3/5 8(2)12	eight numbers: 1, 2, 3, 4, 5, 8, 10, 12
1,2,3/5,8(2)12	the same eight numbers
1 2 3/5 8 10 to 12	the same eight numbers
1,2,3/5,8,10 to 12	the same eight numbers
1 2 3/5 8 10:12	the same eight numbers

`poisson`'s `constraints()` option has syntax `constraints(`*numlist*`)`. Thus, you could type `con-`
`straints(2 4 to 8)`, `constraints(2(2)8)`, etc.

11.1.9   datelist

A *datelist* is a list of dates used with graph options, mainly when the variable being graphed has a date format.

Dates in Stata are recorded as the number of days since 01jan1960, so 0 means 01jan1960, 1 means 02jan1960, and 16,541 means 15apr2005. Similarly, -1 means 31dec1959, -2 means 30dec1959, and $-16,541$ means 18sep1914. That is how dates are recorded if we are talking in terms of daily dates. Stata has weekly, monthly, quarterly, halfyearly, and yearly dates, too. Each is recorded with 1 referring to the first week, month, quarter, halfyear, or year following 31dec1959.

A datelist is either a list of dates, as in

    15apr1973break 17apr1973 20apr1973 23apr1973

or it is a first and last date with an increment in between, as in

    17apr1973(3)23apr1973

or it is a combination:

> 15apr1973 17apr1973(3)23apr1973

Dates specified with spaces, slashes, or commas must be bound in parentheses, as in

> (15 apr 1973) (april 17, 1973)(3)(april 23, 1973)

Evenly spaced calendar dates are not especially useful, but with other time units, even spacing can be useful, as in

> 1999q1(1)2005q1

which means every quarter between 1991q1 and 2005q1, or

> 1999q1(4)2005q1

which means every first quarter.

To interpret a datelist, Stata first looks at the format of the variable and then uses the corresponding date-to-numeric translation function. For instance, if the variable has a %td format (meaning that it is daily), the d() function is used to translate the date. If the variable has a %tq format (meaning that it is quarterly), the q() function is used. The full list is

Format of variable	meaning	translation function	example
%td	daily	d()	15apr1973
%tw	weekly	w()	1980w10
%tm	monthly	m()	2001m2
%tq	quarterly	q()	1999q1
%th	half yearly	h()	2001h1
%ty	yearly	y()	2001

11.1.10   Prefix commands

Stata has a handful of commands that are used to prefix other Stata commands. by varlist:, discussed in section [U] **11.1.2 by varlist:**, is in fact an example of a prefix command. In that section, we demonstrated by using

> by region: summarize marriage_rate divorce_rate

and later,

> by region, sort: summarize marriage_rate divorce_rate

and although we did not, we could also have demonstrated

> by region, sort: summarize marriage_rate divorce_rate, detail

Each of the above runs the summarize command separately on the data for each region.

Note that by itself follows standard Stata syntax:

$$\text{by } \textit{varlist}\big[\textit{, options}\big]: \ \ldots$$

In 'by region, sort: summarize marriage_rate divorce_rate, detail', region is by's varlist and sort is by's option, just as marriage_rate and divorce_rate are summarize's varlist and detail is summarize's option.

by is not the only prefix command, and the full list of such commands is

Prefix command	description
by	run command on subsets of data
statsby	same as by, but collect statistics from each run
rolling	run command on moving subsets, and collect statistics
bootstrap	run command on bootstrap samples
jackknife	run command on jackknife subsets of data
permute	run command on random permutations
simulate	run command on manufactured data
svy	run command and adjust results for survey sampling
stepwise	run command with stepwise variable inclusion/exclusion
xi	run command after expanding factor variables & interactions
capture	run command and capture its return code
noisily	run command and show the output
quietly	run command and suppress the output
version	run command under specified version

The last group—capture, noisily, quietly, and version—have to do with programming Stata and, for historical reasons, capture, noisily, and quietly allow you to omit the colon, so one programmer might code

        quietly regress ...

and another

        quietly: regress ...

All the other prefix commands require the colon.

11.2 Abbreviation rules

Stata allows abbreviations. In this manual, we usually avoid abbreviating commands, variable names, and options to ensure readability:

        . summarize myvar, detail

Experienced Stata users, on the other hand, tend to abbreviate the same command as

        . sum myv, d

As a general rule, command, option, and variable names may be abbreviated to the shortest string of characters that uniquely identifies them.

This rule is violated if the command or option does something that cannot easily be undone; in that case, the command must be spelled out in its entirety.

In addition, a few common commands and options are allowed to have even shorter abbreviations than the general rule would allow.

The general rule is applied, without exception, to variable names.

11.2.1   Command abbreviation

The shortest allowed abbreviation for a command or option can be determined by looking at the command's syntax diagram. This minimal abbreviation is shown by underlining:

regress

rename

replace

rotate

run

If there is no underlining, no abbreviation is allowed. For example, `replace` may not be abbreviated, the underlying reason being that `replace` changes the data.

`regress` can be abbreviated `reg`, `regr`, `regre`, or `regres`, or it can be spelled out in its entirety.

As mentioned above, sometimes very short abbreviations are also allowed. Commands that begin with the letter *d* include `decode`, `describe`, `destring`, `dir`, `discard`, `display`, `do`, and `drop`. This suggests that the shortest allowable abbreviation for `describe` is `desc`. However, since `describe` is such a commonly used command, you may abbreviate it with the single letter d. You may also abbreviate the `list` command with the single letter l.

The other exception to the general abbreviation rule is that commands that alter or destroy data must be spelled out completely. Two commands that begin with the letter *d*, `discard` and `drop`, are destructive in the sense that, once you give one of these commands, there is no way to undo the result. Therefore, both must be spelled out.

The final exceptions to the general rule are commands implemented as ado-files. Such commands may not be abbreviated. Ado-file commands are external, and their names correspond to the names of disk files.

11.2.2   Option abbreviation

Option abbreviation follows the same logic as command abbreviation: You determine the minimum acceptable abbreviation by examining the command's syntax diagram. The syntax diagram for `summarize` reads, in part,

summarize ... , detail format

Option `detail` may be abbreviated d, de, det, ..., `detail`. Similarly, option `format` may be abbreviated f, fo, ..., `format`.

Options `clear` and `replace` occur with many commands. The `clear` option indicates that even though completing this command will result in the loss of all data in memory, and even though the data in memory have changed since the data were last saved on disk, you want to continue. `clear` must be spelled out, as in `use newdata, clear`.

The `replace` option indicates that it's okay to save over an existing dataset. If you type `save mydata` and the file `mydata.dta` already exists, you will receive the message "file mydata.dta already exists", and Stata will refuse to overwrite it. To allow Stata to overwrite the dataset, you would type `save mydata, replace`. `replace` may not be abbreviated.

❏ Technical Note

replace is a stronger modifier than clear and is one you should think about before using. With a mistaken clear, you can lose hours of work, but with a mistaken replace, you can lose days of work.

❏

11.2.3   Variable-name abbreviation

1. Variable names may be abbreviated to the shortest string of characters that uniquely identifies them given the data currently loaded in memory.

   If your dataset contained four variables, state, mrgrate, dvcrate, and dthrate, you could refer to the variable dvcrate as dvcrat, dvcra, dvcr, dvc, or dv. You might type list dv to list the data on dvcrate. You could not refer to the variable dvcrate as d, however, since that abbreviation does not distinguish dvcrate from dthrate. If you were to type list d, Stata would respond with the message "ambiguous abbreviation". (If you wanted to refer to *all* variables that started with the letter *d*, you could type list d*; see [U] **11.4 varlists**.)

2. The character ~ may be used to mean that "zero or more characters go here". For instance, r~8 might refer to the variable rep78, or rep1978, or repair1978, or just r8. (The ~ character is similar to the * character in [U] **11.4 varlists**, except that it adds the restriction "and only one variable matches this specification".)

   In (1), we said that you could abbreviate variables. You could type dvcr to refer to dvcrate, but, if there were more than one variable that started with the letters dvcr, you would receive an error. Note that typing dvcr is the same as typing dvcr~.

11.3   Naming conventions

A name is a sequence of one to thirty-two letters (A–Z and a–z), digits (0–9), and underscores (_).

Programmers: local macro names can have no more than 31 characters in the name; see [U] **18.3.1 Local macros**.

Stata reserves the following names:

_all	double	long	_rc
_b	float	_n	_se
byte	if	_N	_skip
_coef	in	_pi	using
_cons	int	_pred	with

You may not use these reserved names for your variables.

The first character of a name must be a letter or an underscore. We recommend, however, that you not begin your variable names with an underscore. All of Stata's built-in variables begin with an underscore, and we reserve the right to incorporate new _*variables* freely.

Stata respects case; that is, myvar, Myvar, and MYVAR are three distinct names.

All objects in Stata—not just variables—follow this naming convention.

11.4 varlists

A *varlist* is a list of variable names. The variable names in a *varlist* refer either exclusively to new (not yet created) variables or exclusively to existing variables. A *newvarlist* always refers exclusively to new (not yet created) variables.

11.4.1 Lists of existing variables

In lists of existing variable names, variable names may be repeated.

▷ Example 11

If you type `list state mrgrate dvcrate state`, the variable `state` will be listed twice, once in the leftmost column and again in the rightmost column of the list.

◁

Existing variable names may be abbreviated as described in [U] **11.2 Abbreviation rules**. You may also use '*' to indicate that "zero or more characters go here". For instance, if you suffix * to a partial variable name (for example, sta*), you are referring to all variable names that start with that letter combination. If you prefix * to a letter combination (for example, *rate), you are referring to all variables that end in that letter combination. If you put * in the middle (for example, m*rate), you are referring to all variables that begin and end with the specified letters. You may put more than one * in an abbreviation.

▷ Example 12

If the variables `poplt5`, `pop5to17`, and `pop18p` are in our dataset, we may type `pop*` as a shorthand way to refer to all three variables. For instance, `list state pop*` lists the variables `state`, `poplt5`, `pop5to17`, and `pop18p`.

If we had a dataset with variables `inc1990`, `inc1991`, ..., `inc1999` along with variables `incfarm1990`, ..., `incfarm1999`; `pop1990`, ..., `pop1999`; and `ms1990`, ..., `ms1999`; then `*1995` would be a shorthand way of referring to `inc1995`, `incfarm1995`, `pop1995`, and `ms1995`. We could type, for instance, `list *1995`.

In that same dataset, typing `list i*95` would be a shorthand way of listing `inc1995` and `incfarm1995`.

Typing `list i*f*95` would be a shorthand way of listing to `incfarm1995`.

◁

~ is an alternative to *, and really, it means the same thing. The difference is that ~ indicates that if more than one variable matches the specified pattern, Stata will complain rather than substituting all the variables that match the specification.

▷ Example 13

In the previous example, we could have typed `list i~f~95` to list `incfarm1995`. If, however, our dataset also included variable `infant1995`, then `list i*f*95` would list both variables and `list i~f~95` would complain that `i~f~95` is an ambiguous abbreviation.

◁

You may use ? to specify that one character goes here. Remember, * means zero or more characters go here, so ?* can be used to mean one or more characters goes here, ??* can be used to mean two or more characters go here, and so on.

▷ Example 14

In a dataset containing variables rep1, rep2, ..., rep78, rep? would refer to rep1, rep2, ..., rep9, and rep?? would refer to rep10, rep11, ..., rep78.

◁

You may place a dash (−) between two variable names to specify all the variables stored between the two listed variables, inclusive. You can determine storage order using describe; it lists variables in the order in which they are stored.

▷ Example 15

If the dataset contains the variables state, mrgrate, dvcrate, and dthrate, in that order, typing list state-dvcrate is equivalent to typing list state mrgrate dvcrate. In both cases, three variables are listed.

◁

11.4.2   Lists of new variables

In lists of *new variables*, no variable names may be repeated or abbreviated.

You may specify a dash (−) between two variable names that have the same letter prefix and that end in numbers. This form of the dash notation indicates a range of variable names in ascending numerical order.

For example, typing input v1-v4 is equivalent to typing input v1 v2 v3 v4. Typing infile state v1-v3 ssn using rawdata is equivalent to typing infile state v1 v2 v3 ssn using rawdata.

You may specify the storage type before the variable name to force a storage type other than the default. The numeric storage types are byte, int, long, float (the default), and double. The string storage types are str#, where # is replaced with an integer between 1 and 80 (244 for Stata/SE), inclusive, representing the maximum length of the string. See [U] **12 Data**.

For instance, the list var1 str8 var2 var3 specifies that var1 and var3 be given the default storage type, and that var2 be stored as a str8—a string whose maximum length is eight characters.

The list var1 int var2 var3 specifies that var2 be stored as an int. You may use parentheses to bind a list of variable names. The list var1 int(var2 var3) specifies that both var2 and var3 be stored as ints. Similarly, the list var1 str20(var2 var3) specifies that both var2 and var3 be stored as str20s. The different storage types are listed in [U] **12.2.2 Numeric storage types** and [U] **12.4.4 String storage types**.

▷ Example 16

Typing infile str2 state str10 region v1-v5 using mydata reads the state and region strings from the file mydata.raw and stores them as str2 and str10, respectively, along with the variables v1 through v5, which are stored as the default storage type float (unless we have specified a different default with the set type command).

Typing `infile str10(state region) v1-v5 using mydata` would achieve almost the same result, except that the `state` and `region` values recorded in the data would both be assigned to `str10` variables. (We could then use the `compress` command to shorten the strings. See [D] **compress**; it is well worth reading.)

◁

❑ Technical Note

You may append a colon and a *value label name* to numeric variables. (See [U] **12.6 Dataset, variable, and value labels** for a description of value labels.) For instance, `var1 var2:myfmt` specifies that the variable `var2` be associated with the value label stored under the name `myfmt`. This has the same effect as typing the list `var1 var2` and then subsequently giving the command `label values var2 myfmt`.

The advantage of specifying the value label association with the colon notation is that value labels can then be assigned by the current command; see [D] **input** and [D] **infile (free format)**.

❑

▷ Example 17

Typing `infile int(state:stfmt region:regfmt) v1-v5 using mydata, automatic` reads the state and region data from the file `mydata.raw` and stores them as `int`s, along with the variables `v1` through `v5`, which are stored as the default storage type.

In our previous example, both state and region were strings, so how can strings be stored in a numeric variable? See [U] **12.6 Dataset, variable, and value labels** for the complete answer. The colon notation specifies the name of the value label, and the `automatic` option tells Stata to assign unique numeric codes to all character strings. The numeric code for state, which Stata will make up on the fly, will be stored in the `state` variable. The mapping from numeric codes to words will be stored in the value label named `stfmt`. Similarly, regions will be assigned numeric codes, which are stored in `region`, and the mapping will be stored in `regfmt`.

If we were to `list` the data, the `state` and `region` variables would look like strings. `state`, for instance, would appear to contain things like `AL`, `CA`, and `WA`, but actually it would contain only numbers like 1, 2, 3, and 4.

◁

11.4.3 Time-series varlists

Time-series varlists are a variation on varlists of existing variables. When a command allows a time-series varlist, you may include time-series operators. For instance, `L.gnp` refers to the lagged value of variable `gnp`. The time-series operators are

operator	meaning
L.	lag x_{t-1}
L2.	2-period lag x_{t-2}
...	
F.	lead x_{t+1}
F2.	2-period lead x_{t+2}
...	
D.	difference $x_t - x_{t-1}$
D2.	difference of difference $x_t - x_{t-1} - (x_{t-1} - x_{t-2}) = x_t - 2x_{t-1} + x_{t-2}$
...	
S.	"seasonal" difference $x_t - x_{t-1}$
S2.	lag-2 (seasonal) difference $x_t - x_{t-2}$
...	

Time-series operators may be repeated and combined. L3.gnp refers to the third lag of variable gnp. So do LLL.gnp, LL2.gnp, and L2L.gnp. LF.gnp is the same as gnp. DS12.gnp refers to the one-period difference of the 12-period difference. LDS12.gnp refers to the same concept, lagged once.

Note that D1. = S1., but D2. ≠ S2., D3. ≠ S3., and so on. D2. refers to the difference of the difference. S2. refers to the two-period difference. If you wanted the difference of the difference of the 12-period difference of gnp, you would write D2S12.gnp.

Operators may be typed in uppercase or lowercase. Most users would type d2s12.gnp instead of D2S12.gnp.

You may type operators however you wish; Stata internally converts operators to their canonical form. If you typed ld2ls12d.gnp, Stata would present the operated variable as L2D3S12.gnp.

In addition to using *operator#*, Stata understands *operator(numlist)* to mean a set of operated variables. For instance, typing L(1/3).gnp in a varlist is the same as typing 'L.gnp L2.gnp L3.gnp'. The operators can also be applied to a list of variables by enclosing the variables in parentheses; e.g.,

```
. list year L(1/3).(gnp cpi)
```

	year	L.gnp	L2.gnp	L3.gnp	L.cpi	L2.cpi	L3.cpi
1.	1989	.	.	.	.	.	.
2.	1990	5452.8	.	.	100	.	.
3.	1991	5764.9	5452.8	.	105	100	.
4.	1992	5932.4	5764.9	5452.8	108	105	100
		(output omitted)					
8.	1996	7330.1	6892.2	6519.1	122	119	112

In *operator#*, making # zero returns the variable itself. L0.gnp is gnp. Thus the above listing could have been produced by typing list year l(0/3).gnp.

The parentheses notation may be used with any operator. Typing D(1/3).gnp would return the first through third differences.

The parentheses notation may be used in operator lists with multiple operators, such as L(0/3)D2S12.gnp.

Operator lists may include up to one set of parentheses, which may enclose a *numlist*; see [U] **11.1.8 numlist**.

Before you can use time-series operators in varlists, you must set the time variable using the `tsset` command:

```
. list l.gnp
time variable not set
r(111);

. tsset time
  (output omitted)

. list l.gnp
  (output omitted)
```

See [TS] **tsset**. The time variable must take on integer values. In addition, the data must be sorted on the time variable. `tsset` handles this, but later you might encounter

```
. list l.mpg
not sorted
r(5);
```

In that case, type `sort time` or type `tsset` to re-establish the order.

The time-series operators respect the time variable. `L2.gnp` refers to gnp_{t-2}, regardless of missing observations in the dataset. In the following dataset, the observation for 1992 is missing:

```
. list year gnp l2.gnp, separator(0)
```

	year	gnp	L2.gnp
1.	1989	5,452.8	.
2.	1990	5,764.9	.
3.	1991	5,932.4	5,452.8
4.	1993	6,560.0	5,932.4
5.	1994	6,922.4	.
6.	1995	7,237.5	6,560.0

← note, filled in correctly

Operated variables may be used in expressions:

```
. generate gnplag2 = l2.gnp
(3 missing values generated)
```

Stata also understands panel (cross-sectional time-series) data, as well as simple time-series data. If you have cross sections of time series, you indicate this when you `tsset` the data:

```
. tsset country year
```

See [TS] **tsset** and [U] **24.3 Time-series dates**.

11.5   by varlist: construct

by *varlist*: *command*

The by prefix causes *command* to be repeated for each unique set of values of the variables in the *varlist*. *varlist* may contain numeric, string, or a mixture of numeric and string variables. (*varlist* may not contain time-series operators.)

by is an optional prefix to perform a Stata command separately for each group of observations where the values of the variables in the *varlist* are the same.

During each iteration, the values of the system variables _n and _N are set in relation to the first observation in the by-group; see [U] **13.7 Explicit subscripting**. The in *range* qualifier cannot be used with by *varlist*: because ranges specify absolute rather than relative observation numbers.

❏ Technical Note

The inability to combine in and by is not really a constraint since if provides all the functionality of in and quite a bit more. If you wanted to perform *command* for the first three observations in each of the by-groups, you could type

. by *varlist*: *command* if _n<=3

❏

The results of *command* would be the same as if you had formed separate datasets for each group of observations, saved them, used each separately, and issued *command*.

▷ Example 18

We provide some examples using by in [U] **11.1.2 by varlist:** above. We demonstrate the effect of by on _n, _N, and explicit subscripting in [U] **13.7 Explicit subscripting**.

by requires that the data first be sorted. For instance, if we had data on the average January and July temperatures in degrees Fahrenheit for 420 cities located in the Northeast and West and wanted to obtain the averages, by region, across those cities, we might type

```
. use http://www.stata-press.com/data/r9/citytemp3
(City Temperature Data)
. by region: summarize tempjan tempjuly
not sorted
r(5);
```

Stata refused to honor our request since the data are not sorted by region. We must either sort the data by region first (see [D] **sort**) or specify by's sort option (which has the same effect):

```
. by region, sort: summarize tempjan tempjuly
```

| -> region = NE | | | | | |
Variable	Obs	Mean	Std. Dev.	Min	Max
tempjan	164	27.88537	3.543096	16.6	31.8
tempjuly	164	73.35	2.361203	66.5	76.8

| -> region = N Cntrl | | | | | |
Variable	Obs	Mean	Std. Dev.	Min	Max
tempjan	284	21.69437	5.725392	2.2	32.6
tempjuly	284	73.46725	3.103187	64.5	81.4

| -> region = South | | | | | |
Variable	Obs	Mean	Std. Dev.	Min	Max
tempjan	250	46.1456	10.38646	28.9	68
tempjuly	250	80.9896	2.97537	71	87.4

```
-> region = West
    Variable |     Obs        Mean    Std. Dev.       Min        Max
    ---------+-------------------------------------------------------
     tempjan |     256    46.22539    11.25412          13       72.6
    tempjuly |     256    72.10859    6.483131        58.1       93.6
```

◁

▷ Example 19

Using the same data as in the example above, we estimate regressions, by region, of average January temperature on average July temperature. Both temperatures are specified in degrees Fahrenheit.

```
. by region: regress tempjan tempjuly
```

```
-> region = NE
      Source |       SS       df       MS              Number of obs =      164
    ---------+------------------------------           F(  1,   162) =   479.82
       Model | 1529.74026     1  1529.74026            Prob > F      =   0.0000
    Residual | 516.484453   162  3.18817564            R-squared     =   0.7476
    ---------+------------------------------           Adj R-squared =   0.7460
       Total | 2046.22471   163  12.5535258            Root MSE      =   1.7855
```

```
     tempjan |      Coef.   Std. Err.       t    P>|t|     [95% Conf. Interval]
    ---------+--------------------------------------------------------------------
    tempjuly |   1.297424   .0592303      21.90   0.000     1.180461    1.414387
       _cons |  -67.28066   4.346781     -15.48   0.000    -75.86431     -58.697
```

```
-> region = N Cntrl
      Source |       SS       df       MS              Number of obs =      284
    ---------+------------------------------           F(  1,   282) =   115.89
       Model | 2701.97917     1  2701.97917            Prob > F      =   0.0000
    Residual | 6574.79175   282  23.3148644            R-squared     =   0.2913
    ---------+------------------------------           Adj R-squared =   0.2887
       Total | 9276.77092   283  32.7801093            Root MSE      =   4.8285
```

```
     tempjan |      Coef.   Std. Err.       t    P>|t|     [95% Conf. Interval]
    ---------+--------------------------------------------------------------------
    tempjuly |   .9957259   .0924944      10.77   0.000     .8136589    1.177793
       _cons |  -51.45888   6.801344      -7.57   0.000    -64.84673   -38.07103
```

```
-> region = South
      Source |       SS       df       MS              Number of obs =      250
    ---------+------------------------------           F(  1,   248) =    95.17
       Model | 7449.51623     1  7449.51623            Prob > F      =   0.0000
    Residual | 19412.2231   248  78.2750933            R-squared     =   0.2773
    ---------+------------------------------           Adj R-squared =   0.2744
       Total | 26861.7394   249  107.878471            Root MSE      =   8.8473
```

```
     tempjan |      Coef.   Std. Err.       t    P>|t|     [95% Conf. Interval]
    ---------+--------------------------------------------------------------------
    tempjuly |    1.83833   .1884392       9.76   0.000     1.467185    2.209475
       _cons |    -102.74   15.27187      -6.73   0.000    -132.8191   -72.66089
```

```
-> region = West
    Source |       SS       df       MS              Number of obs =     256
-----------+------------------------------           F(  1,    254) =    2.84
     Model | 357.161728      1  357.161728           Prob > F      =  0.0932
  Residual | 31939.9031    254   125.74765           R-squared     =  0.0111
-----------+------------------------------           Adj R-squared =  0.0072
     Total | 32297.0648    255  126.655156           Root MSE      =  11.214

-----------+----------------------------------------------------------------
   tempjan |      Coef.   Std. Err.      t    P>|t|     [95% Conf. Interval]
-----------+----------------------------------------------------------------
  tempjuly |   .1825482   .1083166     1.69   0.093    -.0307648    .3958613
     _cons |    33.0621    7.84194     4.22   0.000     17.61859     48.5056
-----------------------------------------------------------------------------
```

The regressions show that a one-degree increase in the average July temperature in the Northeast corresponds to a 1.3-degree increase in the average January temperature. In the West, however, it corresponds to a 0.18-degree increase, which is only marginally significant.

◁

❑ Technical Note

by has a second syntax that is especially useful when you want to play it safe:

by *varlist*₁ (*varlist*₂): *command*

This says that Stata is to verify that the data are sorted by *varlist*₁ *varlist*₂ and then, assuming that is true, perform *command* by *varlist*₁. For instance,

```
. by subject (time): gen finalval = val[_N]
```

By typing this, we want to create new variable `finalval`, which contains, in each observation, the final observed value of `val` for each subject in the data. The final value will be the last value if, within subject, the data are sorted by time. The above command verifies that the data are sorted by `subject` and `time` and then, if they are, performs

```
. by subject: gen finalval = val[_N]
```

If the data are not sorted properly, an error message will instead be issued. Of course, we could have just typed

```
. by subject: gen finalval = val[_N]
```

after verifying for ourselves that the data were sorted properly, as long as we were careful to look.

by's second syntax can be used with by's `sort` option, so we can also type

```
. by subject (time), sort: gen finalval = val[_N]
```

which is equivalent to

```
. sort subject time
. by subject: gen finalval = val[_N]
```

❑

11.6   File-naming conventions

Some commands require that you specify a *filename*. Filenames are specified in the way natural for your operating system:

Windows	Unix	Macintosh
mydata	mydata	mydata
mydata.dta	mydata.dta	mydata.dta
b:mydata.dta	~friend/mydata.dta	~friend/mydata.dta
"my data"	"my data"	"my data"
"my data.dta"	"my data.dta"	"my data.dta"
myproj\mydata	myproj/mydata	myproj/mydata
"my project\my data"	"my project/my data"	"my project/my data"
C:\analysis\data\mydata	~/analysis/data/mydata	~/analysis/data/mydata
"C:\my project\my data"	"~/my project/my data"	"~/my project/my data"
..\data\mydata	../data/mydata	../data/mydata
"..\my project\my data"	"../my project/my data"	"../my project/my data"

In most cases (the exceptions being copy, dir, ls, erase, rm, and type), Stata automatically provides a file extension if you do not supply one. For instance, if you type use mydata, Stata assumes that you mean use mydata.dta since .dta is the file extension Stata normally uses for data files.

Stata provides fifteen default file extensions that are used by various commands:

.ado	automatically loaded do-files
.dct	ASCII data dictionary
.do	do-file
.dta	Stata-format dataset
.gph	graph image
.log	log file in text format
.mata	Mata source code
.mlib	Mata library
.mmat	Mata matrix
.mo	Mata object file
.out	file saved by outsheet
.raw	ASCII-format dataset
.smcl	log file in SMCL format
.sum	checksum files to verify network transfers
.vrf	impulse–response function datasets

You do not have to name your data files with the .dta extension—if you type an explicit file extension, it will override the default. For instance, if your dataset was stored as myfile.dat, you could type use myfile.dat. If your dataset was stored as simply myfile with no file extension, you could type the period at the end of the filename to indicate that you are explicitly specifying the null extension. You would type use myfile. to use this dataset.

All operating systems allow blanks in filenames, and so does Stata. However, if the filename includes a blank, you must enclose the filename in double quotes. Typing

    . save "my data"

would create the file my data.dta. Typing

    . save my data

would be an error.

❏ Technical Note

Stata also makes use of thirteen other file extensions. These files are of interest only to advanced programmers or are for Stata's internal use. They are

.class	class file for object-oriented programming; see [P] **class**
.hlp	help files
.dlg	dialog resource file
.idlg	dialog resource include file
.ihlp	help include file
.plugin	compiled addition (DLL)
.scheme	control file for a graphics scheme
.style	graphics style file
.key	search's keyword database file
.toc	user-site description file
.pkg	user-site package file
.maint	maintenance file (for Stata's internal use only)
.mnu	menu file (for Stata's internal use only)

❏

11.6.1   A special note for Macintosh users

Have you seen the notation myfolder/myfile before? This notation is called a path and describes the location of a file or folder (also called a directory).

You do not have to use this notation if you do not like it. You could instead restrict yourself to using files only in the current folder. If that turns out to be too restricting, Stata for Macintosh provides enough menus and buttons that you can probably get by. You may, however, find the notation convenient. In case you do, here is the rest of the definition.

The character / is called a path delimiter and delimits folder names and file names in a path. If the path starts with no path delimiter, the path is relative to the current folder.

For example, the path myfolder/myfile refers to the file myfile in the folder myfolder, which is contained in the current folder.

The characters .. refer to the folder containing the current folder. Thus, ../myfile refers to myfile in the folder containing the current folder, and ../nextdoor/myfile refers to myfile in the folder nextdoor in the folder containing the current folder.

If a path starts with a path delimiter, the path is called an absolute path and describes a fixed location of a file or folder name, regardless of what the current folder is. The leading / in an absolute path refers to the root directory, which is the main hard drive from which the operating system is booted. For example, the path /myfolder/myfile refers to the file myfile in the folder myfolder, which is contained in the main hard drive.

The character ~ refers to the user's home directory. Thus, the path ~/myfolder/myfile refers to myfile in the folder myfolder in the user's home directory.

11.6.2   A special note for Unix users

Stata understands ~ to mean your home directory. Stata understands this, even if you do not use csh(1) as your shell.

12   Data

Contents

12.1   Data and datasets

Data form a rectangular table of numeric and string values in which each row is an observation on all the variables and each column contains the observations on a single variable. Variables are designated by *variable names*. Observations are numbered sequentially from 1 to _N. The following example of data contains the first five odd and first five even positive integers, along with a string variable:

```
        odd   even    name
   1.     1      2    Bill
   2.     3      4    Mary
   3.     5      6     Pat
   4.     7      8   Roger
   5.     9     10    Sean
```

The observations are numbered 1 to 5, and the variables are named `odd`, `even`, and `name`. Observations are referred to by number, and variables by name.

A *dataset* is *data* plus labelings, formats, notes, and characteristics.

All aspects of *data* and *datasets* are defined here.

12.2   Numbers

A *number* may contain a sign, an integer part, a decimal point, a fraction part, an e or E, and a signed integer exponent. Numbers may *not* contain commas; for example, the number 1,024 must be typed as 1024 (or 1024. or 1024.0). The following are examples of valid numbers:

```
5
-5
5.2
.5
5.2e+2
5.2e-2
```

❏ **Technical Note**

As a convenience for Fortran users, you may use d or D, as well as e, to indicate exponential notation. Thus, the number 520 may be written 5.2d+2, 5.2D+2, 5.2d+02, or 5.2D+02.

❏

❏ **Technical Note**

Stata also allows numbers to be represented in a hexadecimal/binary format, defined as

$$[+|-]0.0[\langle zeros\rangle]\{X|x\}-3ff$$

or

$$[+|-]1.\langle hexdigit\rangle[\langle hexdigits\rangle]\{X|x\}\{+|-\}\langle hexdigit\rangle[\langle hexdigits\rangle]$$

The lead digit is always 0 or 1; it is 0 only when the number being expressed is zero. A maximum of 13 digits to the right of the hexadecimal point are allowed. The power ranges from $-3ff$ to $+3ff$. The number is expressed in hexadecimal (base 16) digits; the number $aX+b$ means $a \times 2^b$. For instance, 1.0X+3 is 2^3 or 8. 1.8X+3 is 12 because 1.8_{16} is $1 + 8/16 = 1.5$ in decimal and the number is thus $1.5 \times 2^3 = 1.5 \times 8 = 12$.

Stata can also display numbers using this format; see [U] **12.5.1 Numeric formats**. For example,

```
. di 1.81x+2
6.015625

. di %21x 6.015625
+1.8100000000000X+002
```

This hexadecimal format is of special interest to numerical analysts.

❏

12.2.1   Missing values

A number may also take on the special value *missing*, denoted by a single period (.). You specify a missing value anywhere that you may specify a number. Missing values differ from ordinary numbers in one respect: Any arithmetic operation on a missing value yields a missing value.

In fact, there are 27 missing values in Stata: '.', the one just discussed, as well as .a, .b, ..., and .z, which are known as extended missing values. The missing value '.' is known is as the default or system missing value. In any case, some people use extended missing values to indicate why a certain value is unknown—the question was not asked, the person refused to answer, etc. Other people have no use for extended missing values and just use '.'.

Stata's default or system missing value will be returned when you perform an arithmetic operation on missing values or when the arithmetic operation is not defined, such as division by zero, or the logarithm of a nonpositive number.

```
. display 2/0
.

. list
```

	a
1.	.b
2.	.
3.	.a
4.	3
5.	6

```
. generate x = a + 1
(3 missing values generated)
. list
```

	a	x
1.	.b	.
2.	.	.
3.	.a	.
4.	3	4
5.	6	7

Numeric missing values are represented by "large positive values". The ordering is

$$\text{all numbers} < . < .a < .b < \cdots < .z$$

Thus, the expression

$$\text{age} > 60$$

is true if variable age is greater than 60 or is missing. Similarly,

$$\text{gender} \, ! = 0$$

is true if gender is not zero or is missing.

The way to exclude missing values is to ask whether the value is less than '.', and the way to detect missing values is to ask whether the value is greater than or equal to '.'. For instance,

```
. list if age>60 & age<.

. generate agegt60 = 0 if age<=60
. replace agegt60 = 1 if age>60 & age<.

. generate agegt60 = (age>60) if age<.
```

❏ Technical Note

Long-time Stata users: Beware! Before Stata 8, Stata only had a single representation for missing values, the period (.). You could test whether an expression or a variable was missing by typing '...if *exp* ==.' or '...if *exp* !=.'.

Now the statements evaluate to true if *exp* is equal or is not equal to the particular missing value ., excluding the extended missing value cases. Actually, your habits will cause you no harm, assuming that your datasets have no extended missing values in them, but it will be safest to start using the proper syntax as soon as possible: an *exp* is not missing if its value is < ., and it is missing if its value is >= ..

In order to ensure that old programs and do-files continue to work properly, when version is set less than 8, all missing values are treated as being the same. Thus, . == .a == .b == .z, and so '*exp*==.' and '*exp*!=.' work just as they previously did.

❏

▷ Example 1

We have data on the income of husbands and wives recorded in the variables hincome and wincome, respectively. Typing the list command, we see that your data contain

```
. list
```

	hincome	wincome
1.	32000	0
2.	35000	34000
3.	47000	.b
4.	.z	50000
5.	.a	.

The values of wincome in the third and fifth observations are *missing*, as distinct from the value of wincome in the first observation, which is known to be zero.

If we use the generate command to create a new variable, income, that is equal to the sum of hincome and wincome, three missing values would be produced:

```
.  generate income = hincome + wincome
(3 missing values generated)
. list
```

	hincome	wincome	income
1.	32000	0	32000
2.	35000	34000	69000
3.	47000	.b	.
4.	.z	50000	.
5.	.a	.	.

generate produced a warning message that 3 missing values were created, and when we list the data, we see that 47,000 plus *missing* yields *missing*.

◁

❏ Technical Note

Stata stores numeric missing values as the largest 27 numbers allowed by the particular storage type; see [U] **12.2.2 Numeric storage types**. There are two important implications. First, if you sort on a variable that has missing values, the missing values will be placed last, and the sort order of any missing values will follow the rule regarding the properties of missing values stated above.

```
. sort wincome
. list wincome
```

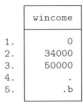

The second implication concerns relational operators and missing values. Do not forget that a missing value will be larger than any numeric value.

```
. list if wincome > 40000
```

	hincome	wincome	income
3.	.z	50000	.
4.	.a	.	.
5.	47000	.b	.

Observations 4 and 5 are listed because '.' and '.b' are both missing and thus are greater than 40,000. Relational operators are discussed in detail in [U] **13.2.3 Relational operators**.

❏

▷ Example 2

In producing statistical output, Stata ignores observations with missing values. Continuing with the example above, if we request summary statistics on hincome and wincome using the summarize command, we obtain

```
. summarize hincome wincome
```

Variable	Obs	Mean	Std. Dev.	Min	Max
hincome	3	38000	7937.254	32000	47000
wincome	3	28000	25534.29	0	50000

Some commands discard the entire observation (known as *casewise deletion*) if one of the variables in the observation is missing. If we use the correlate command to obtain the correlation between hincome and wincome, for instance, we obtain

```
. correlate hincome wincome
(obs=2)
```

	hincome	wincome
hincome	1.0000	
wincome	1.0000	1.0000

Note that the correlation coefficient is calculated over two observations.

◁

12.2.2 Numeric storage types

Numbers can be stored in one of five variable types: byte, int, long, float (the default), or double. bytes are, naturally, stored in 1 byte. ints are stored in 2 bytes, longs and floats in 4 bytes, and doubles in 8 bytes. The table below shows the minimum and maximum values for each storage type.

Storage Type	Minimum	Maximum	Closest to 0 without being 0	bytes
byte	-127	100	± 1	1
int	$-32{,}767$	$32{,}740$	± 1	2
long	$-2{,}147{,}483{,}647$	$2{,}147{,}483{,}620$	± 1	4
float	$-1.70141173319 \times 10^{38}$	$1.70141173319 \times 10^{36}$	$\pm 10^{-36}$	4
double	$-8.9884656743 \times 10^{307}$	$+8.9884656743 \times 10^{308}$	$\pm 10^{-323}$	8

Do not confuse the term *integer*, which is a characteristic of a number, with int, which is a storage type. For instance, the number 5 is an integer, no matter how it is stored; thus, if you read that an argument must be an integer, that does not mean that it must be stored as an int.

12.3 Dates

You can record dates any way you want, but there is one technique that Stata understands, called an elapsed date. An elapsed date is the number of days from January 1, 1960. In this format,

0	means	January 1, 1960
1		January 2, 1960
31		February 1, 1960
365		December 31, 1960
366		January 1, 1961
12,784		January 1, 1995
-1		December 31, 1959
-2		December 30, 1959
$-12,784$		December 31, 1924

Stata understands dates recorded like this from January 1, year 100 (elapsed date $-679{,}350$), to December 31, 9999 (elapsed date 2,936,549), although be careful with dates before Friday, October 15, 1582, when the Gregorian calendar went into effect.

Stata provides functions to convert dates into elapsed dates, formats to print elapsed dates in understandable forms, and other functions to manipulate elapsed dates.

In addition to elapsed dates, Stata provides five other date formats:

weekly	monthly	quarterly	half-yearly	yearly
-1 = 1959 week 52	-1 = Dec. 1959	-1 = 1959 quarter 4	-1 = 2nd half 1959	1959 = 1959
0 = 1960 week 1	0 = Jan. 1960	0 = 1960 quarter 1	0 = 1st half 1960	1960 = 1960
1 = 1960 week 2	1 = Feb. 1960	1 = 1960 quarter 2	1 = 2nd half 1960	1961 = 1961
...	...	...	...	...

For a full discussion of working with date variables, see [U] **24 Dealing with dates**.

12.4   Strings

A *string* is a sequence of printable characters, and is typically enclosed in double quotes. The quotes are not considered a part of the string but delimit the beginning and end of the string. The following are examples of valid strings:

```
"Hello, world"
"String"
"string"
" string"
"string "
""
"x/y+3"
"1.2"
```

All the strings above are distinct; that is, "String" is different from "string", which is different from " string", which is different from "string ". Also note that "1.2" is a string and not a number because it is enclosed in quotes.

All strings in Stata are of varying length, which means that Stata internally records the length of the string and never loses track. There is never a circumstance in which a string cannot be delimited by quotes, but there are rare instances where strings do not have to be delimited by quotes, such as during data input. In those cases, nondelimited strings are stripped of their leading and trailing blanks. Delimited strings are always accepted as is.

The special string "", often called *null string*, is considered by Stata to be a *missing*. No special meaning is given to the string containing a single period, ".".

In addition to double quotes for enclosing strings, Stata also allows compound double quotes: `" and "`. You can type *"string"* or you can type `"string"`, although users seldom type `"string"`. Compound double quotes are of special interest to programmers because they nest and provide a way for a quoted string to itself contain double quotes (either simple or compound). See [U] **18.3.5 Double quotes**.

12.4.1   Strings containing identifying data

String variables often contain identifying information, such as the patient's name or the name of the city or state. Such strings are typically listed but are not used directly in statistical analysis, although the data might be sorted on the string or datasets might be merged on the basis of one or more string variables.

12.4.2   Strings containing categorical data

Occasionally, strings contain information that is to be used directly in analysis, such as the patient's sex, which might be coded "male" or "female". Stata shows a decided preference for such information to be numerically encoded and stored in numeric variables. Stata's statistical routines treat string variables as if every observation records a numeric missing value.

Stata provides two commands for converting string variables into numeric codes and back again: encode and decode; see [U] **23.2 Categorical string variables**. Also see [D] **destring** for information on mapping string variables to numeric and vice versa.

12.4.3 Strings containing numeric data

If a string variable contains the character representation of a number—for instance, myvar contains "1", "1.2", and "-5.2"—you can convert it directly into a numeric variable using the real() function or the destring command; e.g., generate newvar=real(myvar).

Similarly, if you want to convert a numeric variable to its string representation, you can use the string() function or the tostring command; e.g., generate as_str=string(numvar).

See [D] **functions**.

12.4.4 String storage types

Strings are stored in string variables with storage types str1, str2, ..., str244. The storage type merely sets the maximum length of the string, not its actual length; thus, "example" has length 7 whether it is stored as a str7, a str10, or even a str244. On the other hand, an attempt to assign the string "example" to a str6 would result in "exampl".

The maximum length of a string in Stata/SE is 244. String literals may exceed 244 characters, but only the first 244 characters are significant.

The maximum length of a string in Intercooled Stata or Small Stata is 80 characters. Thus, string literals may exceed 80 characters, but only the first 80 characters are significant.

12.5 Formats: controlling how data are displayed

Formats describe how a number or string is to be presented. For instance, how is the number 325.24 to be presented? As 325.2, or 325.24, or 325.240, or 3.2524e+2, or 3.25e+2, or some other way? The *display format* tells Stata exactly how to present such data. You do not have to specify display formats since Stata always makes reasonable assumptions about how to display a variable, but you always have the option.

12.5.1 Numeric formats

A Stata numeric format is formed by

first type	%	to indicate the start of the format
then optionally type	−	if you want the result left-aligned
then optionally type	0	if you want to retain leading zeros (1)
then type	a number w	stating the width of the result
then type	.	
then type	a number d	stating the number of digits to follow the decimal point
then type		
either	e	for scientific notation, e.g., 1.00e+03
or	f	for fixed format, e.g., 1000.0
or	g	for general format; Stata chooses based on the number being displayed
then optionally type	c	to indicate comma format (not allowed with e)

(1) Specifying 0 to mean "include leading zeros" will be honored only with the f format.

For example,

%9.0g	general format, 9 columns wide	
	sqrt(2) =	1.414214
	1,000 =	1000
	10,000,000 =	1.00e+07
%9.0gc	general format, 9 columns wide, with commas	
	sqrt(2) =	1.414214
	1,000 =	1,000
	10,000,000 =	1.00e+07
%9.2f	fixed format, 9 columns wide, 2 decimal places	
	sqrt(2) =	1.41
	1,000 =	1000.00
	10,000,000 =	10000000.00
%9.2fc	fixed format, 9 columns wide, 2 decimal places, with commas	
	sqrt(2) =	1.41
	1,000 =	1,000.00
	10,000,000 =	10,000,000.00
%9.2e	exponential format, 9 columns wide	
	sqrt(2) =	1.41e+00
	1,000 =	1.00e+03
	10,000,000 =	1.00e+07

Stata has three numeric format types: e, f, and g. The formats are denoted by a leading percent sign (%) followed by the string $w.d$, where w and d stand for two integers. The first integer, w, specifies the width of the format. The second integer d specifies the number of digits that are to follow the decimal point. d must be less than w. Finally, a character denotes the format type (e, f, or g), and to that may optionally be appended a c indicating that commas are to be included in the result (c is not allowed with e.)

By default, every numeric variable is given a %w.0g format, where w is large enough to display the largest number of the variable's type. The %w.0g format is a set of formatting rules that present the values in as readable a fashion as possible without sacrificing precision. The g format changes the number of decimal places displayed whenever it improves the readability of the current value.

The default formats for each of the numeric variable types are

byte	%8.0g
int	%8.0g
long	%12.0g
float	%9.0g
double	%10.0g

You can change the format of a variable using the `format` *varname* %*fmt* command.

In addition to %w.0g, allowed is %w.0gc to display numbers with commas. The number one thousand is displayed as 1000 in %9.0g format and as 1,000 in %9.0gc format.

In addition to using %w.0g and %w.0gc, you can use %w.dg and %w.$d$gc, $d > 0$. For example, %9.4g and %9.4gc. The 4 means to display approximately 4 significant digits. For instance, the number 3.14159265 in %9.4g format is displayed as 3.142, 31.4159265 as 31.42, 314.159265 as 314.2, and 3141.59265 as 3142. The format is not exactly a significant-digit format because 31415.9265 is displayed as 31416, not as 3.142e+04.

Under the f format, values are always displayed with the same number of decimal places, even if this results in a loss in the displayed precision. Thus, the f format is similar to the C f format. Stata's f format is also similar to the Fortran F format, but, unlike the Fortran F format, it switches to g whenever a number is too large to be displayed in the specified f format.

In addition to %w.df, the format %w.$d$fc can display numbers with commas.

The e format is similar to the C e and the Fortran E format. Every value is displayed as a leading digit (with a minus sign, if necessary), followed by a decimal point, the specified number of digits, the letter e, a plus sign or a minus sign, and the power of ten (modified by the preceding sign) that multiplies the displayed value. When the e format is specified, the width must exceed the number of digits that follow the decimal point by at least seven to accommodate the leading sign and digit, the decimal point, the e, and the signed power of ten.

▷ Example 3

Below we have a five-observation dataset with three variables: e_fmt, f_fmt, and g_fmt. All three variables have the same values stored in them; only the display format varies. describe shows the display format to the right of the variable type:

```
. use http://www.stata-press.com/data/r9/format
. describe
Contains data from http://www.stata-press.com/data/r9/format.dta
  obs:            5
  vars:           3                              12 Mar 2005 15:18
  size:          80 (99.9% of memory free)
```

variable name	storage type	display format	value label	variable label
e_fmt	float	%9.2e		
f_fmt	float	%10.2f		
g_fmt	float	%9.0g		

```
Sorted by:
```

The formats for each of these variables were set by typing

```
. format e_fmt %9.2e
. format f_fmt %10.2f
```

It was not necessary to set the format for the g_fmt variable since Stata automatically assigned it the %9.0g format. Nevertheless, we could have typed format g_fmt %9.0g if we wished. Listing the data results in

```
. list
```

	e_fmt	f_fmt	g_fmt
1.	2.80e+00	2.80	2.801785
2.	3.96e+06	3962322.50	3962323
3.	4.85e+00	4.85	4.852834
4.	-5.60e-06	-0.00	-5.60e-06
5.	6.26e+00	6.26	6.264982

◁

❑ Technical Note

The discussion above is incomplete. There is one additional format available that will be of interest to numerical analysts. The %21x format displays base 10 numbers in a hexadecimal (base 16) format. The number is expressed in hexadecimal (base 16) digits; the number aX$+b$ means $a \times 2^b$. For example,

```
. display %21x 1234.75
+1.34b0000000000X+00a
```

Thus, the base 10 number 1,234.75 has a base 16 representation of `1.34bX+0a`, meaning

$$\left(1 + 3 \cdot 16^{-1} + 4 \cdot 16^{-2} + 11 \cdot 16^{-3}\right) \times 2^{10}$$

Remember, the hexadecimal—decimal equivalents are

hexadecimal	decimal
0	0
1	1
2	2
3	3
4	4
5	5
6	6
7	7
8	8
9	9
a	10
b	11
c	12
d	13
e	14
f	15

See [U] **12.2 Numbers**.

❏

12.5.2   European numeric formats

The three numeric formats e, f, and g will use ',' to indicate the decimal symbol if you specify their width and depth as w,d rather than $w.d$. For instance, the format %9,0g will display what Stata would usually display as 1.5 as 1,5.

If you use the European specification with fc or gc, the "comma" will be presented as a period. For instance, %9,0gc would display what Stata would usually display as 1,000.5 as 1.000,5.

If this way of presenting numbers appeals to you, consider using Stata's set dp comma command. set dp comma tells Stata to interpret nearly all %$w.d${g|f|e} formats as %w,d{g|f|e} formats. Most of Stata is written using a period to represent the decimal symbol, and that means that, even if you set the appropriate %w,d{g|f|e} format for your data, it will affect only displays of the data. For instance, if you type summarize to obtain summary statistics or regress to obtain regression results, the decimal will still be shown as a period.

set dp comma changes that and affects all of Stata. With set dp comma, it does not matter whether your data are formatted %$w.d${g|f|e} or %w,d{g|f|e}. All results will be displayed using a comma as the decimal character:

```
. use http://www.stata-press.com/data/r9/auto
(1978 Automobile Data)

. set dp comma

. summarize mpg weight foreign
```

Variable	Obs	Mean	Std. Dev.	Min	Max
mpg	74	21,2973	5,785503	12	41
weight	74	3019,459	777,1936	1760	4840
foreign	74	,2972973	,4601885	0	1

```
. regress mpg weight foreign
```

Source	SS	df	MS		
Model	1619,2877	2	809,643849	Number of obs =	74
Residual	824,171761	71	11,608053	F(2, 71) =	69,75
				Prob > F =	0,0000
				R-squared =	0,6627
				Adj R-squared =	0,6532
Total	2443,45946	73	33,4720474	Root MSE =	3,4071

mpg	Coef.	Std. Err.	t	P>|t|	[95% Conf. Interval]	
weight	-,0065879	,0006371	-10,34	0,000	-,0078583	-,0053175
foreign	-1,650029	1,075994	-1,53	0,130	-3,7955	,4954422
_cons	41,6797	2,165547	19,25	0,000	37,36172	45,99768

You can switch the decimal character back to a period by typing `set dp period`.

❏ Technical Note

`set dp comma` makes drastic changes inside Stata, and we mention this because some older, user-written programs may not be able to deal with those changes. If you are using an older user-written program, you might `set dp comma` and then find that the program does not work and instead presents some sort of syntax error.

If, using any program, you do get an unanticipated error, try setting `dp` back to `period`. See [D] **format** for more information.

Also understand that `set dp comma` affects how Stata outputs numbers, not how it inputs them. You must still use the period to indicate the decimal point on all input. Even with `set dp comma`, you type

```
. replace x=1.5 if x==2
```

❏

12.5.3   Date formats

Date formats are really a numeric format because Stata stores dates as the number of days from 01jan1960. See [U] **24 Dealing with dates**.

`%d` is for displaying elapsed dates. The syntax of the `%d` format is

first type	%		to indicate the start of the format
then optionally type	–		if you want the result left-aligned
then type	d		
then optionally type	*other characters*		to indicate how the date is to be displayed

The %d format may be specified as simply %d, or the %d may be followed by up to 11 of the following characters, which specify how the date is to be presented:

c and C	display the century without/with a leading 0
y and Y	display the two-digit year without/with a leading 0
m and M	display Month, first letter capitalized, in three-letter abbreviation (m) or spelled out (M)
l and L	display month, first letter not capitalized, in three-letter abbreviation (l) or spelled out (L)
n and N	display month number 1–12 without/with a leading 0
d and D	display day-within-month number 1–31 without/with a leading 0
j and J	display day-within-year number 1–366 without/with leading 0s
h	display the half of year number 1 or 2
q	display quarter of year number 1, 2, 3, or 4
w and W	display week-of-year number 1–52 without/with a leading 0
_	display a blank
.	display a period
,	display a comma
:	display a colon
−	display a dash
/	display a slash
'	display a close single quote
!c	display character c (code !! to display an exclamation point)

Specifying %d by itself is equivalent to specifying %dD1CY. The first day of January 1999 is displayed as 01jan1999 in this format. For examples of various date formats, see [U] **24 Dealing with dates**.

12.5.4   Time-series formats

Time-series formats—also known as %t formats—are an extension of the date formats coded above. Stata's dates are coded 0 = 01jan1960 and 1 = 02jan1960. Stata also has dates where 0 represents the first week of 1960 (and 1 the second week), 0 represents the first month of 1960 (and 1 the second month), 0 represents the first quarter of 1960 (and 1 the second quarter), 0 represents the first half of 1960 (and 1 the second half), and 1960 represents the year 1960 (and 1961 the next year). %t formats are for displaying these quantities. The %t format is defined as

first type	%	to indicate the start of the format
then optionally type	−	if you want the result left-aligned
then type	t	
then type		a character to indicate how the date is encoded:
type	d	if 0 = 01jan1960, same as %d format
or type	w	if 0 = 1960w1
or type	m	if 0 = 1960m1
or type	q	if 0 = 1960q1
or type	h	if 0 = 1960h1
or type	y	if 1960 = 1960 (it records the year itself)
then optionally type	*other characters*	to indicate how the date is to be displayed

where the optional characters are the same as for the %d format given in the table of the previous section, [U] **12.5.3 Date formats**.

In addition to the above is a %tg format—the g stands for generic. The %tg format is provided merely for completeness; it is equivalent to %9.0g. %tg is provided for users who want to put some sort of %t format on a time variable that is encoded differently than the format Stata understands; whether they do this makes no difference.

The minimal %t formats are %td, %tw, and so on. The default formats for each are

format	default	0 is displayed as	2,000 is displayed as
%td	%tdD1CY	01jan1960	23jun1965
%tw	%twCY!ww	1960w1	1998w25
%tm	%tmCY!mn	1960m1	2126m9
%tq	%tqCY!qq	1960q1	2460q1
%th	%thCY!hh	1960h1	2960h1
%ty	%tyCY	.	2000
%tg	%9.0g	0	2000

There are no mistakes in the table above. For %ty encoded data, the year range is 100–9999, so year 0 displays as missing. More typically, years will be in the range 1900–2100.

For more examples of the %t format, see [U] **24 Dealing with dates**.

12.5.5   String formats

The syntax for a string format is

first type	%	to indicate the start of the format
then optionally type	–	if you want the result left-aligned
then type	a number	indicating the width of the result
then type	s	

For instance, %10s represents a string format of width 10.

For str$w$, the default format is %ws or %9s, whichever is wider. For example, a str10 variable receives a %10s format. Strings are displayed right-justified in the field, unless the minus sign is coded; %-10s would display the string left-aligned.

▷ Example 4

Our automobile data contains a string variable called make.

```
. use http://www.stata-press.com/data/r9/auto
(1978 Automobile Data)

. describe make
```

variable name	storage type	display format	value label	variable label
make	str18	%-18s		Make and Model

```
. list make in 63/67
```

	make
63.	Mazda GLC
64.	Peugeot 604
65.	Renault Le Car
66.	Subaru
67.	Toyota Celica

These values are left-aligned because make has a display format of %-18s. If we want to right-align the values, we could change the format:

```
. format %18s make
. list make in 63/67
```

	make
63.	Mazda GLC
64.	Peugeot 604
65.	Renault Le Car
66.	Subaru
67.	Toyota Celica

◁

12.6   Dataset, variable, and value labels

Labels are strings used to label elements in Stata, such as labels for datasets, variables, and values.

12.6.1   Dataset labels

Associated with every dataset is an 80-character *dataset label*, which is initially set to blanks. You can use the `label data "text"` command to define the dataset label.

▷ Example 5

We have just entered 1980 state data on marriage rates, divorce rates, and median ages. The describe command will describe the data in memory:

```
. describe
Contains data
  obs:            50
  vars:            4
  size:        1,200 (99.8% of memory free)

              storage  display    value
variable name    type   format    label      variable label

state            str8    %9s
median_age       float   %9.0g
marriage_rate    long    %12.0g
divorce_rate     long    %12.0g

Sorted by:
     Note:  dataset has changed since last saved
```

describe shows that there are 50 observations on four variables named state, median_age, marriage_rate, and divorce_rate. state is stored as a str8; median_age is stored as a float; and marriage_rate and divorce_rate are both stored as longs. Each variable's display format (see [U] **12.5 Formats: controlling how data are displayed**) is shown. Finally, the data are not in any particular sort order, and the dataset has changed since it was last saved on disk.

We can label the data by typing label data "1980 state data". We type this and then type describe again:

```
. label data "1980 state data"
. describe
Contains data
  obs:          50                              1980 state data
  vars:          4
  size:      1,200 (99.7% of memory free)
```

variable name	storage type	display format	value label	variable label
state	str8	%9s		
median_age	float	%9.0g		
marriage_rate	long	%12.0g		
divorce_rate	long	%12.0g		

```
Sorted by:
    Note:  dataset has changed since last saved
```

◁

The dataset label is displayed by the describe and use commands.

12.6.2 Variable labels

In addition to the name, every variable has associated with it an 80-character *variable label*. The variable labels are initially set to blanks. You use the label variable *varname* "*text*" command to define a new variable label.

▷ Example 6

We have entered data on four variables: state, median_age, marriage_rate, and divorce_rate. describe portrays the data we entered:

```
. describe
Contains data from states.dta
  obs:          50                              1980 state data
  vars:          4
  size:      1,200 (99.7% of memory free)
```

variable name	storage type	display format	value label	variable label
state	str8	%9s		
median_age	float	%9.0g		
marriage_rate	long	%12.0g		
divorce_rate	long	%12.0g		

```
Sorted by:
    Note:  dataset has changed since last saved
```

We can associate labels with the variables by typing

```
. label variable median_age "Median Age"
. label variable marriage_rate "Marriages per 100,000"
. label variable divorce_rate "Divorces per 100,000"
```

From then on, the result of describe will be

```
. describe
Contains data
    obs:            50                          1980 state data
    vars:            4
    size:         1,200  (99.7% of memory free)

                 storage  display   value
variable name    type     format    label    variable label

state            str8     %9s
median_age       float    %9.0g               Median Age
marriage_rate    long     %12.0g              Marriages per 100,000
divorce_rate     long     %12.0g              Divorces per 100,000

Sorted by:
      Note:  dataset has changed since last saved
```
◁

Whenever Stata produces output, it will use the variable labels rather than the variable names to label the results if there is room.

12.6.3   Value labels

Value labels define a correspondence or mapping between numeric data and the words used to describe what those numeric values represent. Mappings are named and defined by the label define *lblname* # "*string*" # "*string*"... command. The maximum length for the *lblname* is 32 characters. # must be an integer or an extended missing value (.a, .b, ..., .z). The maximum length of *string* is 32,000 characters in Stata/SE and Intercooled Stata and 80 characters in Small Stata. Named mappings are associated with variables by the label values *varname lblname* command.

▷ Example 7

The definition makes value labels sound more complicated than they are in practice. We create a dataset on individuals in which we record a person's sex, coding 0 for males and 1 for females. If our dataset also contained an employee number and salary, it might resemble the following:

```
. describe
Contains data
    obs:             7                          2005 Employee data
    vars:            3
    size:          112  (99.9% of memory free)

                 storage  display   value
variable name    type     format    label    variable label

empno            float    %9.0g               Employee number
sex              float    %9.0g               Sex
salary           float    %8.0fc              Annual salary, exclusive of
                                                bonus

Sorted by:
      Note:  dataset has changed since last saved
```

```
. list
```

	empno	sex	salary
1.	57213	0	24,000
2.	47229	1	27,000
3.	57323	0	24,000
4.	57401	0	24,500
5.	57802	1	27,000
6.	57805	1	24,000
7.	57824	0	22,500

We could create a mapping called sexlabel defining 0 as "Male" and 1 as "Female", and then associate that mapping with the variable sex by typing

```
. label define sexlabel 0 "Male" 1 "Female"
. label values sex sexlabel
```

From then on, our data would appear as

```
. describe
Contains data
  obs:              7                            2005 Employee data
  vars:             3
  size:           112 (99.8% of memory free)
```

variable name	storage type	display format	value label	variable label
empno	float	%9.0g		Employee number
sex	float	%9.0g	sexlabel	Sex
salary	float	%8.0fc		Annual salary, exclusive of bonus

```
Sorted by:
     Note:  dataset has changed since last saved
. list
```

	empno	sex	salary
1.	57213	Male	24,000
2.	47229	Female	27,000
3.	57323	Male	24,000
4.	57401	Male	24,500
5.	57802	Female	27,000
6.	57805	Female	24,000
7.	57824	Male	22,500

Notice not only that the value label is used to produce words when we list the data, but also that the association of the variable sex with the value label sexlabel is shown by the describe command.

◁

❑ Technical Note

Value labels and variables may share the same name. For instance, rather than calling the value label sexlabel in the example above, we could just as well have named it sex. We would then type label values sex sex to associate the value label named sex with the variable named sex.

                                                                                        ❑

▷ Example 8

Stata's encode and decode commands provide a convenient way to go from string variables to numerically coded variables and back again. Let's pretend that, in the example above, rather than coding 0 for males and 1 for females, we created a string variable recording either "male" or "female". Our data look like

```
. describe
Contains data                                       2005 Employee data
  obs:             7
  vars:            3
  size:          126 (99.8% of memory free)

               storage   display    value
variable name   type     format     label     variable label

empno          float    %9.0g                 Employee number
sex            str6     %9s                    Sex
salary         float    %8.0fc                 Annual salary, exclusive of
                                                 bonus

Sorted by:
    Note:  dataset has changed since last saved
. list
```

```
      empno      sex    salary

 1.   57213     male    24,000
 2.   47229   female    27,000
 3.   57323     male    24,000
 4.   57401     male    24,500
 5.   57802   female    27,000

 6.   57805   female    24,000
 7.   57824     male    22,500
```

We now want to create a numerically encoded variable—we will call it gender—from the string variable. We want to do this, say, because we typed anova salary sex to perform a one-way ANOVA of salary on sex, and we were told that there were "no observations". We then remembered that all of Stata's statistical commands treat string variables as if they contain nothing but missing values. The statistical commands work only with numerically coded data.

(*Continued on next page*)

```
. encode sex, generate(gender)

. describe

Contains data
  obs:             7                          2005 Employee data
  vars:            4
  size:          154 (99.8% of memory free)
```

variable name	storage type	display format	value label	variable label
empno	float	%9.0g		Employee number
sex	str6	%9s		Sex
salary	float	%8.0fc		Annual salary, exclusive of bonus
gender	long	%8.0g	gender	Sex

```
Sorted by:
     Note:  dataset has changed since last saved
```

encode adds a new long variable called gender to the data and defines a new value label called
gender. The value label gender maps 1 to the string male and 2 to female, so if we were to list
the data, we could not tell the difference between the gender and sex variables. However, they are
different. Stata's statistical commands know how to deal with gender but do not understand the sex
variable. See [D] **encode**.

❑

❑ Technical Note

Perhaps rather than employee data, our data are on persons undergoing sex-change operations.
As such, there would be two sex variables in our data, sex before the operation and sex after the
operation. Assume that the variables are named presex and postsex. We can associate the *same*
value label to each variable by typing

```
. label define sexlabel 0 "Male" 1 "Female"

. label values presex sexlabel

. label values postsex sexlabel
```

❑

❑ Technical Note

Stata's input commands (input and infile) can switch from the words in a value label back to
the numeric codes. Remember that encode and decode can translate a string to a numeric mapping
and vice versa, so we can map strings to numeric codes either at the time of input or later.

For example,

```
. label define sexlabel 0 "Male" 1 "Female"

. input empno sex:sexlabel salary, label
        empno       sex     salary
 1. 57213 Male 24000
 2. 47229 Female 27000
 3. 57323 0 24000
 4. 57401 Male 24500
 5. 57802 Female 27000
 6. 57805 Female 24000
 7. 57824 Male 22500
 8. end
```

The `label define` command defines the value label sexlabel. `input empno sex:sexlabel salary, label` tells Stata to input three variables from the keyboard (empno, sex, and salary), attach the value label `sexlabel` to the `sex` variable, and look up any words that are typed in the value label to try to convert them to numbers. To prove that it works, we `list` the data that we recently entered:

```
. list
```

	empno	sex	salary
1.	57213	Male	24000
2.	47229	Female	27000
3.	57323	Male	24000
4.	57401	Male	24500
5.	57802	Female	27000
6.	57805	Female	24000
7.	57824	Male	22500

Compare the information we typed for observation 3 with the result listed by Stata. We typed 57323 0 24000. Thus, the value of `sex` in the third observation is 0. When Stata listed the observation, it indicated the value is Male because we told Stata in our `label define` command that zero is equivalent to Male.

Let's now add one more observation to our data:

```
. input, label
          empno      sex      salary
8. 67223 FEmale 23000
'FEmale' cannot be read as a number
8. 67223 Female 23000
9. end
```

At first we typed 67223 FEmale 23000, and Stata responded with "'FEmale' cannot be read as a number". Remember that Stata always respects case, so FEmale is not at all the same thing as Female. Stata prompted us to enter the line again, and we did so, this time correctly.

❑

❑ Technical Note

Coupled with the `automatic` option, Stata can not only go from words to numbers, but can create the mapping as well. Let's input the data again, but this time, rather than type the data in at the keyboard, let's read the data from a file. Assume that we have an ASCII file called `employee.raw` stored on our disk that contains

```
57213 Male 24000
47229 Female 27000
57323 Male 24000
57401 Male 24500
57802 Female 27000
57805 Female 24000
57824 Male 22500
```

The `infile` command can read these data and create the mapping automatically:

```
. label list sexlabel
value label sexlabel not found
r(111);
```

```
. infile empno sex:sexlabel salary using employee, automatic
(7 observations read)
```

Our first command, `label list sexlabel`, is only to prove that we had not previously defined the value label `sexlabel`. Stata `infiled` the data without complaint. We now have

```
. list
```

	empno	sex	salary
1.	57213	Male	24000
2.	47229	Female	27000
3.	57323	Male	24000
4.	57401	Male	24500
5.	57802	Female	27000
6.	57805	Female	24000
7.	57824	Male	22500

Of course, `sex` is just another numeric variable; it does not actually take on the values `Male` and `Female`—it takes on numeric codes that have been automatically mapped to `Male` and `Female`. We can find out what that mapping is by using the `label list` command:

```
. label list sexlabel
sexlabel:
           1 Male
           2 Female
```

We discover that Stata attached the codes 1 to `Male` and 2 to `Female`. Anytime we want to see what our data really look like, ignoring the value labels, we can use the `nolabel` option:

```
. list, nolabel
```

	empno	sex	salary
1.	57213	1	24000
2.	47229	2	27000
3.	57323	1	24000
4.	57401	1	24500
5.	57802	2	27000
6.	57805	2	24000
7.	57824	1	22500

❏

12.6.4   Labels in other languages

A dataset can contain labels—data, variable, and value—in up to 100 languages. To discover the languages available for the dataset in memory, type `label language`. You will see this

```
. label language
Language for variable and value labels

    In this dataset, value and variable labels have been defined in only one
    language:  default

    To create new language:       . label language <name>, new
    To rename current language:   . label language <name>, rename
```

or something like this:

```
. label language
```
Language for variable and value labels

```
        Available languages:
                de
                en
                sp
        Currently set is:                    . label language sp
        To select different language:        . label language <name>
        To create new language:              . label language <name>, new
        To rename current language:          . label language <name>, rename
```

Right now, the example dataset is set with sp (Spanish) labels:

```
. describe
Contains data from /usr/local/stata9/ado/base/a/auto.dta
   obs:           74                        Automóviles, 1978
   vars:          12                        3 Oct 2002 13:53
   size:        3,478 (99.7% of memory free)
```

variable name	storage type	display format	value label	variable label
make	str18	%-18s		Marca y modelo
price	int	%8.0gc		Precio
mpg	int	%8.0g		Consumo de combustible
rep78	int	%8.0g		Historia de reparaciones
headroom	float	%6.1f		Cabeza adelante
trunk	int	%8.0g		Volumen del maletero
weight	int	%8.0gc		Peso
length	int	%8.0g		Longitud
turn	int	%8.0g		Radio de giro
displacement	int	%8.0g		Cilindrada
gear_ratio	float	%6.2f		Relación de cambio
foreign	byte	%8.0g		Extranjero

```
Sorted by:  foreign
```

To create labels in more than one language, you set the new language and then define the labels in the standard way; see [D] **label language**.

12.7 Notes attached to data

A dataset may contain notes, which are nothing more than little bits of text that you define and review with the notes command. Typing note, a colon, and the text defines a note:

```
. note:  Send copy to Bob once verified.
```

You can later display whatever notes you have previously defined by typing notes:

```
. notes
_dta:
   1.  Send copy to Bob once verified.
```

Notes are saved with the data, so once you save your dataset, you can replay this note in the future, too.

You can add more notes:

```
. note: Mary wants a copy, too.
```

```
. notes
_dta:
  1.  Send copy to Bob once verified.
  2.  Mary wants a copy, too.
```

The notes you have added so far are attached to the data generically, which is why Stata prefixes them with _dta when it lists them. You can attach notes to variables:

```
. note state: verify values for Nevada.

. note state: what about the two missing values?

. notes
_dta:
  1.  Send copy to Bob once verified.
  2.  Mary wants a copy, too.
state:
  1.  verify values for Nevada.
  2.  what about the two missing values?
```

When you describe your data, you can see whether notes are attached to the dataset or to any of the variables:

```
. describe
Contains data from states.dta
  obs:           50                          1980 state data
  vars:           4
  size:        1,200 (99.3% of memory free)  (_dta has notes)

                 storage  display   value
variable name    type     format    label    variable label

state            str8     %9s             *
median_age       float    %9.0g                Median Age
marriage_rate    long     %12.0g               Marriages per 100,000
divorce_rate     long     %12.0g               Divorces per 100,000
                                              * indicated variables have notes

Sorted by:
    Note:   dataset has changed since last saved
```

See [D] **notes** for a complete description of this feature.

12.8 Characteristics

Characteristics are an arcane feature of Stata but are of great use to Stata programmers. In fact, the notes command described above was implemented using characteristics.

The dataset itself and each variable within the dataset have associated with them a set of characteristics. Characteristics are named and referred to as *varname*[*charname*], where *varname* is the name of a variable or _dta. The characteristics contain text and are stored with the data in the Stata-format .dta dataset, so they are recalled whenever the data are loaded.

How are characteristics used? The [XT] **xt** commands need to know the name of the variable corresponding to time. These commands allow the variable name to be specified as an option but do not require it. When the user does not specify the time variable, the commands somehow manage to remember it from last time, even from a different Stata session. They do this with characteristics. When the user does not specify the variable name, the commands check the characteristic _dta[tis] for the name of the variable. If the time variable's name is stored there, they continue; if not, they issue an error because they need to know it. When the user specifies the option identifying the time variable, these commands store that name in the characteristic _dta[tis]. This use of characteristics is hidden from the user—no mention is made of how the commands remember the identity of the time variable.

Occasionally, commands identify their use of characteristics explicitly. The xi command (see [R] **xi**) states that it drops the first level of a categorical variable but that, if you wish to control which level is dropped, you can set the variable's omit characteristic. In the documentation, an example is provided where the user types

>   . char agegrp[omit] 3

to set the default omission group to 3. As with the [XT] **xt** commands, if the user saves the data after setting the characteristic, the preferred omission group will be remembered from one session to the next.

As a Stata user, you need only understand how to set and clear a characteristic for the few commands that explicitly reveal their use of characteristics. You set a variable *varname*'s characteristic *charname* to *x* by typing

>   . char *varname*[*charname*] *x*

You set the data's characteristic *charname* to be *x* by typing

>   . char _dta[*charname*] *x*

You clear a characteristic by typing

>   . char *varname*[*charname*]

where *varname* is either a variable name or _dta. You can clear a characteristic, even if it has never been set.

The most important feature of characteristics is that Stata remembers them from one session to the next; they are saved with the data.

❏ Technical Note

Programmers will want to know more. A technical description is found in [P] **char**, but as an overview, you may refer to *varname*'s *charname* characteristic by embedding its name in single quotes and typing '*varname*[*charname*]'; see [U] **18.3.13 Referencing characteristics**.

You can fetch the names of all characteristics associated with *varname* by typing

>   . local *macname* : char *varname*[]

The maximum length of the contents of a characteristic is the same as for macros: 8,681 characters for Small Stata, 67,784 for Intercooled, and $33 * c(\text{max\_k\_theory}) + 200$ for Stata/SE, which for the default setting of 5,000 is 165,200. The association of names with characteristics is by convention. If you, as a programmer, wish to create new characteristics for use in your ado-files, do so, but include at least one capital letter in the characteristic name. The current convention reserves all lowercase names for "official" Stata.

❏

13 Functions and expressions

If you have not read [U] **11 Language syntax**, please do so before reading this entry.

13.1   Overview

Examples of expressions include

```
2+2
miles/gallons
myv+2/oth
(myv+2)/oth
ln(income)
age<25 & income>50000
age<25 | income>50000
age==25
name=="M Brown"
fname + " " + lname
substr(name,1,10)
_pi
val[_n-1]
L.gnp
```

Expressions like those above are allowed anywhere *exp* appears in a syntax diagram. One example is [D] **generate**:

> generate *newvar* = *exp* [*if*] [*in*]

The first *exp* specifies the contents of the new variable, and the optional second expression restricts the subsample over which it is to be defined. Another is [R] **summarize**:

> summarize [*varlist*] [*if*] [*in*]

The optional expression restricts the sample over which summary statistics are calculated.

Algebraic and string expressions are specified in a natural way using the standard rules of hierarchy. You may use parentheses freely to force a different order of evaluation.

▷ Example 1

myv+2/oth is interpreted as myv+(2/oth). If you wanted to change the order of the evaluation, you could type (myv+2)/oth.

◁

13.2   Operators

Stata has four different classes of operators: arithmetic, string, relational, and logical. Each type is discussed below.

13.2.1   Arithmetic operators

The *arithmetic operators* in Stata are + (addition), − (subtraction), * (multiplication), / (division), ^ (raise to a power), and the prefix − (negation). Any arithmetic operation on a missing value or an impossible arithmetic operation (such as division by zero) yields a missing value.

▷ Example 2

The expression -(x+y^(x-y))/(x*y) denotes the formula

$$-\frac{x + y^{x-y}}{x \cdot y}$$

and evaluates to *missing* if x or y is missing or zero.

◁

13.2.2   String operators

The + sign is also used as a string operator for the *concatenation* of two strings. Stata determines by context whether + means addition or concatenation. If + appears between two strings, Stata concatenates them. If + appears between two numeric values, Stata adds them.

▷ Example 3

The expression "this"+"that" results in the string "thisthat", whereas the expression 2+3 results in the number 5. Stata issues the error message "type mismatch" if the arguments on either side of the + sign are not of the same type. Thus, the expression 2+"this" is an error, as is 2+"3".

The expressions on either side of the + can be arbitrarily complex:

substr(string(20+2),1,1) + upper(substr("rf",1+1,1))

The result of the above expression is the string "2F". See [D] **functions** below for a description of the substr(), string(), and upper() functions.

◁

13.2.3   Relational operators

The *relational operators* are > (greater than), < (less than), >= (greater than or equal), <= (less than or equal), == (equal), and != (not equal). Observe that the relational operator for equality is a pair of equal signs. This convention distinguishes relational equality from the =*exp* assignment phrase.

❑ Technical Note

You may use ~ anywhere ! would be appropriate to represent the logical operator "not". Thus, the not-equal operator may also be written as ~=.

❑

Relational expressions are either *true* or *false*. Relational operators may be used on either numeric or string subexpressions; thus, the expression 3>2 is *true*, as is "zebra">"cat". In the latter case, the relation merely indicates that "zebra" comes after the word "cat" in the dictionary. All uppercase letters precede all lowercase letters in Stata's book, so "cat">"Zebra" is also *true*.

Missing values may appear in relational expressions. If x were a numeric variable, the expression x>=. is *true* if x is missing and *false* otherwise. A missing value is greater than any nonmissing value; see [U] **12.2.1 Missing values**.

▷ Example 4

You have data on age and income and wish to list the subset of the data for persons aged 25 years or less. You could type

. list if age<=25

If you wanted to list the subset of data of persons aged exactly 25, you would type

. list if age==25

Note the double equal sign. It would be an error to type list if age=25.

◁

Although it is convenient to think of relational expressions as evaluating to *true* or *false*, they actually evaluate to numbers. A result of *true* is defined as 1 and *false* is defined as 0.

▷ Example 5

The definition of *true* and *false* makes it easy to create indicator, or dummy, variables. For instance,

```
generate incgt10k=income>10000
```

creates a variable that takes on the value 0 when `income` is less than or equal to $10,000, and 1 when `income` is greater than $10,000. Since missing values are greater than all nonmissing values, the new variable `incgt10k` will also take on the value 1 when `income` is *missing*. It would be safer to type

```
generate incgt10k=income>10000 if income<.
```

Now, observations in which `income` is *missing* will also contain *missing* in `incgt10k`. See [U] **25 Dealing with categorical variables** for more examples.

◁

❑ Technical Note

Although you will rarely wish to do so, since arithmetic and relational operators both evaluate to numbers, there is no reason you cannot mix the two types of operators in a single expression. For instance, (2==2)+1 evaluates to 2, since 2==2 evaluates to 1, and $1 + 1$ is 2.

Relational operators are evaluated after all arithmetic operations. Thus, the expression (3>2)+1 is equal to 2, whereas 3>2+1 is equal to 0. Evaluating relational operators last guarantees the *logical* (as opposed to the *numeric*) interpretation. It should make sense that 3>2+1 is *false*.

❑

13.2.4 Logical operators

The *logical operators* are & (and), | (or), and ! (not). The logical operators interpret any nonzero value (including *missing*) as *true* and zero as *false*.

▷ Example 6

If you have data on `age` and `income` and wish to `list` data for persons making more than $50,000 along with persons under the age of 25 making more than $30,000, you could type

```
list if income>50000 | income>30000 & age<25
```

The & takes precedence over the |. If you were unsure, however, you could have typed

```
list if income>50000 | (income>30000 & age<25)
```

In either case, the statement will also `list` all observations for which `income` is *missing*, since *missing* is greater than 50,000.

◁

❑ Technical Note

Like relational operators, logical operators return 1 for *true* and 0 for *false*. For example, the expression 5 & . evaluates to 1. Logical operations, except for !, are performed after all arithmetic and relational operations; the expression 3>2 & 5>4 is interpreted as (3>2) & (5>4) and evaluates to 1.

❑

13.2.5 Order of evaluation, all operators

The order of evaluation (from first to last) of all operators is ! (or ~), ^, - (negation), /, *, - (subtraction), +, != (or ≅), >, <, <=, >=, ==, &, and |.

13.3 Functions

Stata provides mathematical functions, probability and density functions, matrix functions, string functions, functions for dealing with dates and time series, and a set of special functions for programmers. You can find all of these documented in [D] **functions**. Stata's matrix programming language, Mata, provides more functions and those are documented in the *Mata Reference Manual*, or online (type `help mata functions`).

Functions are merely a set of rules; you supply the function with arguments, and the function evaluates the arguments according to the rules that define the function. Since functions are essentially subroutines that evaluate arguments and cause no action on their own, functions must be used in conjunction with a Stata command. Functions are indicated by the function name, an open parenthesis, an expression or expressions separated by commas, and a close parenthesis.

For example,

```
. display sqrt(4)
2
```

or

```
. display sqrt(2+2)
2
```

demonstrates the simplest use of a function. In this case, we have used the mathematical function, `sqrt()`, which takes a single number (or expression) as its argument and returns its square root. Note that the function was used with the Stata command `display`. If we had simply typed

```
. sqrt(4)
```

Stata would have returned the error message

```
unrecognized command:  sqrt
r(199);
```

Functions can operate on variables, as well. For example, suppose you wanted to generate a random variable that has observations drawn from a log-normal distribution. You could type

```
. set obs 5
obs was 0, now 5
. generate y = uniform()
. replace y = invnormal(y)
(5 real changes made)
. replace y = exp(y)
(5 real changes made)
. list
```

	y
1.	.686471
2.	2.380994
3.	.2814537
4.	1.215575
5.	.2920268

You could have saved yourself quite a bit of typing by just typing

    . generate y = exp(invnormal(uniform()))

Functions accept expressions as arguments.

All functions are defined over a specified domain and return values within a specified range. Whenever an argument is outside of a function's domain, the function will return a missing value or issue an error message, whichever is most appropriate. For example, if you supplied the log() function with an argument of zero, the log(0) would return a missing value because zero is outside of the natural logarithm function's domain. If you supplied the log() function with a string argument, Stata would issue a "type mismatch" error because log() is a numerical function and is undefined for strings. If you supply an argument that evaluates to a value that is outside of the function's range, the function will return a missing value. Whenever a function accepts a string as an argument, the string must be enclosed in double quotes, unless you provide the name of a variable that has a string storage type.

13.4   System variables (_variables)

Expressions may also contain _variables (pronounced "underscore variables"), which are built-in system variables that are created and updated by Stata. They are called _variables because their names all begin with the underscore '_' character.

The _variables are

[eqno]_b[varname] (synonym: [eqno]_coef[varname]) contains the value (to machine precision) of the coefficient on varname from the most recently fitted model (such as ANOVA, regression, Cox, logit, probit, multinomial logit, and the like). See [U] **13.5 Accessing coefficients and standard errors** below for a complete description.

_cons is always equal to the number 1 when used directly and refers to the intercept term when used indirectly, as in _b[_cons].

_n contains the number of the current observation.

_N contains the total number of observations in the dataset.

_pi contains the value of π to machine precision.

_rc contains the value of the return code from the most recent capture command.

[eqno]_se[varname] contains the value (to machine precision) of the standard error of the coefficient on varname from the most recently fitted model (such as ANOVA, regression, Cox, logit, probit, multinomial logit, and the like). See [U] **13.5 Accessing coefficients and standard errors** below for a complete description.

13.5   Accessing coefficients and standard errors

After fitting a model, you can access the coefficients and standard errors and use them in subsequent expressions. Also see [R] **predict** (and [U] **20 Estimation and postestimation commands**) for an easier way to obtain predictions, residuals, and the like.

13.5.1   Simple models

First, let's consider estimation methods that yield a single estimated equation with a one-to-one correspondence between coefficients and variables such as cnreg, logit, ologit, oprobit, probit, regress, and tobit. _b[*varname*] (synonym _coef[*varname*]) contains the coefficient on *varname* and _se[*varname*] contains its standard error, and both are recorded to machine precision. Thus, _b[age] refers to the calculated coefficient on the age variable after typing, say, regress response age sex, and _se[age] refers to the standard error on the coefficient. _b[_cons] refers to the constant and _se[_cons] to its standard error. Thus, you might type

```
. regress response age sex
. generate asif = _b[_cons] + _b[age]*age
```

13.5.2   ANOVA and MANOVA models

In ANOVA there is no simple relationship between the coefficients and the variables. For continuous variables in the model, _b[*varname*] refers to the coefficient. This works just as it does in simple models. For categorical variables, you must specify the level as well as the variable. _b[drug[2]] refers to the coefficient on the second level of drug. For interactions, _b[drug[2]*disease[1]] refers to the coefficient on the second level of drug and the first level of disease. Standard errors are obtained similarly using _se[]. Thus, you might type

```
. anova outcome sex age drug sex*age drug*age, continuous(age)
. generate age_effect = _b[age]*age
. replace age_effect = age_effect + _b[sex[1]*age] if sex==1
. replace age_effect = age_effect + _b[sex[2]*age] if sex==2
```

You access the coefficients and standard errors after manova in the same manner as with anova—using _b[] and _se[]—but, because manova is a multiple-equation estimator, the _b[] and _se[] must be preceded by an equation number in square brackets; see [U] **13.5.3 Multiple-equation models**.

13.5.3   Multiple-equation models

The syntax for referring to coefficients and standard errors in multiple-equation models is the same as in the simple-model case, except that _b[] and _se[] are preceded by an equation number in square brackets. There are, however, numerous alternatives in how you may type requests. The way that you are supposed to type requests is

[*eqno*] _b[*varname*]

[*eqno*] _se[*varname*]

but you may substitute _coef[] for _b[]. In fact, you may omit the _b[] altogether, and most Stata users do:

[*eqno*] [*varname*]

You may also omit the second pair of square brackets:

[*eqno*] *varname*

There are two ways to specify the equation number *eqno*: either as an absolute equation number or as an "indirect" equation number. In the absolute form, the number is preceded by a '#' sign. Thus, [#1]displ refers to the coefficient on displ in the first equation (and [#1]_se[displ] refers to its standard error). You can even use this form for simple models, such as regress, if you prefer. regress estimates a single equation, so [#1]displ refers to the coefficient on displ, just as _b[displ] does. Similarly, [#1]_se[displ] and _se[displ] are equivalent. The logic works both ways—in the multiple equation context, _b[displ] refers to the coefficient on displ in the first equation and _se[displ] refers to its standard error. _b[*varname*] (_se[*varname*]) is just another way of saying [#1]*varname* ([#1]_se[*varname*]).

Equations may also be referenced indirectly. [res]displ refers to the coefficient on displ in the equation named res. Equations are often named after the corresponding dependent variable name if there is such a concept in the fitted model, so [res]displ might refer to the coefficient on displ in the equation for variable res.

In the case of multinomial logit (mlogit), however, equations are named after the levels of the single dependent categorical variable. In multinomial logit, there is one dependent variable, and there is an equation corresponding to each of the outcomes (values taken on) recorded in that variable, except for the one that is arbitrarily labeled the base. [res]displ would be interpreted as the coefficient on displ in the equation corresponding *to the outcome* res. If outcome res is the base outcome, Stata treats [res]displ as zero (and Stata does the same thing for [res]_se[displ]).

Continuing with the multinomial logit case, the outcome variable must be numeric, although it need not be an integer. [res]displ would only be understood if there were a value label associated with the numeric outcome variable and res were one of the labels. If your data are not labeled, you may refer to the numeric value directly by omitting the '#'. [1]displ refers to the coefficient on displ in the equation corresponding to the outcome 1, which may be different from [#1]displ. [1.2]displ would be the coefficient on displ in the equation corresponding to outcome 1.2. [1.2]_cons refers to the constant in the equation corresponding to outcome 1.2. [1.2]_se[_cons] refers to the standard error on the constant.

Thus, you might type

```
. mlogit outcome displ weight
. generate cont_din1 = [1]displ*displ
```

For every observation in your data, cont_din1 would contain the coefficient on displ in the equation corresponding to the outcome 1 multiplied by displ, or the contribution of displ in determining outcome 1.

13.6   Accessing results from Stata commands

Most Stata commands—not just estimation commands—save results so that you can access them in subsequent expressions. You do that by referring to e(*name*), r(*name*), s(*name*), or c(*name*).

```
. summarize age
. generate agedev = age-r(mean)
. regress mpg weight
. display "The number of observations used is " e(N)
```

Most commands are categorized as r-class, meaning that they save results in r(). The returned results—such as r(mean)—are available immediately following the command, and if you are going to refer to them, you need to refer to them soon because the next command will probably replace what is in r().

e-class commands are Stata's estimation commands—commands that fit models. Results in e() remain available until the next model is fitted.

s-class commands are parsing commands—commands used by programmers to interpret commands you type. Very few commands save anything in s().

There are no c-class commands. c() contains values that are always available, such as c(current_date) (today's date), c(pwd) (the current directory), C(N) (the number of observations), and so on. There are lots of c() values and they are documented in [P] **creturn**.

Every command of Stata is designated r-class, e-class, or s-class, or, if the command saves nothing, n-class. r stands for return as in returned results, e stands for estimation as in estimation results, s stands for string, and, admittedly, this last acronym is weak, n stands for null.

You can find out what is stored where by looking in the *Saved Results* section for the particular command in the *Reference* manual. If you know the class of a command—and it is easy enough to guess—you can also see what is stored by typing 'return list', 'ereturn list', or 'sreturn list':

See [R] **saved results** and [U] **18.8 Accessing results calculated by other programs**.

13.7   Explicit subscripting

Individual observations on variables can be referenced by subscripting the variables. Explicit subscripts are specified by following a variable name with square brackets that contain an expression. The result of the subscript expression is truncated to an integer, and the value of the variable for the indicated observation is returned. If the value of the subscript expression is less than 1 or greater than _N, a missing value is returned.

13.7.1   Generating lags and leads

When you type something like

    . generate y = x

Stata interprets it as if you typed

    . generate y = x[_n]

which means that the first observation of y is to be assigned the value from the first observation of x, the second observation of y is to be assigned the value from the second observation on x, and so on. If you instead typed

    . generate y = x[1]

you would set each observation of y equal to the first observation on x. If you typed

    . generate y = x[2]

you would set each observation of y equal to the second observation on x. If you typed

    . generate y = x[0]

Stata would merely copy a missing value into every observation of y because observation 0 does not exist. Exactly the same would happen if you typed

    . generate y = x[100]

and you had fewer than 100 observations in your data.

When you type the square brackets, you are specifying explicit subscripts. Explicit subscripting combined with the _*variable* _n can be used to create lagged values on a variable. The lagged value of a variable x can be obtained by typing

    . generate xlag = x[_n-1]

If you are really interested in lags and leads, you probably have time-series data and would be better served by using the time-series operators, such as L.x. Time-series operators can be used with varlists and expressions and they are safer because they account for gaps in the data; see [U] **11.4.3 Time-series varlists** and [U] **13.8 Time-series operators**. Even so, it is important that you understand how the above works.

The built-in underscore variable _n is understood by Stata to mean the observation number of the current observation. That is why

. generate y = x[_n]

results in observation 1 of x being copied to observation 1 of y and similarly for the rest of the observations. Consider

. generate xlag = x[_n-1]

_n-1 evaluates to the observation number of the previous observation. For the first observation, _n-1 = 0 and therefore xlag[1] is set to missing. For the second observation, _n-1 = 1 and xlag[2] is set to the value of x[1], and so on.

Similarly, the lead of x can be created by

. generate xlead = x[_n+1]

In this case, the last observation on the new variable xlead will be *missing* because _n+1 will be greater than _N (_N is the total number of observations in the dataset).

13.7.2   Subscripting within groups

When a command is preceded by the by *varlist*: prefix, subscript expressions and the underscore variables _n and _N are evaluated relative to the subset of the data currently being processed. For example, consider the following (admittedly not very interesting) data:

. list

	bvar	oldvar
1.	1	1.1
2.	1	2.1
3.	1	3.1
4.	2	4.1
5.	2	5.1

To see how _n, _N, and explicit subscripting work, let's create three new variables demonstrating each and then list their values:

. generate small_n = _n
. generate big_n = _N
. generate newvar = oldvar[1]
. list

	bvar	oldvar	small_n	big_n	newvar
1.	1	1.1	1	5	1.1
2.	1	2.1	2	5	1.1
3.	1	3.1	3	5	1.1
4.	2	4.1	4	5	1.1
5.	2	5.1	5	5	1.1

small_n (which is equal to _n) goes from 1 to 5, and big_n (which is equal to _N) is 5. This should not be surprising; there are 5 observations in the data, and _n is supposed to count observations, whereas _N is the total number. newvar, which we defined as oldvar[1], is 1.1. Indeed, we see that the first observation on oldvar is 1.1.

Now, let's repeat those same three steps, only this time preceding each step with the prefix by bvar:. First, we will drop the old values of small_n, big_n, and newvar so that we start fresh:

```
. drop small_n big_n newvar
. by bvar, sort: generate small_n=_n
. by bvar: generate big_n =_N
. by bvar: generate newvar=oldvar[1]
. list
```

	bvar	oldvar	small_n	big_n	newvar
1.	1	1.1	1	3	1.1
2.	1	2.1	2	3	1.1
3.	1	3.1	3	3	1.1
4.	2	4.1	1	2	4.1
5.	2	5.1	2	2	4.1

The results are different. Remember that we claimed that _n and _N are evaluated relative to the subset of data in the by-group. Thus, small_n (_n) goes from 1 to 3 for bvar = 1 and from 1 to 2 for bvar = 2. big_n (_N) is 3 for the first group and 2 for the second. Finally, newvar (oldvar[1]) is 1.1 and 4.1.

▷ Example 7

You now know enough to do some amazing things.

Suppose that you have data on individual states and you have another variable in your data called region that divides the states into the four census regions. You have a variable x in your data, and you want to make a new variable called avgx to include in your regressions. This new variable is to take on the average value of x for the region in which the state is located. Thus, for California you will have the observation on x and the observation on the average value in the region, avgx. Here's how:

```
. by region, sort: generate avgx=sum(x)/_n
. by region: replace avgx=avgx[_N]
```

First, by region, we generate avgx equal to the running sum of x divided by the number of observations so far. The , sort ensures that the data are in region order. We have, in effect, created the running average of x within region. It is the last observation of this running average, the overall average within the region, that interests us. So, by region, we replace every avgx observation in a region with the last observation within the region, avgx[_N].

Here is what we will see when we type these two commands:

```
. by region, sort: generate avgx=sum(x)/_n
. by region: replace avgx=avgx[_N]
(46 real changes made)
```

In our example, there are no missing observations on x. If there had been, we would have obtained the wrong answer. When we created the running average, we typed

```
. by region, sort: generate avgx=sum(x)/_n
```

The problem is not with the sum() function. When sum() encounters a missing, it adds zero to the sum. The problem is with _n. Let's assume that the second observation in the first region has recorded a missing for x. When Stata processes the third observation in that region, it will calculate the sum of two elements (remember that one is missing) and then divide the sum by 3 when it should be divided by 2. There is an easy solution:

```
. by region: generate avgx=sum(x)/sum(x<.)
```

Rather than divide by _n, we divide by the total number of nonmissing observations seen on x so far, namely, the sum(x<.).

If our goal were simply to obtain the mean, we could have more easily accomplished it by typing egen avgx=mean(x), by(region); see [D] **egen**. egen, however, is written in Stata, and the above is how egen's mean() function works. The general principles are worth understanding.

◁

▷ Example 8

You have some patient data recording vital signs at various times during an experiment. The variables include patient, an ID number or name of the patient; time, a variable recording the date or time or epoch of the vital-sign reading; and vital, a vital sign. You probably have more than one vital sign, but one is enough to illustrate the concept. Each observation in your data represents a patient-time combination.

Let's assume that you have 1,000 patients and, for every observation on the same patient, you want to create a new variable called orig that records the patient's initial value of this vital sign.

```
. sort patient time
. by patient: generate orig=vital[1]
```

Observe that vital[1] refers not to the first reading on the first patient, but to the first reading on the current patient, because we are performing the generate command by patient.

◁

▷ Example 9

Let's do one more example with this patient data. Suppose that we want to create a new dataset from our patient data that records not only the patient's identification, the time of the reading of the first vital sign, and the first vital sign reading itself, but also the time of the reading of the last vital sign and its value. We want one observation per patient. Here's how:

```
. sort patient time
. by patient: generate lasttime=time[_N]
. by patient: generate lastvital=vital[_N]
. by patient: drop if _n!=1
```

◁

13.8   Time-series operators

Time-series operators allow you to refer to the lag of `gnp` by typing `L.gnp`, the second lag by typing `L2.gnp`, etc. There are also operators for lead (`F`), difference `D`, and seasonal difference `S`.

Time-series operators can be used with varlists and with expressions. See [U] **11.4.3 Time-series varlists** if you have not read it already. This section has to do with using time-series operators in expressions such as with `generate`. You do not have to create new variables; you can use the time-series operated variables directly.

13.8.1   Generating lags and leads

In a time-series context, referring to `L2.gnp` is better than referring to `gnp[_n-2]` because there might be missing observations. Pretend that observation 4 contains data for $t = 25$ and observation 5 data for $t = 27$. `L2.gnp` will still produce correct answers; `L2.gnp` for observation 5 will be the value from observation 4 because the time-series operators look at t to find the relevant observation. The more mechanical `gnp[_n-2]` just goes two observations back, which, in this case, would not produce the desired result.

Time-series operators can be used with varlists or with expressions, so you can type

        . regress val L.gnp r

or

        . generate gnplagged = L.gnp
        . regress val gnplagged

Before you can type either one, however, you must use the `tsset` command to tell Stata the identity of the time variable; see [TS] **tsset**. Once you have `tsset` the data, anyplace you see an *exp* in a syntax diagram, you may type time-series operated variables, so you can type

        . summarize r if F.gnp<gnp

or

        . generate grew = 1 if gnp>L.gnp & L.gnp<.
        . replace grew = 0 if grew>=. & L.gnp<.

or

        . generate grew = (gnp>L.gnp) if L.gnp<.

13.8.2   Operators within groups

Stata also understands panel or cross-sectional time-series data. For instance, if you type

        . tsset country time

you are declaring that you have time-series data. The time variable is `time`, and you have time-series data for separate countries.

Once you have `tsset` both cross-sectional and time identifiers, you proceed just as you would if you had a simple time series.

        . generate grew = (gnp>L.gnp) if L.gnp<.

would produce correct results. The `L.` operator will not confuse the observation at the end of one panel with the beginning of the next.

13.9 Label values

If you have not read [U] **12.6 Dataset, variable, and value labels**, please do so. You may use labels in an expression in place of the numeric values with which they are associated. To use a label in this way, type the label in double quotes followed by a colon and the name of the value label.

▷ Example 10

If the value label yesno associates the label yes with 1 and no with 0, then "yes":yesno (said aloud as the value of yes under yesno) is evaluated as 1. If the double-quoted label is not defined in the indicated value label, or if the value label itself is not found, a missing value is returned. Thus, the expression "maybe":yesno is evaluated as *missing*.

```
. list
```

	name	answer
1.	Sribney	no
2.	Gaines	no
3.	Hilbe	yes
4.	DeLeon	no
5.	Cain	no
6.	Willis	yes
7.	Schroeder	no
8.	Cox	no
9.	Roman	no
10.	Hardin	yes
11.	Lancaster	yes
12.	Johnson	no

```
. list if answer=="yes":yesno
```

	name	answer
3.	Hilbe	yes
6.	Willis	yes
10.	Hardin	yes
11.	Lancaster	yes

In the above example, the variable answer is not a string variable; it is a numeric variable that has the associated value label yesno. Since yesno associates yes with 1 and no with 0, we could have typed list if answer==1 instead of what we did type. We could not have typed list if answer=="yes" because answer is not a string variable. If we had, we would have received the error message "type mismatch".

◁

13.10   Precision and problems therein

Examine the following short Stata session:

```
. drop _all
. input x y
              x          y
  1. 1 1.1
  2. 2 1.2
  3. 3 1.3
  4. end
. count if x==1
      1
. count if y==1.1
      0
. list
```

	x	y
1.	1	1.1
2.	2	1.2
3.	3	1.3

We created a dataset containing two variables, x and y. The first observation has x equal to 1 and y equal to 1.1. When we asked Stata to count the number of times that the variable x took on the value 1, we were told that it occurred once. Yet when we asked Stata to count the number of times y took on the value 1.1, we were told zero—meaning that it never occurred. What has gone wrong? When we list the data, we see that the first observation has y equal to 1.1.

Despite appearances, Stata has not made a mistake. Stata stores numbers internally in binary form, and the number 1.1 has no exact binary representation—that is, there is no finite string of binary digits that is exactly equal to 1.1.

❏ Technical Note

The number 1.1 in binary form is 1.0001100110011 ..., where the period represents the binary point. The problem binary computers have with storing numbers like 1/10 is much like the problem we base-10 users have in precisely writing 1/11, which is 0.0909090909 ....

❏

The number that appears as 1.1 in the listing above is actually 1.1000000238419, which is off by roughly 2 parts in 10^8. Unless we tell Stata otherwise, it stores all numbers as floats, which are also known as *single-precision* or *4-byte reals*. On the other hand, Stata performs all internal calculations in double, which is also known as *double-precision* or *8-byte reals*. This is what leads to the difficulty.

In the above example, we compared the number 1.1, stored as a float, with the number 1.1 stored as a double. The double-precision representation of 1.1 is more accurate than the single-precision representation, but what is important is that it is also different. Those two numbers are not equal.

There are a number of ways around this problem. The problem with 1.1 apparently not equaling 1.1 would never arise if the storage precision and the precision of the internal calculations were the same. Thus, you could store all your data as doubles. This takes more computer memory, however, and it is unlikely that (1) your data are really that accurate and (2) the extra digits would meaningfully affect any calculated result, even if the data were that accurate.

❏ Technical Note

This is unlikely to affect any calculated result because Stata performs all internal calculations in double precision. This is all rather ironic, since the problem would also not arise if we had designed Stata to use single precision for its internal calculations. Stata would be less accurate, but the problem would have been completely disguised from the user, making this entry unnecessary.

❏

Another solution is to use the `float()` function. `float(x)` rounds x to its `float` representation. If we had typed `count if y==float(1.1)` in the above example, we would have been informed that there is one such value.

14 Matrix expressions

Contents

14.1   Overview

Stata has two matrix programming languages, one that might be called Stata's older matrix language and another that is called Mata. Stata's Mata is the new one, and there is an uneasy relationship between the two.

Below we discuss Stata's older language and leave the newer one to another manual—the *Mata Reference Manual* ([M])—or you can learn about the newer one by typing `help mata`.

We admit that the newer language is better in almost every way than the older language, but the older one still has a use because it is the one that Stata truly and deeply understands. Even when Mata wants to talk to Stata, matrixwise, it is the older language that Mata must use, so you must learn to use the older language as well as the new.

This is not nearly as difficult, or messy, as you might imagine because Stata's older language is remarkably easy to use, and really, there is not much to learn. Just remember that for heavy-duty programming, it will be worth your time to learn Mata, too.

14.1.1   Definition of a matrix

Stata's definition of a matrix includes a few details that go beyond the mathematics. To Stata, a matrix is a named entity containing an $r \times c$ ($0 < r \leq$ matsize, $0 < c \leq$ matsize) rectangular array of double-precision numbers (including missing values) that is bordered by a row and a column of names.

```
. matrix list A

A[3,2]
     c1  c2
r1    1   2
r2    3   4
r3    5   6
```

159

In this case, we have a 3×2 matrix named A containing elements 1, 2, 3, 4, 5, and 6. Row 1, column 2 (written $A_{1,2}$ in math and A[1,2] in Stata) contains 2. The columns are named c1 and c2 and the rows r1, r2, and r3. These are the default names Stata comes up with when it cannot do better. The names do not play a role in the mathematics, but they are of great help when it comes to labeling the output.

The names are operated on just as the numbers are. For instance,

```
. matrix B=A'*A
. matrix list B
symmetric B[2,2]
     c1  c2
c1   35
c2   44  56
```

We defined $\mathbf{B} = \mathbf{A}'\mathbf{A}$. Note that the row and column names of $\mathbf{B}$ are the same. Multiplication is defined for any $a \times b$ and $b \times c$ matrices, the result being $a \times c$. Thus, the row and column names of the result are the row names of the first matrix and the column names of the second matrix. We formed $\mathbf{A}'\mathbf{A}$, using the transpose of $\mathbf{A}$ for the first matrix—which also interchanged the names—and so obtained the names shown.

14.1.2   matsize

Matrices are limited to being no larger than matsize $\times$ matsize. The default value of matsize is 400 for Stata/SE and 200 for Intercooled Stata, but you can reset this using the set matsize command; see [R] **matsize**.

The maximum value of matsize is 800 for Intercooled Stata, so matrices are not suitable for holding large amounts of data. This restriction does not prove a limitation because terms that appear in statistical formulas are of the form $(\mathbf{X}'\mathbf{W}\mathbf{Z})$ and Stata provides a command, matrix accum, for efficiently forming such matrices; see [U] **14.6 Creating matrices by accumulating data** below. The maximum value of matsize is 11,000 for Stata/SE, so performing matrix operations directly on large amounts of data is more feasible. Note that the matsize limit does not apply to Mata matrices; see the *Mata Reference Manual*.

14.2   Row and column names

Matrix rows and columns always have names. Stata is smart about setting these names when the matrix is created, and the matrix commands and operators manipulate these names throughout calculations, so the names typically are set correctly at the conclusion of matrix calculations.

For instance, consider the matrix calculation $\mathbf{b} = (\mathbf{X}'\mathbf{X})^{-1}\mathbf{X}'\mathbf{y}$ performed on real data:

```
. use http://www.stata-press.com/data/r9/auto
(1978 Automobile Data)
. matrix accum XprimeX = weight foreign
(obs=74)
. matrix vecaccum yprimeX = mpg weight foreign
. matrix b = invsym(XprimeX)*yprimeX'
. matrix list b
b[3,1]
                mpg
 weight  -.00658789
foreign  -1.6500291
  _cons   41.679702
```

Note that these names were produced without our ever having given a special command to place the names on the result. When we formed matrix XprimeX, Stata produced the result

```
. matrix list XprimeX
symmetric XprimeX[3,3]
              weight     foreign       _cons
  weight   7.188e+08
 foreign       50950          22
   _cons      223440          22          74
```

matrix accum forms $\mathbf{X}'\mathbf{X}$ matrices from data and sets the row and column names to the variable names used. The names are correct in the sense that, for instance, the (1,1) element is the sum across the observations of squares of weight and the (2,1) element is the sum of the product of weight and foreign.

Similarly, matrix vecaccum forms $\mathbf{y}'\mathbf{X}$ matrices, and it also sets the row and column names to the variable names used, so matrix vecaccum yprimeX = mpg weight foreign resulted in

```
. matrix list yprimeX
yprimeX[1,3]
         weight     foreign       _cons
 mpg    4493720         545        1576
```

The final step, matrix b = invsym(XprimeX)*yprimeX', manipulated the names, and, if you think carefully, you can derive for yourself the rules. invsym() (inversion) is much like transposition, so row and column names must be swapped. In this case, however, the matrix was symmetric, so that amounted to leaving the names as they were. Multiplication amounts to taking the column names of the first matrix and the row names of the second. The final result is

```
. matrix list b
b[3,1]
                    mpg
  weight   -.00658789
 foreign   -1.6500291
   _cons    41.679702
```

and the interpretation is $\mathrm{mpg} = -.00659\,\mathrm{weight} - 1.65\,\mathrm{foreign} + 41.68 + e$.

Researchers realized long ago that using matrix notation simplifies the description of complex calculations. What they may not have realized is that, corresponding to each mathematical definition of a matrix operator, there is a definition of the operator's effect on the names that can be used to carry the names forward through long and complex matrix calculations.

14.2.1 The purpose of row and column names

Mostly, matrices in Stata are used in programming estimators, and Stata uses row and column names to produce pretty output. For instance, say that we wrote code—interactively or in a program—that produced the following coefficient vector b and covariance matrix V:

```
. matrix list b
b[1,3]
          weight  displacement          _cons
 y1    -.00656711      .00528078     40.084522
. matrix list V
symmetric V[3,3]
                     weight  displacement          _cons
       weight      1.360e-06
 displacement       -.0000103      .00009741
       _cons       -.00207455      .01188356      4.0808455
```

We could now produce standard estimation output by coding two more lines:

```
. ereturn post b V
. ereturn display
```

	Coef.	Std. Err.	z	P>|z|	[95% Conf. Interval]	
weight	-.0065671	.0011662	-5.63	0.000	-.0088529	-.0042813
displacement	.0052808	.0098696	0.54	0.593	-.0140632	.0246248
_cons	40.08452	2.02011	19.84	0.000	36.12518	44.04387

Stata's `ereturn` command knew to produce this output because of the row and column names on the coefficient vector and variance matrix. Moreover, in most cases, we do nothing special in our code that produces b and V to set the row and column names because, given how matrix names work, they work themselves out.

In addition, sometimes row and column names help us detect programming errors. Assume that we wrote code to produce matrices b and V but made a mistake. Sometimes our mistake will result in the wrong row and column names. Rather than the b vector we previously showed you, we might produce

```
. matrix list b
b[1,3]
         weight         c2       _cons
y1   -.00656711       42.23   40.084522
```

If we posted our estimation results now, Stata would refuse because it can tell by the names that there is a problem:

```
. ereturn post b V
name conflict
r(507);
```

Understand, however, that Stata follows the standard rules of matrix algebra; the names are just along for the ride. Matrices are summed by position, meaning that a directive to form $\mathbf{C} = \mathbf{A} + \mathbf{B}$ results in $C_{11} = A_{11} + B_{11}$, regardless of the names, and it is not an error to sum matrices with different names:

```
. matrix list a
symmetric a[3,3]
              c1          c2          c3
   mpg      14419
weight    1221120   1.219e+08
 _cons        545       50950          22
. matrix list b
symmetric b[3,3]
                    c1          c2          c3
displacement   3211055
        mpg     227102       22249
      _cons      12153        1041          52
. matrix c = a + b
. matrix list c
symmetric c[3,3]
                    c1          c2          c3
displacement   3225474
        mpg    1448222   1.219e+08
      _cons      12698       51991          74
```

Matrix row and column names are used to label output; they do not affect how matrix algebra is performed.

14.2.2   Three-part names

Row and column names have three parts: *equation_name:ts_operator.subname*.

In the examples shown so far, the first two parts have been blank; the row and column names consisted of *subnames* only. This is typical. Run any single-equation model (such as those produced by `regress`, `probit`, `logistic`, etc.), and if you fetch the resulting matrices, you will find that they have row and column names of the *subname* form.

Those who work with time-series data will find matrices with row and column names of the form *ts_operator.subname*. For example,

```
. matrix list example1
symmetric example1[3,3]
                                    L.
                 rate           rate         _cons
  rate     3.0952534
L.rate      .0096504      .00007742
 _cons    -2.8413483     -.01821928     4.8578916
```

We obtained this matrix by running a linear regression on `rate` and `L.rate` and then fetching the covariance matrix. Think of the row and column name `L.rate` no differently than you think of `rate` or, in the previous examples, `r1`, `r2`, `c1`, `c2`, `weight`, and `foreign`.

Equation names are used to label partitioned matrices and, in estimation, occur in the context of multiple equations. Here is a matrix with *equation_names* and *subnames*:

```
. matrix list example2
symmetric example2[5,5]
                      mpg:          mpg:          mpg:          mpg:          mpg:
                   foreign         displ         _cons       foreign         _cons
   mpg:foreign   1.6483972
     mpg:displ     .004747      .00003876
     mpg:_cons   -1.4266352    -.00905773     2.4341021
weight:foreign  -51.208454     -4.665e-19     15.224135     24997.727
 weight:_cons    15.224135      2.077e-17    -15.224135    -7431.7565     7431.7565
```

Here is an example with all three parts filled in:

```
. matrix list example3
symmetric example3[5,5]
                      val:          val:          val:       weight:       weight:
                                      L.
                      rate          rate         _cons       foreign         _cons
     val:rate     2.2947268
   val:L.rate     .00385216      .0000309
    val:_cons    -1.4533912     -.0072726     2.2583357
weight:foreign  -163.86684      7.796e-17     49.384526     25351.696
 weight:_cons    49.384526     -1.566e-16    -49.384526    -7640.237      7640.237
```

`val:L.rate` is a column name, just as, in the previous section, `c2` and `foreign` were names.

Say that this last matrix is the variance matrix produced by a program we wrote and that our program also produced a coefficient vector `b`:

```
. matrix list b
b[1,5]
                      val:          val:          val:       weight:       weight:
                                      L.
                      rate          rate         _cons       foreign         _cons
   y1     4.5366753     -.00316923     20.68421    -1008.7968     3324.7059
```

Here is the result of posting and displaying the results:

```
. ereturn post b example3
. ereturn display
```

		Coef.	Std. Err.	z	P>|z|	[95% Conf. Interval]	
val							
rate							
	—	4.536675	1.514836	2.995	0.003	1.567652	7.505698
	L1	-.0031692	.0055591	-0.570	0.569	-.0140648	.0077264
_cons		20.68421	1.502776	13.764	0.000	17.73882	23.6296
weight							
foreign		-1008.797	159.2222	-6.336	0.000	-1320.866	-696.7271
_cons		3324.706	87.40845	38.036	0.000	3153.388	3496.023

The equation names are used to separate one equation from the next.

14.2.3   Setting row and column names

You reset row and column names using the `matrix rownames` and `matrix colnames` commands.

Before resetting the names, use `matrix list` to verify that the names are not set correctly; often, they already are. When you enter a matrix by hand, however, the row names are unimaginatively set to r1, r2, ..., and the column names to c1, c2, ...

```
. matrix a = (1,2,3\4,5,6)
. matrix list a
a[2,3]
     c1  c2  c3
r1    1   2   3
r2    4   5   6
```

Regardless of the current row and column names, `matrix rownames` and `matrix colnames` reset them:

```
. matrix colnames a = foreign alpha _cons
. matrix rownames a = one two
. matrix list a
a[2,3]
       foreign    alpha    _cons
one          1        2        3
two          4        5        6
```

You may set the *ts_operator* as well as the *subname*,

```
. matrix colnames a = foreign l.rate _cons
. matrix list a
a[2,3]
                    L.
       foreign    rate    _cons
one          1       2        3
two          4       5        6
```

and you may set equation names:

```
. matrix colnames a = this:foreign this:l.rate that:_cons

. matrix list a

a[2,3]
          this:      this:      that:
                       L.
       foreign       rate      _cons
  one      1          2          3
  two      4          5          6
```

See [P] **matrix rownames** for more information.

14.2.4   Obtaining row and column names

matrix list displays the matrix with its row and column names. In a programming context, you can fetch the row and column names into a macro using

$$\text{local } \dots : \text{ rowfullnames } \textit{matname}$$
$$\text{local } \dots : \text{ colfullnames } \textit{matname}$$
$$\text{local } \dots : \text{ rownames } \textit{matname}$$
$$\text{local } \dots : \text{ colnames } \textit{matname}$$
$$\text{local } \dots : \text{ roweq } \textit{matname}$$
$$\text{local } \dots : \text{ coleq } \textit{matname}$$

rowfullnames and colfullnames return the full names (*equation_name:ts_operator.subname*) listed one after the other.

rownames and colnames omit the equations and return *ts_operator.subname*, listed one after the other.

roweq and coleq return the equation names, listed one after the other.

See [P] **macro** and [P] **matrix define** for more information.

14.3   Vectors and scalars

Stata does not have vectors as such—they are considered special cases of matrices and are handled by the matrix command.

Stata does have scalars, although they are not strictly necessary because they, too, could be handled as special cases. See [P] **scalar** for a description of scalars.

14.4   Inputting matrices by hand

You input matrices using

matrix input *matname* = (...)

or

matrix *matname* = (...)

In either case, you enter the matrices by row. You separate one element from the next using commas (,) and one row from the next using backslashes (\). If you omit the word input, you are using the expression parser to input the matrix:

```
. matrix a = (1,2\3,4)
. matrix list a
a[2,2]
     c1  c2
r1    1   2
r2    3   4
```

This has the advantage that you can use expressions for any of the elements:

```
. matrix b = (1, 2+3/2 \ cos(_pi), _pi)
. matrix list b
b[2,2]
          c1          c2
r1         1         3.5
r2        -1   3.1415927
```

The disadvantage is that the matrix must be small, say, no more than 50 elements (regardless of the value of `matsize`).

`matrix input` has no such restriction, but you may not use subexpressions for the elements:

```
. matrix input c  = (1,2\3,4)
. matrix input d = (1, 2+3/2 \ cos(_pi), _pi)
invalid syntax
r(198);
```

Either way, after inputting the matrix, you will probably want to set the row and column names; see [U] **14.2.3 Setting row and column names** above.

14.5   Accessing matrices created by Stata commands

Some Stata commands—including all estimation commands—leave behind matrices that you can subsequently use. After executing an estimation command, type `ereturn list` to see what is available:

```
. use http://www.stata-press.com/data/r9/auto
(1978 Automobile Data)

. probit foreign mpg weight
 (output omitted )

. ereturn list
scalars:
                 e(N) =  74
              e(ll_0) = -45.03320955699139
                e(ll) = -26.8441890057987
              e(df_m) =  2
              e(chi2) =  36.37804110238538
              e(r2_p) =  .4039023807124769

macros:
            e(title) : "Probit regression"
           e(depvar) : "foreign"
             e(cmd) : "probit"
         e(crittype) : "log likelihood"
          e(predict) : "probit_p"
       e(properties) : "b V"
        e(estat_cmd) : "probit_estat"
         e(chi2type) : "LR"
```

```
matrices:
                    e(b) :  1 x 3
                    e(V) :  3 x 3

functions:
            e(sample)
```

Most estimation commands leave behind e(b) (the coefficient vector) and e(V) (the variance–covariance matrix of the estimator):

```
. matrix list e(b)

e(b)[1,3]
            mpg      weight      _cons
y1   -.10395033   -.00233554   8.275464
```

You can refer to e(b) and e(V) in any matrix expression:

```
. matrix myb = e(b)

. matrix list myb

myb[1,3]
            mpg      weight      _cons
y1   -.10318674    -.0023264   8.234735

. matrix c = e(b)*invsym(e(V))*e(b)'

. matrix list c

symmetric c[1,1]
            y1
y1   22.440544
```

14.6   Creating matrices by accumulating data

In programming estimators, matrices of the form $\mathbf{X'X}$, $\mathbf{X'Z}$, $\mathbf{X'WX}$, and $\mathbf{X'WZ}$ often occur, where $\mathbf{X}$ and $\mathbf{Z}$ are data matrices. matrix accum, matrix glsaccum, matrix vecaccum, and matrix opaccum produce such matrices; see [P] **matrix accum**.

We recommend that you not load the data into a matrix and use the expression parser directly to form such matrices, although see [P] **matrix mkmat** if that is your interest. If that is your interest, be sure to read the *Technical Note* at the end of [P] **matrix mkmat**. There is much to recommend learning how to use the matrix accum commands.

14.7   Matrix operators

You can create new matrices or replace existing matrices by typing

$$\text{matrix } \textit{matname} = \textit{matrix\_expression}$$

For instance,

```
. matrix A = invsym(R*V*R')
. matrix IAR = I(rowsof(A)) - A*R
. matrix beta = b*IAR' + r*A'
. matrix C = -C'
. matrix D = (A, B \ B', A)
. matrix E = (A+B)*C'
. matrix S = (S+S')/2
```

The following operators are provided:

Operator	Symbol
Unary operators	
negation	−
transposition	'
Binary operators	
(lowest precedence)	
row join	\
column join	,
addition	+
subtraction	−
multiplication	*
division by scalar	/
Kronecker product	#
(highest precedence)	

Parentheses may be used to change the order of evaluation.

Note in particular that , and \ are operators; (1,2) creates a 1×2 matrix (vector), and (A,B) creates a rowsof(A) $\times$ colsof(A)+colsof(B) matrix, where rowsof(A) = rowsof(B). (1\2) creates a 2×1 matrix (vector), and (A\B) creates a rowsof(A)+rowsof(B) $\times$ colsof(A) matrix, where colsof(A) = colsof(B). Thus, expressions of the form

```
matrix R = (A,B)*Vinv*(A,B)'
```

are allowed.

14.8   Matrix functions

In addition to the functions listed below, see [P] **matrix svd** for singular value decomposition, [P] **matrix symeigen** for eigenvalues and eigenvectors of symmetric matrices, and [P] **matrix eigenvalues** for eigenvalues of nonsymmetric matrices. For a full description of the matrix functions, see [D] **functions**.

Matrix functions returning matrices:

cholesky(M)	I(n)	sweep(M,i)
corr(M)	inv(M)	invsym(M)
diag(v)	J(r,c,z)	vec(M)
e(*name*)	matuniform(r, c)	vecdiag(M)
get(*systemname*)	nullmat(*matname*)	
hadamard(M, N)	return(*name*)	

Matrix functions returning scalars:

colnumb(M,s)	el(s,i,j)	rownumb(M,s)
colsof(M)	issymmetric(M)	rowsof(M)
det(M)	matmissing(M)	trace(M)
diag0cnt(M)	mreldif(X,Y)	

14.9 Subscripting

1. In matrix and scalar expressions, you may refer to *matname*$[r,c]$, where r and c are scalar expressions, to obtain a single element of *matname* as a scalar.

   Examples:
   ```
   matrix A = A / A[1,1]
   generate newvar = oldvar / A[2,2]
   ```

2. In matrix expressions, you may refer to *matname*$[s_r,s_c]$, where s_r and s_c are string expressions, to obtain a submatrix with a single element. The element returned is based on searching the row and column names.

   Examples:
   ```
   matrix B = V["price","price"]
   generate sdif = dif / sqrt(V["price","price"])
   ```

3. In matrix expressions, you may mix these two syntaxes and refer to *matname*$[r,s_c]$ or to *matname*$[s_r,c]$.

   Example:
   ```
   matrix b = b * R[1,"price"]
   ```

4. In matrix expressions, you may use *matname*$[r_1..r_2,c_1..c_2]$ to refer to submatrices; r_1, r_2, c_1, and c_2 may be scalar expressions. If r_2 evaluates to missing, it is taken as referring to the last row of *matname*; if c_2 evaluates to missing, it is taken as referring to the last column of *matname*. Thus, *matname*$[r_1...,c_1...]$ is allowed.

   Examples:
   ```
   matrix S = Z[1..4, 1..4]
   matrix R = Z[5..., 5...]
   ```

5. In matrix expressions, you may refer to *matname*$[s_{r1}..s_{r2},s_{c1}..s_{c2}]$ to refer to submatrices where s_{r1}, s_{r2}, s_{c1}, and s_{c2}, are string expressions. The matrix returned is based on looking up the row and column names.

   If the string evaluates to an equation name only, all the rows or columns for the equation are returned.

   Examples:
   ```
   matrix S = Z["price".."weight", "price".."weight"]
   matrix L = D["mpg:price".."mpg:weight", "mpg:price".."mpg:weight"]
   matrix T1 = C["mpg:", "mpg:"]
   matrix T2 = C["mpg:", "price:"]
   ```

6. In matrix expressions, any of the above syntaxes may be combined.

   Examples:
   ```
   matrix T1 = C["mpg:", "price:weight".."price:displ"]
   matrix T2 = C["mpg:", "price:weight"...]
   matrix T3 = C["mpg:price", 2..5]
   matrix T4 = C["mpg:price", 2]
   ```

7. When defining an element of a matrix, use

$$\texttt{matrix } \textit{matname}[i,j] = \textit{expression}$$

where i and j are scalar expressions. The matrix *matname* must already exist.

Example:
```
matrix A = J(2,2,0)
matrix A[1,2] = sqrt(2)
```

8. To replace a submatrix within a matrix, use the same syntax. If the expression on the right evaluates to a scalar or 1×1 matrix, the element is replaced. If it evaluates to a matrix, the submatrix with top-left element at (i,j) is replaced. The matrix *matname* must already exist.

Example:
```
matrix A = J(4,4,0)
matrix A[2,2] = C'*C
```

14.10   Using matrices in scalar expressions

Scalar expressions are documented as *exp* in the Stata manuals:

```
generate newvar = exp if exp ...
replace  newvar = exp if exp ...
regress ... if exp ...
if exp {... }
while exp {... }
```

Most importantly, scalar expressions occur in `generate` and `replace`, in the `if` *exp* modifier allowed on the end of many commands, and in the `if` and `while` commands for program control.

You will rarely need to refer to a matrix in any of these situations except when using the `if` and `while` commands.

In any case, you may refer to matrices in any of these situations, but the expression cannot require evaluation of matrix expressions returning matrices. Thus, you could refer to `trace(A)` but not to `trace(A+B)`.

It can be difficult to predict when an evaluation of an expression requires evaluating a matrix; even experienced users can be surprised. If you get the error message "matrix operators that return matrices not allowed in this context", r(509), you have encountered such a situation.

The solution is to split the line in two. For instance, you would change

```
if trace(A+B)==0 {
        ...
}
```

to

```
matrix AplusB = A+B
if trace(AplusB)==0 {
        ...
}
```

or even to

```
matrix Trace = trace(A+B)
if Trace[1,1]==0 {
        ...
}
```

15 Printing and preserving output

Contents

15.1   Overview

Stata can record your session into a file called a log file but does not start a log automatically; you must tell Stata to record your session. By default, the resulting log file contains what you type and what Stata produces in response, recorded in a format called Stata Markup and Control Language (SMCL); see [P] **smcl**. The file can be printed or converted to ASCII text for incorporation into documents you create with your word processor.

To start a log:
  Your session is now being recorded
  in file *filename*.smcl.
	. log using *filename*

To temporarily stop logging:
  Temporarily stop:
  Resume:
	. log off
	. log on

To stop logging and close the file:
  You can now print *filename*.smcl or type:
  to create *filename*.log that you can
  load into your word processor.
	. log close
	. translate *filename*.smcl *filename*.log

Alternative ways to start logging:
  append to an existing log:
  replace an existing log:
	. log using *filename*, append
	. log using *filename*, replace

Using the GUI:
  To start a log:
  To temporarily stop logging:
  To resume:
  To stop logging and close the file:
  To print previous or current log:
	click the **Log** button
	click the **Log** button, and choose **Suspend**
	click the **Log** button, and choose **Resume**
	click the **Log** button, and choose **Close**
	select **File > View**, choose file,
	    right-click on the Viewer, and select **Print**

In addition, `cmdlog` will produce logs containing solely what you typed—logs that, while not containing your results, are sufficient to recreate the session.

To start a command-only log:	. cmdlog using *filename*
To stop logging and close the file:	. cmdlog close
To recreate your session:	. do *filename*.txt

15.1.1 Starting and closing logs

With great foresight, you begin working in Stata and type `log using session` (or click the **Log** button) before starting your work:

```
. log using session
```

```
        log:  C:\example\session.smcl
   log type:  smcl
  opened on:  17 Mar 2005, 12:35:08
. use http://www.stata-press.com/data/r9/census
(1980 Census data by state)
. tabulate reg [freq=pop]
     Census │
     region │       Freq.      Percent         Cum.
 ───────────┼───────────────────────────────────────
         NE │    49135283        21.75        21.75
    N Cntrl │    58865670        26.06        47.81
      South │    74734029        33.08        80.89
       West │    43172490        19.11       100.00
 ───────────┼───────────────────────────────────────
      Total │   225907472       100.00
. summarize medage
    Variable │       Obs        Mean    Std. Dev.       Min        Max
 ────────────┼─────────────────────────────────────────────────────────
      medage │        50       29.54     1.693445       24.2       34.7
. log close
        log:  C:\example\session.smcl
   log type:  smcl
  closed on:  17 Mar 2005, 12:35:38
```

There is now a file named `session.smcl` on your disk. If you were to look at it in a text editor or word processor, you would see something like this:

```
{smcl}
{com}{sf}{ul off}{txt}{.-}
        log:  {res}C:\example\session.smcl
  {txt}log type:  {res}smcl
  {txt}opened on:  {res}17 Mar 2005, 12:35:08
{txt}
{com}. use http://www.stata-press.com/data/r9/census
{txt}(1980 Census data by state)

{com}. tabulate reg [freq=pop]

     {txt}Census {c |}
     region {c |}      Freq.      Percent        Cum.
{hline 12}{c +}{hline 35}
        NE {c |}{res}    49135283        21.75        21.75
{txt}    N Cntrl {c |}{res}    58865670        26.06        47.81
  (output omitted )
```

What you are seeing is SMCL, which Stata understands. Here is the result of typing the file using Stata's `type` command:

```
. type session.smcl
```

```
        log:  C:\example\session.smcl
   log type:  smcl
  opened on:  17 Mar 2005, 12:35:08
. use http://www.stata-press.com/data/r9/census
(1980 Census data by state)

. tabulate reg [freq=pop]

    Census  |
    region  |      Freq.        Percent        Cum.
------------+-----------------------------------
        NE  |   49135283         21.75        21.75
    N Cntrl |   58865670         26.06        47.81
     South  |   74734029         33.08        80.89
      West  |   43172490         19.11       100.00
------------+-----------------------------------
     Total  |  225907472        100.00

. summarize medage

    Variable |      Obs        Mean     Std. Dev.       Min        Max
-------------+-----------------------------------------------------
      medage |       50       29.54      1.693445       24.2       34.7

. log close
        log:  C:\example\session.smcl
   log type:  smcl
  closed on:  17 Mar 2005, 12:35:38
```

```
. _
```

What you will see is a perfect copy of what you previously saw. If you use Stata to print the file, you will get a perfect printed copy, too.

SMCL files can be translated to ASCII text, which is a format more useful for inclusion into a word processing document. If you type translate *filename*.smcl *filename*.log, Stata will translate *filename*.smcl to ASCII and store the result in *filename*.log:

```
. translate session.smcl session.log
```

The resulting file session.log looks like this:

```
---------------------------------------------------------------------------
        log:  C:\example\session.smcl
   log type:  smcl
  opened on:  17 Mar 2005, 12:35:08
. use http://www.stata-press.com/data/r9/census
(1980 Census data by state)

. tabulate reg [freq=pop]

    Census  |
    region  |      Freq.        Percent        Cum.
------------+-----------------------------------
        NE  |   49135283         21.75        21.75
    N Cntrl |   58865670         26.06        47.81
     South  |   74734029         33.08        80.89
(output omitted )
```

When you use translate to create *filename*.log from *filename*.smcl, *filename*.log must not already exist:

```
. translate session.smcl session.log
file session.log already exists
r(602);
```

If the file does already exist and you wish to overwrite the existing copy, you can specify the `replace` option:

```
. translate session.smcl session.log, replace
```

See [R] **translate** for more information.

If you prefer, you can skip the SMCL and create ASCII text logs directly, either by specifying that you want the log in `text` format,

```
. log using session, text
```

or by specifying that the file to be created be a `.log` file:

```
. log using session.log
```

15.1.2 Appending to an existing log

Stata never lets you accidentally write over an existing log file. If you have an existing log file and you want to continue logging, you have three choices:

• create a new log file

• append the new log onto the existing log file by typing log using *logname*, `append`

• replace the existing log file by typing log using *logname*, `replace`

For example, if you have an existing log file named `session.smcl`, you might type

```
. log using session, append
```

to append the new log to the end of the existing log file, `session.smcl`.

15.1.3 Temporarily suspending and resuming logging

Once you have started logging your session, you can turn logging on and off. When you turn logging off, Stata temporarily stops recording your session but leaves the log file open. When you turn logging back on, Stata continues to record your session, appending the additional record to the end of the file.

For instance, say that the first time something interesting happens, you type log using `results` (or click on **Log** and open `results.smcl`). You then retype the command that produced the interesting result (or double-click the command in the Review window, or use the PageUp key to retrieve the command; see [U] **10 Keyboard use**). You now have a copy of the interesting result saved in the log file.

You are now reasonably sure that nothing interesting will occur, at least for a while. Rather than type `log close`, however, you type `log off`, or you click on **Log** and choose **Suspend**. From now on, nothing goes into the file. The next time something interesting happens, you type `log on` (or click on **Log** and choose **Resume**) and reissue the (interesting) command. After that, you type `log off`. You keep working like this—toggling the log on and off.

15.2   Placing comments in logs

Stata treats lines starting with a '*' as comments and ignores them. Thus, if you are working interactively and wish to make a comment, you can type '*' followed by your comment:

```
. * check that all the spells are completed
. _
```

Stata ignores your comment but, if you have a log going, the comment now appears in the file.

❏ Technical Note

log can be combined with #review (see [U] **10 Keyboard use**) to bail you out when you have not adequately planned ahead. Say that you have been working in front of your computer, and you now realize that you have done what you wanted to do. Unfortunately, you are not sure exactly what it is you have done. Did you make a mistake? Could you reproduce the result? Unfortunately, you have not been logging your output. Typing #review will allow you to look over what commands you have issued, and, combined with log, will allow you to make a record. You can also see the commands that you have issued in the Review window. You can save those commands to a file by right-clicking on the Review window and selecting **Save Review Contents...**.

Type log using *filename*. Type #review 100. Stata will list the last 100 commands you gave, or however many it has stored. Since log is making a record, that list will also be stored in the file. Finally, type log close.

❏

15.3   Logging only what you type

Log files record everything that happens during a session, both what you type and what Stata produces in response.

Stata can also produce command log files—files that contain only what you type. These files are perfect for later going back and creating a Stata do-file.

cmdlog creates command log files, and its basic syntax is

cmdlog using *filename* [, replace]	creates *filename*.txt
cmdlog off	temporarily suspends command logging
cmdlog on	resumes command logging
cmdlog close	closes the command log file

See [R] **log** for all the details.

Command logs are plain ASCII text files. If you typed

```
. cmdlog using session
(cmdlog C:\example\session.txt opened)
. use http://www.stata-press.com/data/r9/census
(Census Data)
. tabulate reg [freq=pop]
  (output omitted )
. summarize medage
  (output omitted )
. cmdlog close
(cmdlog C:\example\session.txt closed)
```

file `mycmds.txt` would contain

```
use census
tabulate reg [freq=pop]
summarize medage
```

You can create both kinds of logs—full session logs and command logs—simultaneously, if you wish. A command log file can later be used as a do-file; see [R] **do**.

15.4   The log-button alternative

The capabilities of the `log` command (but not the `cmdlog` command) are available from Stata's GUI interface; just click on the **Log** button or select **Log** from the **File** menu.

You can use the Viewer to view logs, even logs that are in the process of being created. Just select **File > View**. If you are currently logging, the filename to view will already be filled in with the current log file, and all you need to do is click **OK**. Periodically, you can click the **Refresh** button to bring the Viewer up to date.

You can also use the Viewer to view previous logs.

You can access the Viewer by selecting **File > View**, or you can use the `view` command:

```
. view myoldlog.smcl
```

15.5   Printing logs

You print logs from the Viewer. Select **File > View**, or type '`view` *logfilename*' from the command line to load the log into the Viewer, and then right-click on the Viewer and select **Print**.

Alternatively, you can print logs by other means; see [R] **translate**.

16 Do-files

Contents

16.1 Description

Rather than typing commands at the keyboard, you can create a text file containing commands and instruct Stata to execute the commands stored in that file. Such files are called *do-files* since the command that causes them to be executed is do.

A do-file is a standard ASCII text file that is executed by Stata when you type do *filename*. You can use any text editor or the built-in Do-file Editor to create do-files; see [GS] **14 Using the Do-file Editor**.

▷ Example 1

You can use do-files to create a batch-like environment in which you place all the commands you want to perform in a file and then instruct Stata to do that file. For instance, assume that you use your text editor or word processor to create a file called myjob.do that contains these three lines:

```
──────────────────────────────────────────── top of myjob.do ────────
use http://www.stata-press.com/data/r9/census5
tabulate region
summarize marriage_rate divorce_rate median_age if state!="Nevada"
──────────────────────────────────────────── end of myjob.do ────────
```

You then enter Stata and instruct Stata to do the file:

```
. do myjob

. use http://www.stata-press.com/data/r9/census5
(1980 Census data by state)
```

177

```
. tabulate region

    Census
    region |      Freq.       Percent         Cum.
-----------+-----------------------------------
        NE |          9         18.00        18.00
    N Cntrl |         12         24.00        42.00
     South |         16         32.00        74.00
      West |         13         26.00       100.00
-----------+-----------------------------------
     Total |         50        100.00

. summarize marriage_rate divorce_rate median_age if state !="Nevada"

    Variable |        Obs         Mean     Std. Dev.         Min          Max
-------------+--------------------------------------------------------------
  marriage_r~e |         49     .0106791     .0021746     .0074654     .0172704
  divorce_rate |         49     .0054268     .0015104     .0029436      .008752
   median_age |         49     29.52653     1.708286         24.2         34.7
```

You typed only do myjob to produce this output. Since you did not specify the file extension, Stata assumed you meant do myjob.do; see [U] **11.6 File-naming conventions**.

◁

16.1.1 Version

We recommend that the first line in your do-file declare the Stata release you used when you wrote the do-file; myjob.do would better read as

```
─────────────────────────────────── top of myjob.do ───────────
version 9
use http://www.stata-press.com/data/r9/census
tabulate region
summarize marriage_rate divorce_rate median_age if state!="Nevada"
─────────────────────────────────── end of myjob.do ───────────
```

We admit that we do not always follow our own advice, as you will see many examples in this manual that do not include the version 9 line.

If you intend to keep the do-file, however, you should include this line since it ensures that your do-file will continue to work with future versions of Stata. Stata is under constant development, and sometimes things change in surprising ways.

For instance, in Stata 3.0, a new syntax for specifying the weights was introduced. If you had an old do-file written for Stata 2.1 that analyzed weighted data and did not have version 2.1 at the top, you would find that today's Stata would flag some of its lines as syntax errors. If you had the version 2.1 line, it would work just as it used to.

In Stata 4.0, we updated the random-number generator uniform()—the new one is better in that it has a longer period. If you wrote a do-file back in the days of Stata 3.1 that made a bootstrap calculation of variance and did not include version 3.1 at the top, it would now produce different (but equivalent) results. If you had included the line, it would produce the same results that it used to.

When running an old do-file that includes a version statement, you need not worry about setting the version back. Stata automatically restores the previous value of version when the do-file completes.

16.1.2 Comments and blank lines in do-files

You may freely include blank lines in your do-file. In the previous example, the do-file could just as well have read

─────────────────────────── top of myjob.do ───────────
```
version 9

use http://www.stata-press.com/data/r9/census
tabulate region
summarize marriage_rate divorce_rate median_age if state!="Nevada"
```
─────────────────────────── end of myjob.do ───────────

There are four ways to include comments in a do-file.

1. Begin the line with a '*'; Stata ignores such lines.

2. Place the comment in /* */ delimiters.

3. Place the comment after two forward slashes; i.e., //. Everything after the // to the end of the current line is considered a comment (unless the // is part of http://...).

4. Place the comment after three forward slashes, i.e., ///. Everything after the /// to the end of the current line is considered a comment. However, when you use ///, the next line joins with the current line. /// lets you split very long lines across multiple lines in the do-file.

❏ Technical Note

The /* */, //, and /// comment indicators can be used in do-files and ado-files only; you may not use them interactively. You can, however, use the '*' comment indicator interactively.

❏

myjob.do then might read

─────────────────────────── top of myjob.do ───────────
```
* a sample analysis job
version 9

use http://www.stata-press.com/data/r9/census

/* obtain the summary statistics: */
tabulate region
summarize marriage_rate divorce_rate median_age if state!="Nevada"
```
─────────────────────────── end of myjob.do ───────────

or, equivalently,

─────────────────────────── top of myjob.do ───────────
```
// a sample analysis job
version 9

use http://www.stata-press.com/data/r9/census

// obtain the summary statistics:
tabulate region
summarize marriage_rate divorce_rate median_age if state!="Nevada"
```
─────────────────────────── end of myjob.do ───────────

The style of comment indicator you use is up to you. One advantage of the /* */ method is that it can be put at the end of lines:

```
———————————————————————————————— top of myjob.do ——————
* a sample analysis job
version 9

use http://www.stata-press.com/data/r9/census

tabulate region                    /* obtain summary statistics */
summarize marriage_rate divorce_rate median_age if state!="Nevada"
———————————————————————————————————————— end of myjob.do ——————
```

In fact, /* */ can be put anywhere, even in the middle of a line:

```
———————————————————————————————— top of myjob.do ——————
* a sample analysis job
version 9

use /* confirm this is latest */ http://www.stata-press.com/data/r9/census

tabulate region                    /* obtain summary statistics */
summarize marriage_rate divorce_rate median_age if state!="Nevada"
———————————————————————————————————————— end of myjob.do ——————
```

You can achieve the same results using the // and /// methods:

```
———————————————————————————————— top of myjob.do ——————
// a sample analysis job
version 9

use http://www.stata-press.com/data/r9/census

tabulate region                    // obtain summary statistics
summarize marriage_rate divorce_rate median_age if state!="Nevada"
———————————————————————————————————————— end of myjob.do ——————
```

or

```
———————————————————————————————— top of myjob.do ——————
// a sample analysis job
version 9

use /// confirm this is latest
http://www.stata-press.com/data/r9/census

tabulate region                    // obtain summary statistics
summarize marriage_rate divorce_rate median_age if state!="Nevada"
———————————————————————————————————————— end of myjob.do ——————
```

16.1.3 Long lines in do-files

When you use Stata interactively, you press *Enter* to end a line and tell Stata to execute it. If you need to type a line that is wider than the screen, you just do it, letting it wrap or scroll.

You can follow the same procedure in do-files—if your editor or word processor will let you—but you can do better. You can change the end-of-line delimiter to ';' using #delimit, you can comment out the line break using /* */ comment delimiters, or you can use the /// line-join indicator.

▷ Example 2

In the following fragment of a do-file, we temporarily change the end-of-line delimiter:

── fragment of example.do ────────────

```
use mydata
#delimit ;
summarize weight price displ headroom rep78 length turn gear_ratio
          if substr(company,1,4)=="Ford" |
             substr(company,1,2)=="GM", detail ;
gen byte ford = substr(company,1,4)=="Ford" ;
#delimit cr
gen byte gm = substr(company,1,2)=="GM"
```

── fragment of example.do ────────────

Once we change the line delimiter to semicolon, all lines, even short ones, must end in semicolons. Stata treats carriage returns as no different from blanks. We can change the delimiter back to carriage return by typing #delimit cr.

The #delimit command is allowed only in do-files—it is not allowed interactively. You need not remember to set the delimiter back to carriage return at the end of a do-file because Stata will reset it automatically.

◁

▷ Example 3

The other way around long lines is to comment out the carriage return using /* */ comment brackets or to use the /// line-join indicator. Thus, our code fragment could also read

── fragment of example.do ────────────

```
use mydata
summarize weight price displ headroom rep78 length turn gear_ratio /*
     */  if substr(company,1,4)=="Ford" |     /*
     */     substr(company,1,2)=="GM", detail
gen byte ford = substr(company,1,4)=="Ford"
gen byte gm = substr(company,1,2)=="GM"
```

── fragment of example.do ────────────

or

── fragment of example.do ────────────

```
use mydata
summarize weight price displ headroom rep78 length turn gear_ratio ///
          if substr(company,1,4)=="Ford" |    ///
             substr(company,1,2)=="GM", detail
gen byte ford = substr(company,1,4)=="Ford"
gen byte gm = substr(company,1,2)=="GM"
```

── fragment of example.do ────────────

◁

16.1.4 Error handling in do-files

A do-file stops executing when (1) the end of the file is reached, (2) an exit is executed, or (3) an error (nonzero *return code*) occurs. If an error occurs, the remaining commands in the do-file are not executed.

If you press *Break* while executing a do-file, Stata responds as though an error has occurred, stopping the do-file. This happens because the return code is nonzero; see [U] **8 Error messages and return codes** for an explanation of return codes.

▷ Example 4

Here is what happens when we execute a do-file and then press *Break*:

```
. do myjob2
. version 9
. use census
(Census data)
. tabulate region

     Census
     region |      Freq.     Percent        Cum.
—Break—
r(1);

end of do-file
—Break—
r(1);

. _
```

When we pressed *Break*, Stata responded by typing —Break— and showed a return code of 1. Stata seemingly repeated itself, typing first "end of do-file", and then —Break— and the return code of 1 again. Do not worry about the repeated messages. The first message indicates that Stata was stopping the tabulate because you pressed *Break*, and the second message indicates that Stata is stopping the do-file for the same reason.

◁

▷ Example 5

Let's try our example again, but this time, let's introduce an error. We change the file myjob2.do to read

―――――――――――――――――――――――――――――――――――――― top of myjob2.do ――――――――
```
use http://www.stata-press.com/data/r9/censas
tabulate region
summarize marriage_rate divorce_rate median_age if state!="Nevada"
```
―― end of myjob2.do ――――――――

Note our subtle typographical error. We typed use censas when we meant use census. We assume that there is no file called censas.dta, so now we have an error. Here is what happens when you instruct Stata to do the file:

```
. do myjob2
. version 9
. use censas
file censas.dta not found
r(601);

end of do-file
r(601);

. _
```

When Stata was told to use censas, it responded with "file censas.dta not found" and a return code of 601. Stata then typed "end of do-file" and repeated the return code of 601. The repeated message occurred for the same reason it did when we pressed *Break* in the previous example. The use resulted in a return code of 601, so the do-file itself resulted in the same return code. The important thing to understand is that Stata stopped executing the file because there was an error.

◁

❏ Technical Note

We can tell Stata to continue executing the file even if there are errors by typing do *filename*, nostop. Here is the result:

```
. do myjob2, nostop
. version 9
. use censas
file censas.dta not found
r(601);
. tabulate region
no variables defined
r(111);
summarize marriage_rate divorce_rate median_age if state!="Nevada"
no variables defined
r(111);
end of do-file
. _
```

None of the commands worked because the do-file's first command failed. That is why Stata ordinarily stops. However, if our file had contained anything that could work, it would have worked. In general, we do not recommend coding in this manner, as unintended consequences can result when errors do not stop execution.

❏

16.1.5 Logging the output of do-files

You log the output of do-files just as you would an interactive session; see [U] **15 Printing and preserving output**.

Many users include the commands to start and stop the logging in the do-file itself:

```
──────────────────────────────── top of myjob3.do ────────────
version 9
log using myjob3, replace
* a sample analysis job
use http://www.stata-press.com/data/r9/census
tabulate region                    // obtain summary statistics
summarize marriage_rate divorce_rate median_age if state!="Nevada"
log close
──────────────────────────────── end of myjob3.do ────────────
```

We chose to open with log using myjob3, replace, the important part being the replace option. Had we omitted the option, we could not easily rerun our do-file. If myjob3.smcl had already existed and log was not told that it is okay to replace the file, the do-file would have stopped and instead reported that "file myjob3.smcl already exists". We could get around that, of course, by erasing the log file before running the do-file.

16.1.6   Preventing —more— conditions

Assume that you are running a do-file and logging the output so that you can look at it later. In that case, Stata's feature of pausing every time the screen is full is just an irritation: It means that you have to sit and watch the do-file run so you can clear the —more—.

The way around this is to include the line `set more off` in your do-file. Setting more to `off`, as explained in [U] **7 —more— conditions**, prevents Stata from ever issuing a —more—.

16.2   Calling other do-files

Do-files may call other do-files. For instance, say that you wrote `makedata.do`, which infiles your data, generates a few variables, and saves `step1.dta`. Say that you wrote `anlstep1.do`, which performed a little analysis on `step1.dta`. You could then create a third do-file,

```
                                                ── top of master.do ─────
     version 9
     do makedata
     do anlstep1
                                                ── end of master.do ─────
```

and so, in effect, combine the two do-files.

Do-files may call other do-files, which, in turn, call other do-files, and so on. Stata allows do-files to be nested 64 deep.

Be not confused: `master.do` above could call 1,000 do-files one after the other, and still the level of nesting would be only two.

16.3   Running a do-file (Stata for Windows)

1. You can execute do-files by typing `do` followed by the filename, as we did above.

2. You can execute do-files by selecting **File > Do...**.

3. You can use the Do-file Editor to compose, save, and execute do-files; see [GSW] **14 Using the Do-file Editor**. To use the Do-file Editor, click on the **Do-file Editor** button, or type `doedit` in the Command window.

4. You can double-click the icon for the do-file to launch Stata and run the do-file. When the do-file completes, Stata will prompt you for the next command just as if you had started Stata the normal way. If you want Stata to exit instead, include `exit, STATA clear` as the last line of your do-file.

5. You can run the do-file in batch mode. See [GSW] **A.9 Executing Stata in background (batch) mode** for details, but the short explanation is that you open a DOS window and type

        C:\data> "C:\Program Files\Stata9\wstata" /s do myjob

   or

        C:\data> "C:\Program Files\Stata9\wstata" /b do myjob

to run in batch mode, assuming that you have installed Stata in the folder C:\Program Files\Stata9. /b and /s determine the kind of log produced, but put that aside for a second. What is important is that when you start Stata in these ways, Stata will run in the background. When the do-file completes, the Stata icon on the taskbar will flash. You can then click on it to close Stata. If you want to stop the do-file before it completes, click on the Stata icon on the taskbar, and Stata will ask you if you want to cancel the job.

To log the output, you can start the log before executing the do-file or you can include the log using and log close in your do-file.

When you run Stata in these ways, Stata takes the following actions:

  a. Stata automatically opens a log. If you specified /s, Stata will open a SMCL log; if you specified /b, Stata will open an ASCII text log. If your do-file is named *xyz*.do, the log will be called *xyz*.smcl (/s) or *xyz*.log (/b) in the same directory.

  b. If your do-file explicitly opens another log, Stata will save two copies of the output.

  c. Stata ignores —more— conditions and anything else that would cause the do-file to stop were it running interactively.

16.4 Running a do-file (Stata for Macintosh)

1. You can execute do-files by typing do followed by the filename, as we did above.

2. You can execute do-files by selecting **File > Do...**.

3. With Stata running, you can go to the Desktop and double-click the do-file.

4. You can use the Do-file Editor to compose, save, and execute do-files; see [GSM] **14 Using the Do-file Editor**. Click the **Do-file Editor** button, or type doedit in the Command window.

5. If Stata is not running, you can double-click on the icon for the do-file to launch Stata and run the do-file. When the do-file completes, Stata will prompt you for the next command just as if you had started Stata the normal way. If you want Stata to exit instead, include exit, STATA clear as the last line of your do-file.

6. You can run the do-file in batch mode. See [GSM] **A.12 Executing Stata in background (batch) mode** for details, but the short explanation is that you open a Terminal window and type

       % /Applications/Stata/Stata.app/Contents/MacOS/Stata -s do myjob

   or

       % /Applications/Stata/Stata.app/Contents/MacOS/Stata -b do myjob

   to run in batch mode, assuming that you have installed Intercooled Stata in the folder /Applications/Stata. -b and -s determine the kind of log produced, but put that aside for a second. What is important is that when you start Stata in these ways, Stata will run in the background. When the do-file completes, the Stata icon on the Dock will bounce until you put Stata into the foreground. You can then exit Stata. If you want to stop the do-file before it completes, right-click on the Stata icon on the Dock, and select **Quit**.

To log the output, you can start the log before executing the do-file or you can include the log using and log close in your do-file.

When you run Stata in these ways, Stata takes the following actions:

  a. Stata automatically opens a log. If you specified -s, Stata will open a SMCL log; if you specified -b, Stata will open an ASCII text log. If your do-file is named *xyz*.do, the log will be called *xyz*.smcl (-s) or *xyz*.log (-b) in the same directory.

b. If your do-file explicitly opens another log, Stata will save two copies of the output.

c. Stata ignores —more— conditions and anything else that would cause the do-file to stop were it running interactively.

16.5   Running a do-file (Stata for Unix)

1. You can execute do-files by typing `do` followed by the filename, as we did above.

2. You can execute do-files by selecting **File > Do...**.

3. You can use the Do-file Editor to compose, save, and execute do-files; see [GSW] **14 Using the Do-file Editor**. Click on the **Do-file Editor** button or type `doedit` in the Command window.

4. At the Unix prompt, you can type

   $ `xstata do` *filename*

   or

   $ `stata  do` *filename*

   to launch Stata and run the do-file. When the do-file completes, Stata will prompt you for the next command just as if you had started Stata the normal way. If you want Stata to exit instead, include `exit, STATA clear` as the last line of your do-file.

To log the output, you can start the log before executing the do-file or you can include the `log using` and `log close` in your do-file.

5. At the Unix prompt, you can type

   $ `stata -s do` *filename* `&`

   or

   $ `stata -b do` *filename* `&`

   to run the do-file in the background. Note that the above two examples both involve the use of `stata`, not `xstata`. Type `stata`, even if you usually use the GUI version of Stata, `xstata`. The examples differ only in that one specifies the `-s` option and the other the `-b` option, which determines the kind of log that will be produced. In the above examples, Stata takes the following actions:

   a. Stata automatically opens a log. If you specified `-s`, Stata will open a SMCL log; if you specified `-b`, Stata will open an ASCII text log. If your do-file is named *xyz*.`do`, the log will be called *xyz*.`smcl` (`-s`) or *xyz*.`log` (`-b`) in the current directory (the directory from which you issued the `stata` command).

   b. If your do-file explicitly opens another log, Stata will save two copies of the output.

   c. Stata ignores —more— conditions and anything else that would cause the do-file to stop were it running interactively.

To reiterate, one way to run a do-file in the background and obtain an ASCII text log is by typing

   $ `stata -b do myfile &`

Another way uses standard redirection:

   $ `stata < myfile.do > myfile.log &`

The first way is slightly more efficient. Either way, Stata knows it is in the background and ignores —more— conditions and anything else that would cause the do-file to stop if it were running interactively. However, if your do-file contains either the `#delimit` command or the comment characters (`/*` at the end of one line and `*/` at the beginning of the next), the second method will not work. We recommend that you use the first method: `stata -b do myfile &`.

The choice between `stata -b do myfile &` and `stata -s do myfile &` is more personal. We prefer obtaining SMCL logs (`-s`) because they look better when printed, and, in any case, they can always be converted to ASCII text format using `translate`; see [R] **translate**.

16.6   Programming with do-files

This is an advanced topic, and we are going to refer to concepts not yet explained; see [U] **18 Programming Stata** for further information.

16.6.1   Argument passing

Do-files accept arguments, just as Stata programs do; this is described in [U] **18 Programming Stata** and [U] **18.4 Program arguments**. In fact, the logic Stata follows when invoking a do-file is the same as when invoking a program: the local macros are saved, and new ones are defined. Arguments are stored in the local macros '1', '2', and so on. When the do-file completes, the previous definitions are restored, just as with programs.

Thus, if you wanted your do-file to

1. use a dataset of your choosing,

2. tabulate a variable named `region`, and

3. summarize variables `marriage_rate` and `divorce_rate`,

you could write the do-file

```
───────────────────────────────────── top of myxmpl.do ───────────
    use '1'
    tabulate region
    summarize marriage_rate divorce_rate
───────────────────────────────────── end of myxmpl.do ───────────
```

and you could run this do-file by typing, for instance,

```
. do myxmpl census
  (output omitted )
```

The first command—`use '1'`—would be interpreted as `use census` because `census` was the first argument you typed after `do myxmpl`.

An even better version of the do-file would read

```
───────────────────────────────────── top of myxmpl.do ───────────
    args dsname
    use 'dsname'
    tabulate region
    summarize marriage_rate divorce_rate
───────────────────────────────────── end of myxmpl.do ───────────
```

The `args` command merely assigns a better name to the argument passed. `args dsname` does not verify that what we type following `do myxmpl` is a filename—we would have to use the `syntax` command if we wanted to do that—but substituting 'dsname' for '1' does make the code more readable.

If our program were to receive two arguments, we could refer to them as '1' and '2', or we could put an 'args dsname other' at the top of our do-file and then refer to 'dsname' and 'other'.

To learn more about argument passing, see [U] **18.4 Program arguments**.

16.6.2   Suppressing output

There is an alternative to typing do *filename*; it is run *filename*. run works in the same way as do, except that neither the instructions in the file nor any of the output caused by those instructions is shown on the screen or in the log file.

For instance, using the above `myxmpl.do`, typing `run myxmpl census` results in

```
. run myxmpl census

.
```

All the instructions were executed, but none of the output was shown.

This is not useful in this case, but if the do-file contained only the definitions of Stata programs—see [U] **18 Programming Stata**—and you merely wanted to load the programs without seeing the code, run would be useful.

17 Ado-files

Contents

17.1   Description

Stata is programmable, and even if you never write a Stata program, Stata's programmability is still important. Many of Stata's features are implemented as Stata programs, and new features are implemented every day, both by StataCorp and by others.

1. You can obtain additions from the *Stata Journal*. You subscribe to the printed journal, but the software additions are available free over the Internet.

2. You can obtain additions from the Stata listserver, Statalist, where an active group of users advises each other on how to use Stata, and often, in the process, trades programs. Visit the Stata web site, *http://www.stata.com*, for instructions on how to subscribe; subscribing to the listserver is free.

3. The Boston College Statistical Software Components Archive (SSC) is a distributed database making available a large and constantly growing number of Stata programs. You can browse and search the archive, and you can find links to the archive from *http://www.stata.com*. Importantly, Stata knows how to access the archive and other places, as well. You can search for additions using Stata's `search, net` command; see [R] **search**. You can immediately install materials you find using `search, net` by using the hyperlinks that will be displayed by `search` in the Results window, or by using the `net` command. A specialized command, `ssc`, has a number of options available to help you find and install the user-written commands that are available from this site; see [R] **ssc**.

4. You can write your own additions to Stata.

This chapter is written for people who want to consume ado-files. All users should read it. If you later decide you want to write ado-files, see [U] **18.11 Ado-files**.

17.2   What is an ado-file?

An ado-file defines a Stata command, but not all Stata commands are defined by ado-files.

When you type `summarize` to obtain summary statistics, you are using a command built into Stata.

When you type `ci` to obtain confidence intervals, you are running an ado-file. The results of using a built-in command or an ado-file are indistinguishable.

An ado-file is an ASCII text file that contains a Stata program. When you type a command that Stata does not know, it looks in certain places for an ado-file of that name. If Stata finds it, Stata loads and executes it, so it appears to you as if the ado-command is just another command built into Stata.

We just told you that Stata's `ci` command is implemented as an ado-file. That means that, somewhere, there is a file named `ci.ado`.

Ado-files usually come with help files. When you type `help ci` (or select **Help > Stata Command**, and type `ci`), Stata looks for `ci.hlp`, just as it looks for `ci.ado` when you use the `ci` command. A help file is also an ASCII text file that tells Stata's help system what to display.

17.3   How can I tell if a command is built in or an ado-file?

You can use the `which` command to determine if a file is built in or implemented as an ado-file. For instance, `logistic` is an ado-file, and here is what happens when you type `which logistic`:

```
. which logistic
C:\Program Files\Stata9\ado\base\l\logistic.ado
*! version 3.2.8  22mar2005
```

`logit` is a built-in command:

```
. which logit
built-in command:  logit
```

17.4   Can I look at an ado-file?

Certainly. When you type `which` followed by an ado-command, Stata reports where the file is stored:

```
. which logistic
C:\Program Files\Stata9\ado\base\l\logistic.ado
*! version 3.2.8  22mar2005
```

Ado-files are just ASCII text files containing the Stata program, so you can type them or view them in Stata's Viewer (or even look at them in your editor or word processor):

```
. type "C:\Program Files\Stata9\ado\base\l\logistic.ado"
*! version 3.2.8  22mar2005
program logistic, prop(or svyb svyj svyr swml) byable(onecall)
        version 6.0, missing
  (output omitted )
end
```

or

```
. viewsource logistic.ado
*! version 3.2.8  22mar2005
program logistic, prop(or svyb svyj svyr swml) byable(onecall)
        version 6.0, missing
  (output omitted )
end
```

The `type` command displays the contents of a file. The `viewsource` command searches for a file along the ado directories and displays the file in the Viewer. You can also look at the corresponding help file in raw form if you wish. If there is a help file, it is stored in the same place as the ado-file:

```
. type "C:\Program Files\Stata9\ado\base\l\logistic.hlp", asis
{smcl}
* 09mar2005...
{cmd:help logistic}{right:dialog:  {bf:{dialog logistic}}{space 15}}
{right:also see:  {help logistic postestimation}}
{hline}
  (output omitted )
```

or

```
. viewsource logistic.hlp
{smcl}
* 09mar2005...
{cmd:help logistic}{right:dialog:  {bf:{dialog logistic}}{space 15}}
{right:also see:  {help logistic postestimation}}
{hline}
  (output omitted )
```

17.5   Where does Stata look for ado-files?

Stata looks for ado-files in seven places, which can be categorized in three ways:

I. The official ado directories:
  1. (UPDATES), the official updates directory containing updated ado-files from StataCorp
  2. (BASE), the official base directory containing the ado-files shipped with your version of Stata

II. Your personal ado directories:
  3. (SITE), the directory for ado-files your site might have installed
  4. (PLUS), the directory for ado-files you personally might have installed
  5. (PERSONAL), the directory for ado-files you personally might have written
  6. (OLDPLACE), the directory where Stata users used to save their personally written ado-files

III. The current directory:
  7. (.), the ado-files you have written just this instant or for just this project

The location of these directories varies from computer to computer, but Stata's `sysdir` command will tell you where they are on your computer:

```
. sysdir
    STATA:  C:\Program Files\Stata9\
  UPDATES:  C:\Program Files\Stata9\ado\updates\
     BASE:  C:\Program Files\Stata9\ado\base\
     SITE:  C:\Program Files\Stata9\ado\site\
     PLUS:  C:\ado\plus\
 PERSONAL:  C:\ado\personal\
 OLDPLACE:  C:\ado\
```

17.5.1   Where are the official ado directories?

These are the directories listed as BASE and UPDATES by sysdir:

```
. sysdir
    STATA:  C:\Program Files\Stata9\
  UPDATES:  C:\Program Files\Stata9\ado\updates\
     BASE:  C:\Program Files\Stata9\ado\base\
     SITE:  C:\Program Files\Stata9\ado\site\
     PLUS:  C:\ado\plus\
 PERSONAL:  C:\ado\personal\
 OLDPLACE:  C:\ado\
```

1. BASE contains the ado-files we originally shipped to you.

2. UPDATES contains any updates you might have installed since then. You can install these updates using the update command or by selecting **Help > Official Updates**; see [U] **17.8 How do I install official updates?**.

17.5.2   Where is my personal ado directory?

These are the directories listed as PERSONAL, PLUS, SITE, and OLDPLACE by sysdir:

```
. sysdir
    STATA:  C:\Program Files\Stata9\
  UPDATES:  C:\Program Files\Stata9\ado\updates\
     BASE:  C:\Program Files\Stata9\ado\base\
     SITE:  C:\Program Files\Stata9\ado\site\
     PLUS:  C:\ado\plus\
 PERSONAL:  C:\ado\personal\
 OLDPLACE:  C:\ado\
```

1. PERSONAL is for ado-files you personally have written. Store your private ado-files here; see [U] **17.7 How do I add my own ado-files?**.

2. PLUS is for ado-files you personally installed but did not write. Such ado-files are usually obtained from the SJ, but they are sometimes found in other places, too. You find and install such files using Stata's net command, or you can select **Help > SJ and User-written Programs**; see [U] **17.6 How do I install an addition?**.

3. SITE is really the opposite of a personal ado directory—it is a public directory corresponding to PLUS. If you are on a networked computer, the site administrator can install ado-files here, and all Stata users will then be able to use them just as if each found and installed them in their PLUS directory for themselves. Site administrators find and install the ado-files just as you would, using Stata's net command, but they specify an option when they install something that tells Stata to write the files into SITE rather than PLUS; see [R] **net**.

4. OLDPLACE is for old-time Stata users. Prior to Stata 6, all "personal" ado-files, whether personally written or just personally installed, were written in the same directory—OLDPLACE. So that the old-time Stata users do not have to go back and rearrange what they have already done, Stata still looks in OLDPLACE.

17.6   How do I install an addition?

Additions come in three flavors:

1. User-written additions, which you might find in the SJ, etc.

2. Ado-files you have written.

> See [U] **17.7 How do I add my own ado-files?**. If you have an ado-file obtained from the Stata listserver or a friend, treat it as belonging to this case.

3. Official updates provided by StataCorp.

> See [U] **17.8 How do I install official updates?**.

User-written additions you might find in the *Stata Journal* (SJ), etc., are obtained over the Internet. To access them on the Internet,

1. select **Help > SJ and User-written Programs**, and click on one of the links.

or

2. type `net from http://www.stata.com`.

What to do next will be obvious, but, in case it is not, see **Using the Internet** in the *Getting Started* manual for your computer. Also see [U] **28 Using the Internet to keep up to date** and [R] **net**.

17.7 How do I add my own ado-files?

You write a Stata program (see [U] **18 Programming Stata**), store it in a file ending in `.ado`, perhaps write a help file, and copy everything to the directory `sysdir` lists as PERSONAL:

```
. sysdir
   STATA:  C:\Program Files\Stata9\
 UPDATES:  C:\Program Files\Stata9\ado\updates\
    BASE:  C:\Program Files\Stata9\ado\base\
    SITE:  C:\Program Files\Stata9\ado\site\
    PLUS:  C:\ado\plus\
PERSONAL:  C:\ado\personal\
OLDPLACE:  C:\ado\
```

In this case, we would copy the files to `C:\ado\personal`.

While you are writing your ado-file, it is sometimes convenient to store the pieces in the current directory. Do that if you wish; you can move them to your personal ado directory when the program is debugged.

17.8 How do I install official updates?

Updates are available over the Internet:

1. Select **Help > Official Updates**, and then click on *http://www.stata.com*

or

2. type `update query`.

What to do next should be obvious, but, in case it is not, see **Using the Internet** in the *Getting Started* manual for your computer. Also see [U] **28 Using the Internet to keep up to date** and [R] **net**.

The official updates include bug fixes and new features but do not change the syntax of an existing command or change the way Stata works.

Once you have installed the updates, you can enter Stata and type `help whatsnew` (or select **Help > What's New?**) to learn about what has changed.

18 Programming Stata

Contents

This is an advanced topic. Some Stata users live productive lives without ever programming Stata. After all, you do not need to know how to program Stata to input data, create new variables, and fit models. On the other hand, programming Stata is not difficult—at least if the problem is not difficult—and Stata's programmability is one of its best features. The real power of Stata is not revealed until you program it.

If you are uncertain whether to read this chapter, we recommend that you start reading and then bail out when it gets too arcane for you. You will learn things about Stata that you may find useful even if you never write a Stata program.

If you want even more, we offer courses over the Internet on Stata programming; see [U] **3.7 Net-Courses**.

18.1   Description

When you type a command that Stata does not recognize, Stata first looks in its memory for a program of that name. If Stata finds it, Stata executes the program.

There is no Stata command named `hello`,

```
. hello
unrecognized command
r(199);
```

but there could be if you defined a program named `hello`, and after that, the following might happen when you typed `hello`:

```
. hello
hi there
. _
```

This would happen if, beforehand, you had typed

```
. program hello
  1. display "hi there"
  2. end

. _
```

That is how programming works in Stata. A program is defined by

> `program` *progname*
>   *Stata commands*
> `end`

and it is executed by typing *progname* at Stata's dot prompt.

18.2   Relationship between a program and a do-file

Stata treats programs the same way it treats do-files. Below we will discuss passing arguments, consuming results from Stata commands, and other topics, but realize that everything we say applies equally to do-files and programs.

Programs and do-files differ in the following ways:

1. You invoke a do-file by typing do *filename*. You invoke a program by simply typing the program's name.

2. Programs must be defined (loaded) before they are used, whereas all that is required to run a do-file is that the file exist. There are ways to make programs load automatically, however, so this difference is of little importance.

3. When you type do *filename*, Stata displays the commands it is executing and the results. When you type *progname*, Stata shows only the results, not the display of the underlying commands. This is an important difference in outlook: In a do-file, how it does something is as important as what it does. In a program, the how is no longer important. You might think of a program as a new feature of Stata.

Let us now mention some of the similarities:

1. Arguments are passed to programs and do-files in the same way.

2. Programs and do-files both contain Stata commands. Any Stata command you put in a do-file can be put in a program.

3. Programs may call other programs. Do-files may call other do-files. Programs may call do-files (this rarely happens), and do-files may call programs (this often happens). Stata allows programs (and do-files) to be nested up to 64 deep.

Now, here is the interesting thing: programs are typically defined in do-files (or in a variant of do-files called ado-files; we will get to that later).

You can define a program interactively, and that is useful for pedagogical purposes, but in real applications, you will compose your program in a text editor and store its definition in a do-file.

You have already seen your first program:

```
program hello
        display "hi there"
end
```

You could type those commands interactively, but if the body of the program were more complicated, that would be inconvenient. So instead, suppose that you typed the commands into a do-file:

```
──────────────────────────────────────── top of hello.do ────────
program hello
        display "hi there"
end
──────────────────────────────────────── end of hello.do ────────
```

Now, returning to Stata, you type

```
. do hello

. program hello
  1.          display "hi there"
  2. end

.
end of do-file
```

Do you see that typing do hello did nothing but load the program? Typing do hello is the same as typing out the program's definition because that is all the do-file contains. Understand that the do-file was executed, but the statements in the do-file only defined the program hello; they did not execute it. Now that the program is loaded, we can execute it interactively:

```
. hello
hi there
```

So, that is one way you could use do-files and programs together. If you wanted to create new commands for interactive use, you could

1. Write the command as a `program` ... `end` in a do-file.

2. `do` the do-file before you use the new command.

3. Use the new command during the rest of the session.

There are more convenient ways to do this that would automatically load the do-file, but put that aside. The above method would work.

Another way we could use do-files and programs together is to put the definition of the program and its execution together into a do-file:

```
──────────────────────────────────── top of hello.do ──────────
program hello
        display "hi there"
end
hello
──────────────────────────────────── end of hello.do ──────────
```

Here is what would happen if we executed this do-file:

```
. do hello

. program hello
  1.        display "hi there"
  2. end
. hello
hi there

.
end of do-file
```

Do-files and programs are often used in such combinations. Why? Say that program `hello` is long and complicated and you have a problem where you need to do it twice. That would be a good reason to write a program. Moreover, you may wish to carry forth this procedure as a step of your analysis and, being cautious, do not want to perform this analysis interactively. You never intended program `hello` to be used interactively—it was just something you needed in the midst of a do-file—so you defined the program and used it there.

Anyway, there are lots of variations on this theme, but understand that few people actually sit in front of Stata and interactively type `program` and then compose a program. They instead do that in front of their text editor. They compose the program in a do-file and then execute the do-file.

There is one other (minor) thing to know: Once a program is defined, Stata does not allow you to redefine it:

```
. program hello
hello already defined
r(110);
```

Thus, in our most recent do-file that defines and executes `hello`, we could not rerun it in the same Stata session:

```
. do hello

. program hello
hello already defined
r(110);

end of do-file
r(110);
```

That problem is solved by typing `program drop hello` before redefining it. We could do that interactively, or we could modify our do-file:

```
─────────────────────────────────────────── top of hello.do ───────────
program drop hello
program hello
        display "hi there"
end
hello
─────────────────────────────────────────── end of hello.do ───────────
```

There is a problem with this solution. We can now rerun our do-file, but the first time we tried to run it in a Stata session, it would fail:

```
. do hello

. program drop hello
hello not found
r(111);

end of do-file
r(111);
```

The way around this conundrum is to modify the do-file:

```
─────────────────────────────────────────── top of hello.do ───────────
capture program drop hello
program hello
        display "hi there"
end
hello
─────────────────────────────────────────── end of hello.do ───────────
```

`capture` in front of a command makes Stata indifferent to whether the command works; see [P] **capture**. In real do-files containing programs, you will often see `capture program drop` before the program's definition.

To learn about the `program` command itself, see [P] **program**. It manipulates programs. `program` can define programs, drop programs, and show you a directory of programs that you have defined.

A program can contain any Stata command, but certain Stata commands are of special interest to program writers; see the *Programming* heading in the subject table of contents in the *Quick Reference and Index*.

18.3   Macros

Before we can begin programming, we must discuss macros. which are the variables of Stata programs.

A *macro* is a string of characters, called the *macroname*, that stands for another string of characters, called the *macro contents*.

Macros can be local or global. We will start with local macros because they are the most commonly used, but nothing really distinguishes one from the other at this stage.

18.3.1 Local macros

Local macro names can be up to 31 (not 32) characters in length.

One sets the contents of a local macro with the `local` command. In fact, we can do this interactively. We will begin by experimenting with macros in this way to learn about them. If we type

```
. local shortcut "myvar thisvar thatvar"
```

then 'shortcut' is a synonym for "myvar thisvar thatvar". Note the single quotes around `shortcut`. We said that sentence exactly the way we meant to because

if you type	'shortcut'
which is to say	left-single-quote shortcut right-single-quote
Stata hears	myvar thisvar thatvar

To access the contents of the macro, we use a left single-quote (located at the upper left on most keyboards), the macro name, and a right single-quote (located under the " on the right side of most keyboards).

The single quotes bracketing the macroname `shortcut` are called the macro-substitution characters. `shortcut` means shortcut. 'shortcut' means myvar thisvar thatvar.

So, if you typed

```
. list 'shortcut'
```

the effect would be exactly as if you typed

```
. list myvar thisvar thatvar
```

Macros can be used literally anywhere. For instance, if we also defined

```
. local cmd "list"
```

we could type

```
. 'cmd' 'shortcut'
```

to mean `list myvar thisvar thatvar`.

As another example, consider the definitions

```
. local prefix "my"
. local suffix "var"
```

Then

```
. 'cmd' 'prefix''suffix'
```

would mean `list myvar`.

18.3.2 Global macros

Let's put aside why Stata has two kinds of macros—local and global—and focus right now on how global macros work.

Global macros can have names that are up to 32 (not 31) characters in length. You set the contents of a global macro using the `global` rather than the `local` command:

```
. global shortcut "alpha beta"
```

You obtain the contents of a global macro by prefixing its name with a dollar sign: $shortcut is equivalent to "alpha beta".

In the previous section, we defined a local macro named shortcut, which is a different macro. 'shortcut' is still "myvar thisvar thatvar".

Local and global macros may have the same names, but even if they do, they are unrelated and are still distinguishable.

Global macros are just like local macros except that you set their contents with global rather than local, and you substitute their contents by prefixing them with a $ rather than enclosing them in ' '.

18.3.3   The difference between local and global macros

The difference between local and global macros is that local macros are private and global macros are public.

Say that you have written a program

```
program myprog
        code using local macro alpha
end
```

The local macro alpha in myprog is private in that no other program can modify or even look at alpha's contents. To make this point absolutely clear, assume that your program looks like this:

```
program myprog
        code using local macro alpha
        mysub
        more code using local macro alpha
end
program mysub
        code using local macro alpha
end
```

Note that myprog calls mysub and that both programs use a local macro named alpha. Even so, the local macros in each program are different. mysub's alpha macro may contain one thing, but that has nothing to do with what myprog's alpha macro contains. Even when mysub begins execution, its alpha macro is different from myprog's. It is not that mysub's inherits myprog's alpha macro contents but is then free to change it. It is that myprog's alpha and mysub's alpha are entirely different things.

When you write a program using local macros, you need not worry that some other program has been written using local macros with the same names. Local macros are just that: local to your program.

Global macros, on the other hand, are available to all programs. If both myprog and mysub use the global macro beta, they are using the same macro. Whatever the contents of $beta are when mysub is invoked, those are the contents when mysub begins execution, and, whatever the contents of $beta are when mysub completes, those are the contents when myprog regains control.

18.3.4 Macros and expressions

From now on, we are going to use local and global macros according to whichever is convenient; understand that whatever is said about one applies to the other.

Consider the definitions

```
. local one 2+2
. local two = 2+2
```

(which we could just as well have illustrated using the global command). In any case, note the equal sign in the second macro definition and the lack of the equal sign in the first. Formally, the first should be

```
. local one "2+2"
```

but Stata does not mind if we omit the double quotes in the local (global) statement.

local one 2+2 (with or without double quotes) copies the string 2+2 into the macro named one.

local two = 2+2 evaluates the expression 2+2, producing 4, and stores 4 in the macro named two.

That is, you type

local *macname contents*

if you want to copy *contents* to *macname*, and you type

local *macname* = *expression*

if you want to evaluate *expression* and store the result in *macname*.

In the second form, *expression* can be numeric or string. 2+2 is a numeric expression. As an example of a string expression,

```
. local res = substr("this",1,2) + "at"
```

stores that in res.

Since the expression can be either numeric or string, what is the difference between the following statements?

```
. local a "example"
. local b = "example"
```

Both statements store example in their respective macros. The first does so by a simple copy operation, whereas the second evaluates the expression "example", which is a string expression because of the double quotes that, in this case, evaluates to itself.

There is, however, a difference. Stata's expression parser is limited to handling strings of 244 characters in Stata/SE and 80 characters in Intercooled Stata or Small Stata. Strings longer than that are truncated.

The copy operation of the first syntax is not limited—it can copy up to the maximum length of a macro, which is currently 67,784 characters for Intercooled Stata and 3,400 for Small Stata. For Stata/SE, the limit is $33 * c(\texttt{max\_k\_theory}) + 200$ characters, which for the default setting of 5,000 is 165,200 characters.

To a programmer, the length limit for string expressions may seem limiting, but it turns out it is not, due to another feature discussed in [U] **18.3.6 Extended macro functions**.

There are some other issues of using macros and expressions that look a little strange to programmers coming from other languages, at least the first time they see them. For instance, say that the macro 'i' contains 5. How would you increment i so that it contains $5 + 1 = 6$? The answer is

```
local i = 'i' + 1
```

Do you see why the single quotes are on the right but not the left? Remember, 'i' refers to the contents of the local macro named i, which, we just said, is 5. Thus, after expansion, the line reads

```
local i = 5 + 1
```

which is the desired result.

There is a another way to increment local macros that will be more familiar to some programmers, especially C programmers:

```
local ++i
```

As C programmers would expect, local ++i is more efficient (executes more quickly) than local i = i+1, but in terms of outcome, it is equivalent. You can decrement a local macro using

```
local --i
```

local --i is equivalent to local i = i-1, but executes more quickly. Finally, note that

```
local i++
```

will *not* increment the local macro i, but instead redefines the local macro i to contain ++. There is, however, a context in which i++ (and i--) do work as expected; see [U] **18.3.7 Macro expansion operators and function**.

18.3.5   Double quotes

Consider another local macro, 'answ', which might contain yes or no. In a program that was supposed to do something different based on answ's content, you might code

```
if "'answ'" == "yes" {
        ...
}
else {
        ...
}
```

Note the odd-looking "'answ'", and now think about the line after substitution. The line reads either

```
if "yes" == "yes" {
```

or

```
if "no" == "yes" {
```

either of which is the desired result. Had we omitted the double quotes, the line would have read

```
if no == "yes" {
```

(assuming 'answ' contains no), and that is not at all the desired result. As the line reads now, no would not be a string, but would be interpreted as a variable in the data.

The key to all of this is to think of the line after substitution.

Double quotes are used to enclose strings: "yes", "no", "my dir\my file", "'answ'" (meaning the contents of local macro answ, treated as a string), and so on. Double quotes are used with macros,

```
local a "example"
if "'answ'" == "yes" {
        ...
}
```

and double quotes are used by lots of Stata commands,

```
. regress lnwage age ed if sex=="female"
. gen outa = outcome if drug=="A"
. use "person file"
```

Do not omit the double quotes just because you are using a "quoted" macro:

```
. regress lnwage age ed if sex=="`x'"
. gen outa = outcome if drug=="`firstdrug'"
. use "`filename'"
```

Stata has two sets of double-quote characters, of which "" is one. The other is `""'. They both work the same way:

```
. regress lnwage age ed if sex==`"female"'
. gen outa = outcome if drug==`"A"'
. use `"person file"'
```

No rational user would use `""' (called compound double quotes) instead of "" (called simple double quotes), but smart programmers do use them:

```
local a `"example"'
if `"`answ'"' == `"yes"' {
        ...
}
```

Why is `"example"' better than "example", `"`answ'"' better than `"answ"', and `"yes"' better than "yes"? The answer is that only `"`answ'"' is better than "`answ'"; `"example"' and `"yes"' are no better—and no worse—than "example" and "yes".

`"`answ'"' is better than "`answ'" because the macro answ might itself contain (simple or compound) double quotes. The really great thing about compound double quotes is that they nest. Say that `answ' contained the string "I "think" so". Then,

Stata would find	if "`answ'"=="yes"
confusing because it would expand to	if "I "think" so"=="yes"
Stata would not find	if `"`answ'"'==`"yes"'
confusing because it would expand to	if `"I "think" so"'==`"yes"'

Open and close double quote in the simple form look the same; open quote is " and so is close quote. Open and close double quote in the compound form are distinguishable; open quote is `" and close quote is "', and so Stata can pair the close with the corresponding open double quote. `"I "think" so"' is easy for Stata to understand, whereas "I "think" so" is a hopeless mishmash. (If you disagree, consider what "A"B"C" might mean. Is it the quoted string A"B"C, or is it quoted string A, followed by B, followed by quoted string C?)

Since Stata can distinguish open from close quotes, even nested compound double quotes are understandable: `"I `"think"' so"'. (What does "A"B"C" mean? Either it means `"A`"B"'C"' or it means `"A"'B`"C"'.)

Yes, compound double quotes make you think your vision is stuttering, especially when combined with the macro substitution `' characters. That is why we rarely use them, even when writing programs. You do not have to use exclusively one or the other style of quotes. It is perfectly acceptable to code

```
local a "example"
if `"`answ'"' == "yes" {
        ...
}
```

using compound double quotes where it might be necessary (`‘"‘answ’"’`) and using simple double quotes in other places (such as `"yes"`). It is also acceptable to use simple double quotes around macros (e.g., `"‘answ’"`) if you are certain that the macros themselves do not contain double quotes or (more likely) if you do not care what happens if they do.

In some instances, careful programmers should use compound double quotes. Later you will learn that Stata's `syntax` command interprets standard Stata syntax and so makes it easy to write programs that understand things like

```
. myprog mpg weight if strpos(make,"VW")!=0
```

`syntax` works—we are getting ahead of ourselves—by placing the if *exp* typed by the user in the local macro `if`. Thus, `‘if’` will contain "if `strpos(make,"VW")!=0`" in this case. Now, say that you are at a point in your program where you want to know whether the user specified an if *exp*. It would be natural to code

```
if ‘"‘if’"’ != "" {
        // the if exp was specified
        ...
}
else {
        // it was not
        ...
}
```

Note that we used compound double quotes around the macro `‘if’`. The local macro `‘if’` might contain double quotes, so we placed compound double quotes around it.

18.3.6   Extended macro functions

In addition to allowing =*exp*, `local` and `global` provide *extended functions*. The use of an extended function is denoted by a colon (`:`) following the macro name, as in

```
local       lbl : variable label myvar
local filenames : "." files "*.dta"
local       xi : word ‘i’ of ‘list’
```

Some macro extended functions access a piece of information. In the first example, the variable label associated with variable `myvar` will be stored in macro `lbl`. Other macro extended functions perform operations to gather the information. In the second example, macro `filenames` will contain the names of all the `.dta` datasets in the current directory. Still other macro extended functions perform an operation on their arguments and return the result. In the third example, `xi` will contain the `‘i’`th word (element) of `‘list’`. See [P] **macro** for a complete list of the macro extended functions.

Another useful source of information is `c()`, documented in [P] **creturn**:

```
local  today "‘c(current_date)’"
local curdir "‘c(pwd)’"
local   newn = c(N)+1
```

`c()` refers to a prerecorded list of values, which may be used directly in expressions or which may be quoted and the result substituted anywhere. `c(current_date)` returns today's date in the form "*dd MON yyyy*". Thus, the first example stores in macro `today` that date. `c(pwd)` returns the current directory, such as `C:\data\proj`. Thus, the second example stores in macro `curdir` the current directory. `c(N)` returns the number of observations of the data in memory. Thus, the third example stores in macro `newn` that number, plus one.

Note the use of quotes with c(). We could just as well have coded the first two examples as

```
local  today = c(current_date)
local curdir = c(pwd)
```

c() is a Stata function in the same sense that sqrt() is a Stata function. Thus, we can use c() directly in expressions. It is a special property of macro expansion, however, that you may use the c() function inside macro-expansion quotes. The same is not true of sqrt().

In any case, whenever you need a piece of information, whether it be about the dataset or about the environment, look in [P] **macro** and [P] **creturn**. It is likely to be in one place or the other, and sometimes, it is in both. You can obtain the current directory using

```
local curdir = c(pwd)
```

or using

```
local curdir : pwd
```

When information is in both, it does not matter which source you use.

18.3.7   Macro increment and decrement functions

We mentioned incrementing macros in [U] **18.3.4 Macros and expressions**. The construct

```
command that makes reference to 'i'
local ++i
```

occurs so commonly in Stata programs that it is convenient (and faster when executed) to collapse both lines of code into one, and to increment (or decrement) i at the same time that it is referenced. Stata allows this:

```
while ('++i' < 1000) {
        ...
}
while ('i++' < 1000) {
        ...
}
while ('--i' > 0) {
        ...
}
while ('i--' > 0) {
        ...
}
```

Above we have chosen to illustrate this using Stata's while command, but understand that ++ and -- can be used anyplace in any context, just so long as it is enclosed in macro-substitution quotes.

When the ++ or -- appears before the name, the macro is first incremented or decremented, and then the result is substituted.

When the ++ or -- appears after the name, the current value of the macro is substituted and then the macro is incremented or decremented.

❏ Technical Note

Do not use the inline ++ or -- operators when a part of the line might not be executed. Consider

```
if ('i'==0) local j = 'k++'
```

versus

```
if (`i'==0) {
        local j = `k++'
}
```

The first will not do what you expect because macros are expanded before the line is interpreted. Thus, the first will result in k being incremented in all cases, whereas the second increments k only when `i'==0.

❑

18.3.8 Macro expressions

Typing

> *command that makes reference to* `=exp`

is equivalent to

> local *macroname = exp*
> *command that makes reference to* `macroname`

although the former runs faster and is easier to type. When you use `=exp` within some larger command, *exp* is evaluated by Stata's expression evaluator, and the results are inserted as a literal string into the larger command. Then the command is executed. For example,

```
summarize u4
summarize u`=2+2'
summarize u`=4*(cos(0)==1)'
```

all do the same thing. Note that *exp* can be any valid Stata expression and, thus, may include references to variables, matrices, scalars, or even other macros. In the last case, just remember to enclose the submacros in quotes:

```
replace `var' = `group'[`=`j'+1']
```

In addition, typing

> *command that makes reference to* `:extended macro function`

is equivalent to

> local *macroname* : *extended macro function*
> *command that makes reference to* `macroname`

Thus, one might code

```
format y `:format x'
```

to assign to variable ty y the same format as the variable x.

❑ Technical Note

There is another macro expansion operator, . (called dot), which is used in conjunction with Stata's class system; see [P] **class** for more information.

There is also a macro expansion function, macval(), which is for use when expanding a macro— `macval(*name*)`—which confines the macro expansion to the first level of *name*, thereby suppressing the expansion of any embedded references to macros within *name*. Only two or three Stata users have or will ever need this, but, if you suspect you are one of them, see [P] **macro** and then see [P] **file** for an example.

❑

18.3.9   Advanced local macro manipulation

This section is really an aside to help test your understanding of macro substitution. The tricky examples illustrated below sometimes occur in real programs.

1. Say that you have macros x1, x2, x3, and so on. Obviously, 'x1' refers to the contents of x1, 'x2' to the contents of x2, etc. What does 'x'i'' refer to? Suppose that 'i' contains 6.

   The rule is to expand the inside first:

$$
\begin{array}{ll}
\text{'x'i''} & \text{expands to}\quad \text{'x6'} \\
\text{'x6'} & \text{expands to}\quad \text{the contents of local macro x6}
\end{array}
$$

   So, there you have a vector of macros.

2. We have already shown adjoining expansions: 'alpha''beta' expands to myvar if 'alpha' contains my and 'beta' contains var. What does 'alpha''gamma''beta' expand to when gamma is undefined?

   Stata does not mind if you reference a nonexisting macro. A nonexisting macro is treated as a macro with no contents. If local macro gamma does not exist, then

$$
\text{'gamma'}\quad \text{expands to}\quad \text{nothing}
$$

   It is not an error. Thus, 'alpha''gamma''beta' expands to myvar.

3. You clear a local macro by setting its contents to nothing:

$$
\begin{array}{ll}
 & \text{local } \textit{macname} \\
\text{or} & \text{local } \textit{macname} \text{ ""} \\
\text{or} & \text{local } \textit{macname} = \text{ ""}
\end{array}
$$

18.3.10   Advanced global macro manipulation

Global macros are rarely used, and when they are used, it is typically for communication between programs. You should never use a global macro where a local macro would suffice.

1. Constructions like xi are expanded sequentially. If $x contained this and $i 6, then xi expands to this6. If $x was undefined, then xi is just 6 because undefined global macros, like undefined local macros, are treated as containing nothing.

2. You can nest macro expansion by including braces, so assuming that $i contains 6, ${x$i} expands to ${x6}, which expands to the contents of $x6 (which would be nothing if $x6 is undefined).

3. You can mix global and local macros. Assume that local macro j contains 7. Then, ${x'j'} expands to the contents of $x7.

4. You also use braces to force the contents of global macros to run up against the succeeding text. For instance, assume that the macro drive contains "b:". If drive were a local macro, you could type

   'drive'myfile.dta

   to obtain b:myfile.dta. Because drive is a global macro, however, you must type

   ${drive}myfile.dta

You could not type

```
$drive myfile.dta
```

because that would expand to b: myfile.dta. You could not type

```
$drivemyfile.dta
```

because that would expand to .dta.

5. Because Stata uses $ to mark global-macro expansion, printing a real $ is sometimes tricky. To display the string $22.15 using the display command, you can type display "\$22.15", although you can get away with display "$22.15" because Stata is rather smart. Stata would not be smart about display "$this" if you really wanted to display $this and not the contents of the macro this. You would have to type display "\$this". Another alternative would be to use the SMCL code for a dollar sign when you wanted to display it: display "{c S|}this"; see [P] smcl.

6. Real dollar signs can also be placed into the contents of macros, thus postponing substitution. First, let's understand what happens when we do not postpone substitution; consider the following definitions:

```
global baseset "myvar thatvar"
global bigset "$baseset thisvar"
```

$bigset is equivalent to "myvar thatvar thisvar". Now, say that we redefine the macro baseset:

```
global baseset "myvar thatvar othvar"
```

The definition of bigset has not changed—it is still equivalent to "myvar thatvar thisvar". It has not changed because bigset used the definition of baseset that was current at the time it was defined. bigset no longer knows that its contents are supposed to have any relation to baseset.

Instead, let us assume that we had defined bigset as

```
global bigset "\$baseset thisvar"
```

at the outset. Then $bigset is equivalent to "$baseset thisvar", which in turn is equivalent to "myvar thatvar othvar thisvar". Since bigset explicitly depends upon baseset, anytime we change the definition of baseset, we will automatically change the definition of bigset as well.

18.3.11 Constructing Windows filenames using macros

Stata uses the \ character to tell its parser not to expand macros.

Windows uses the \ character as the directory path separator.

Mostly, there is no problem using a \ in a filename. However, if you are writing a program that contains a Windows path in macro path and a filename in fname, do not assemble the final result as

```
`path'\`fname'
```

because Stata will interpret the \ as an instruction to not expand `fname'. Instead, assemble the final result as

```
`path'/`fname'
```

Stata understands / as a directory separator on all platforms.

18.3.12   Accessing system values

Stata programs often need access to system parameters and settings, such as the value of π, the current date and time, or the current working directory.

System values are accessed via Stata's c-class values. The syntax works much the same as if you were referencing a local macro. For example, a reference to the c-class value for π, `‘c(pi)’`, will expand to a literal string containing 3.141592653589793 and could be used to do

```
. display sqrt(2*‘c(pi)’)
2.5066283
```

You could also access the current time

```
. display "‘c(current_time)’"
11:34:57
```

C-class values are designed to provide one all-encompassing way to access system parameters and settings, including system directories, system limits, string limits, memory settings, properties of the data currently in memory, output settings, efficiency settings, network settings, debugging settings, etc.

See [P] **creturn** for a complete detailed list of what is available. Typing

```
. creturn list
```

will give you the whole list of current settings.

18.3.13   Referencing characteristics

Characteristics—see [U] **12.8 Characteristics**—are like macros associated with variables. They have names of the form *varname*[*charname*]—such as mpg[comment]—and you quote their names just as you do macro names to obtain their contents:

To substitute the value of *varname*[*charname*], type	‘*varname*[*charname*]’
For example,	‘mpg[comment]’

You set the contents using the char command:

> char *varname*[*charname*] [["]*text*["]]

This is similar to the local and global commands, except that there is no =*exp* variation. You clear a characteristic by setting its contents to nothing just as you would with a macro:

Type	char *varname*[*charname*]
or	char *varname*[*charname*] ""

What is unique about characteristics is that they are saved with the data, meaning that their contents survive from one session to the next, and they are associated with variables in the data, so if you ever drop a variable, the associated characteristics disappear, too. (In addition, _dta[*charname*] is associated with the data but not with any variable in particular.)

All the standard rules apply: characteristics may be referenced by quotation in any context, and the characteristic's contents are substituted for the quoted characteristic name. As with macros, referencing a nonexistent characteristic is not an error; it merely substitutes to nothing.

18.4   Program arguments

When you invoke a program or do-file, what you type following the program or do-file name are the arguments. For instance, if you have a program called xyz and type

. xyz mpg weight

then mpg and weight are the program's arguments, mpg being the first argument and weight the second.

Program arguments are passed to programs via local macros:

Macro	Contents
'0'	what the user typed exactly as the user typed it, odd spacing, double quotes, and all
'1'	the first argument (first word of '0')
'2'	the second argument (second word of '0')
'3'	the third argument (third word of '0')
...	...
'*'	the arguments '1', '2', '3', ..., listed one after the other and with a single blank in between; similar to but different from '0' because odd spacing and double quotes are removed

That is, what the user types is passed to you in three different ways:

1. It is passed in '0' exactly as the user typed it, meaning quotes, odd spacing, and all.

2. It is passed in '1', '2', ... broken out into arguments on the basis of blanks (but with quotes used to force binding; we will get to that).

3. It is passed in '*' as "'1' '2' '3' ...", which is a crudely cleaned up version of '0'.

You will probably not use all three forms in one program.

We recommend that you ignore '*', at least for receiving arguments; it is included so that old Stata programs will continue to work.

Operating directly with '0' takes considerable programming sophistication, although Stata's syntax command makes interpreting '0' according to standard Stata syntax easy. That will be covered in [U] **18.4.4 Parsing standard Stata syntax** below.

The easiest way to receive arguments, however, is to deal with the positional macros '1', '2', ...

At the start of this section, we imagined an xyz program invoked by typing xyz mpg weight. In that case, '1' would contain mpg, '2' would contain weight, and '3' would contain nothing.

Let's write a program to report the correlation between two variables. Of course, Stata already has a command that can do this—correlate—and, in fact, we will implement our program in terms of correlate. It is silly, but all we want to accomplish right now is to show how Stata passes arguments to a program.

Here is our program:

```
program xyz
        correlate '1' '2'
end
```

Once the program is defined, we can try it:

```
. use http://www.stata-press.com/data/r9/auto
(1978 Automobile Data)

. xyz mpg weight
(obs=74)
                 |    mpg    weight
          -------+------------------
             mpg |  1.0000
          weight | -0.8072   1.0000
```

See how this works? We typed `xyz mpg weight`, which invoked our `xyz` program with '1' being `mpg` and '2' being `weight`. Our program gave the command `correlate '1' '2'`, and that expanded to `correlate mpg weight`.

Stylistically, this is not a good example of the use of positional arguments, but realistically, there is nothing wrong with it. The stylistic problem is that if `xyz` is really to report the correlation between two variables, it ought to allow standard Stata syntax, and that is not a difficult thing to do. Realistically, the program works.

Positional arguments, however, play an important role, even for programmers who care about style. When we write a subroutine—a program to be called by another program and not intended for direct human use—we often pass information using positional arguments.

Stata forms the positional arguments '1', '2', ... by taking what the user typed following the command (or do-file), parsing it on white space with double quotes used to force binding, and stripping the quotes. What that means is that the arguments are formed on the basis of words, but double-quoted strings are kept together as a single argument but with the quotes removed.

Let's create a program to illustrate these concepts. Although we would not normally define programs interactively, this program is short enough that we will:

```
. program listargs
  1.   display "The 1st argument you typed is:  '1'"
  2.   display "The 2nd argument you typed is:  '2'"
  3.   display "The 3rd argument you typed is:  '3'"
  4.   display "The 4th argument you typed is:  '4'"
  5.   end
```

The `display` command simply types the double-quoted string following it; see [P] **display**.

Let's try our program:

```
. listargs
The 1st argument you typed is:
The 2nd argument you typed is:
The 3rd argument you typed is:
The 4th argument you typed is:
```

We type `listargs`, and the result shows us what we already know—we typed nothing after the word `listargs`. There are no arguments. Let's try it again, this time adding `this is a test`:

```
. listargs this is a test
The 1st argument you typed is:  this
The 2nd argument you typed is:  is
The 3rd argument you typed is:  a
The 4th argument you typed is:  test
```

We learn that the first argument is 'this', the second is 'is', and so on. Blanks always separate arguments. You can, however, override this feature by placing double quotes around what you type:

```
. listargs "this is a test"
The 1st argument you typed is:  this is a test
The 2nd argument you typed is:
The 3rd argument you typed is:
The 4th argument you typed is:
```

This time we typed only one argument, 'this is a test'. When we place double quotes around what we type, Stata interprets whatever we type inside the quotes to be a single argument. In this case, '1' contains 'this is a test' (note that the double quotes were removed).

We can use double quotes more than once:

```
. listargs "this is" "a test"
The 1st argument you typed is:  this is
The 2nd argument you typed is:  a test
The 3rd argument you typed is:
The 4th argument you typed is:
```

The first argument is 'this is' and the second argument is 'a test'.

18.4.1 Named positional arguments

Positional arguments can be named: in your code, you do not have to refer to '1', '2', '3', ...; you can instead refer to more meaningful names, such as n, a, and b; numb, alpha, and beta; or whatever else you find convenient. You want to do this because programs coded in terms of '1', '2', ... are hard to read and therefore are more likely to contain errors.

You obtain better-named positional arguments using the args command:

```
program progname
        args argnames
        ...
end
```

For instance, if your program received four positional arguments and you wanted to call them varname, n, oldval, and newval, you would code

```
program progname
        args varname n oldval newval
        ...
end
```

varname, n, oldval, and newval become new local macros, and args simply copies '1', '2', '3', and '4' to them. It does not change '1', '2', '3', and '4'—you can still refer to the numbered macros if you wish—and it does not verify that your program receives the right number of arguments. If our example above were invoked with just two arguments, 'oldval' and 'newval' would contain nothing. If it were invoked with five arguments, the fifth argument would still be out there, stored in local macro '5'.

Let's make a command to create a dataset containing n observations on x ranging from a to b. Such a command would be useful, for instance, if we wanted to graph some complicated mathematical function and experiment with different ranges. It is convenient if we can type the range of x over which we wish to make the graph rather than concocting the range by hand. (In fact, Stata already has such a command—range—but it will be instructive to write our own.)

Before writing this program, we had better know how to proceed, so here is how you could create a dataset containing n observations with x ranging from a to b:

1. drop _all
   to clear whatever data are in memory.

2. set obs n
   to make a dataset of n observations on no variables; if n were 100, we would type set obs 100.

3. gen x = (_n-1)/(n-1)*($b-a$)+a
   because the built-in variable _n is 1 in the first observation, 2 in the second, and so on; see [U] **13.4 System variables (_variables)**.

So, the first version of our program might read

```
program rng                              // arguments are n a b
        drop _all
        set obs `1'
        generate x = (_n-1)/(_N-1)*(`3'-`2')+`2'
end
```

The above is just a direct translation of what we just said. '1' corresponds to n, '2' corresponds to a, and '3' corresponds to b. This program, however, would be far more understandable if we changed it to read

```
program rng
        args n a b
        drop _all
        set obs `n'
        generate x = (_n-1)/(_N-1)*(`b'-`a')+`a'
end
```

18.4.2 Incrementing through positional arguments

Some programs contain k arguments, where k varies, but it does not much matter because the same thing is done to each argument. One such program is summarize: type summarize mpg to obtain summary statistics on mpg, and type summarize mpg weight to obtain first summary statistics on mpg and then summary statistics on weight.

```
program ...
        local i = 1
        while "`i''" != "" {
                logic stated in terms of `i''
                local ++i
        }
end
```

Equivalently, if the logic that uses `i'' contains only one reference to `i'',

```
program ...
        local i = 1
        while "`i''" != "" {
                logic stated in terms of `i++''
        }
end
```

Note the tricky construction `i'', which then itself is placed in double quotes—"`i''"—for the while loop. To understand it, say i contains 1 or, equivalently, `i' is 1. Then `i'' is `1' is the name of the first variable. "`i''" is the name of the first variable in quotes. The while asks if the name of the variable is nothing and, if it is not, executes. Now `i' is 2, and "`i''" is the name of the second variable, in quotes. If that name is not "", we continue. If the name is "", we are done.

Say that you were writing a subroutine that was to receive k variables, but the code that processes each variable needs to know (while it is processing) how many variables were passed to the subroutine. You need first to count the variables (and so derive k) and then, knowing k, pass through the list again.

```
program progname
        local k = 1                     // count the number of arguments
        while "`‘k’’`" != "" {
                local ++k
        }
        local --k                  // k contains one too many
                                   // now pass through again
        local i = 1
        while `i' <= `k' {
                code in terms of ‘‘i’’ and ‘k’
                local ++i
        }
end
```

In the above example, we have used `while`, Stata's all-purpose looping command. Stata has two other looping commands, `foreach` and `forvalues` and they sometimes produce code that is more readable and executes more quickly. We direct you to read [P] **foreach** and [P] **forvalues**, but emphasize that, at this point, there is nothing they can do that `while` cannot do. Above we coded

```
local i = 1
while `i' <= `k' {
        code in terms of ‘‘i’’ and ‘k’
        local ++i
}
```

to produce logic that looped over the values ‘i’ = 1 to ‘k’. We could have instead coded

```
forvalues i = 1(1)`k' {
        code in terms of ‘‘i’’ and ‘k’
}
```

Similarly, at the beginning of this subsection, we said that you could use the following code in terms of `while` to loop over the arguments received:

```
program ...
        local i = 1
        while "`‘i’’`" != "" {
                logic stated in terms of ‘‘i’’
                local ++i
        }
end
```

Equivalent to the above would be

```
program ...
        foreach x of local 0 {
                logic stated in terms of ‘x’
        }
end
```

See [P] **foreach** and [P] **forvalues**.

You can combine `args` and incrementing through an unknown number of positional arguments. For instance, say that you were writing a subroutine that was to receive (1) `varname`, the name of some variable; (2) `n`, which is some sort of count; and (3) at least one and maybe 20 variable names. Perhaps you are to sum the variables, divide by `n`, and store the result in the first variable. What the program does is irrelevant; here is how we could receive the arguments:

```
program progname
        args varname n
        local i 3
        while "`i'" != "" {
                logic stated in terms of `i'
                local ++i
        }
end
```

18.4.3   Using macro shift

Another way to code the repeat-the-same-process problem for each argument is

```
program ...
        while "`1'" != "" {
                logic stated in terms of `1'
                macro shift
        }
end
```

macro shift shifts '1', '2', '3', ..., one to the left: what was '1' disappears, what was '2' becomes '1', what was '3' becomes '2', and so on.

The outside while loop continues the process until macro '1' contains nothing.

macro shift is an older construct that we no longer advocate using. Instead, we recommend that you use the techniques described in the previous subsection, that is, references to `i' and foreach/forvalues.

There are two reasons we make this recommendation: (1) macro shift destroys the positional macros '1', '2', which must then be reset using tokenize should you wish to pass through the argument list again, and (more importantly) (2) if the number of arguments is large (which in Stata/SE is more likely), macro shift can be incredibly slow.

❏ Technical Note

macro shift can do one thing that would be difficult to do by other means.

'*', the result of listing the contents of the numbered macros one after the other with a single blank in between, changes with macro shift. Say that your program received a list of variables and that the first variable was the dependent variable and the rest were independent variables. You want to save the first variable name in 'lhsvar' and all the rest in 'rhsvars'. You could code

```
program progname
        local lhsvar "`1'"
        macro shift 1
        local rhsvars "`*'"
        ...
end
```

Getting ahead of ourselves, suppose that a single macro contains a list of variables and you want to split the contents of the macro in two. Perhaps 'varlist' is the result of a syntax command (see [U] **18.4.4 Parsing standard Stata syntax**), and you now wish to split 'varlist' into 'lhsvar' and 'rhsvars'. tokenize will reset the numbered macros:

```
        program progname
                ...
                tokenize 'varlist'
                local lhsvar "'1'"
                macro shift 1
                local rhsvars "'*'"
                ...
        end
```

❏

18.4.4   Parsing standard Stata syntax

Let us now switch to '0' from the positional arguments '1', '2', ...

You can parse '0' (what the user typed) according to standard Stata syntax with a single command. Remember that standard Stata syntax is

$$\left[\,\text{by } \textit{varlist}\text{:}\,\right]\quad \textit{command} \quad \left[\,\textit{varlist}\,\right]\quad \left[\,\text{=}\textit{exp}\,\right]\quad \left[\,\text{using } \textit{filename}\,\right]\quad \left[\,\textit{if}\,\right]\quad \left[\,\textit{in}\,\right]\quad \left[\,\textit{weight}\,\right]$$

$$\left[\,\text{, } \textit{options}\,\right]$$

See [U] **11 Language syntax**.

The syntax command parses standard syntax. You code what amounts to the syntax diagram of your command in your program, and then syntax looks at '0' (it knows to look there) and compares what the user typed with what you are willing to accept. Then one of two things happens: either syntax stores the pieces in an easily processable way or, if what the user typed does not match what you specified, syntax issues the appropriate error message and stops your program.

Consider a program that is to take two or more variable names along with an optional if *exp* and in *range*. The program would read

```
        program ...
                syntax varlist(min=2) [if] [in]
                ...
        end
```

You will have to read [P] **syntax** to learn how to specify the syntactical elements, but the command is certainly readable, and it will not be long until you are guessing correctly about how to fill it in. And yes, the square brackets really do indicate optional elements, and you just use them with syntax in the natural way.

The one syntax command you code encompasses the entire parsing process. In this case, if what the user typed matches "two or more variables and an optional if and in", syntax defines new local macros:

'varlist'	the two or more variable names
'if'	the if *exp* specified by the user (or nothing)
'in'	the in *range* specified by the user (or nothing)

To see that this works, experiment with the following program:

```
        program tryit
                syntax varlist(min=2) [if] [in]
                display "varlist now contains |'varlist'|"
                display '"if now contains |'if'|"'
                display "in now contains |'in'|"
        end
```

Below we experiment:

```
. tryit mpg weight
varlist now contains |mpg weight|
if now contains ||
in now contains ||

. tryit mpg weight displ if foreign==1
varlist now contains |mpg weight displ|
if now contains |if foreign==1|
in now contains ||

. tryit mpg wei in 1/10
varlist now contains |mpg weight|
if now contains ||
in now contains |in 1/10|

. tryit mpg
too few variables specified
r(102);
```

Note that in our third try we abbreviated the weight variable as `wei`, yet, after parsing, `syntax` unabbreviated the variable for us.

If this program were next going to step through the variables in the varlist, the positional macros '1', '2', ... could be reset by coding

```
tokenize `varlist'
```

See [P] **tokenize**. `tokenize `varlist'` resets '1' to be the first word of '`varlist`', '2' to be the second word, and so on.

18.4.5   Parsing immediate commands

Immediate commands are described in [U] **19 Immediate commands**—they take numbers as arguments. By convention, when you name immediate commands, you should make the last letter of the name *i*. Assume that `mycmdi` takes as arguments two numbers, the first of which must be a positive integer, and also allows the options `alpha` and `beta`. The basic structure is

```
program mycmdi
        gettoken n 0 : 0, parse(" ,")          /* get first number */
        gettoken x 0 : 0, parse(" ,")          /* get second number */
        confirm integer number `n'             /* verify first is integer */
        confirm number `x'                     /* verify second is number */

        if `n'<=0 error 2001                   /* check that n is positive */
        place any other checks here

        syntax [, Alpha Beta]                  /* parse remaining syntax */
        make calculation and display output
end
```

See [P] **gettoken**.

18.4.6   Parsing nonstandard syntax

If you wish to interpret nonstandard syntax and positional arguments are not adequate for you, you know that you face a formidable programming task. The key to the solution is the `gettoken` command.

`gettoken` can pull a single token from the front of a macro according to the parsing characters you specify and, optionally, define another macro or redefine the initial macro to contain the remaining (unparsed) characters. That is,

Say that '0' contains	"this is what the user typed"
After gettoken,	
new macro 'token' could contain	"this"
and '0' could still contain	"this is what the user typed"
or	
new macro 'token' could contain	"this"
and new macro 'rest' could contain	" is what the user typed"
and '0' could still contain	"this is what the user typed"
or	
new macro 'token' could contain	"this"
and '0' could contain	" is what the user typed"

A simplified syntax of `gettoken` is

> `gettoken` *emname1* [*emname2*] : *emname3* [, <u>parse</u>(*pchars*) <u>quotes</u>
>
> <u>match</u>(*lmacname*) `bind`]

where *emname1*, *emname2*, *emname3*, and *lmacname* are the names of local macros. (Stata provides a way to work with global macros, but, in practice, that is seldom necessary; see [P] **gettoken**.)

`gettoken` pulls the first token from *emname3* and stores it in *emname1*, and if *emname2* is specified, stores the remaining characters from *emname3* in *emname2*. Any of *emname1*, *emname2*, and *emname3* may be the same macro. Typically, `gettoken` is coded

> `gettoken` *emname1* : 0 [, *options*]
>
> `gettoken` *emname1* 0 : 0 [, *options*]

since '0' is the macro containing what the user typed. The first coding is used for token lookahead, should that be necessary, and the second is used for committing to taking the token.

`gettoken`'s options are

`parse("`*string*`")` for specifying parsing characters
the default is `parse(" ")`, meaning to parse on white space
it is common to specify `parse('"" "')`, meaning to parse on white space and double quote
(`'"" "'` is the string double-quote-space in compound double quotes)

`quotes` to specify that outer double quotes *not* be stripped

`match(`*lmacname*`)` to bind on parentheses and square brackets
lmacname will be set to contain "(", "[", or nothing, depending on whether *emname1* was bound on parentheses or brackets, or if `match()` turned out to be irrelevant
emname1 will have the outside parentheses or brackets removed

`gettoken` binds on double quotes whenever a (simple or compound) double quote is encountered at the beginning of *emname3*. Specifying `parse('"" "')` ensures that double-quoted strings are isolated.

quote specifies that double quotes not be removed from the source in defining the token. For instance, in parsing `""this is" a test"`, the next token is "`this is`" if quote is not specified and is "`"this is"`" if quote is specified.

`match()` specifies that parentheses and square brackets be matched in defining tokens. The outside level of parentheses or brackets is stripped. In parsing "`(2+3)/2`", the next token is "`2+3`" if `match()` is specified. In practice, `match()` might be used with expressions, but it is more likely to be used to isolate bound varlists and time-series varlists.

18.5   Scalars and matrices

In addition to macros, scalars and matrices are provided for programmers; see [U] **14 Matrix expressions**, [P] **scalar** and [P] **matrix**.

As far as scalar calculations go, you can use macros or scalars. Remember, macros can hold numbers. Stata's scalars are, however, slightly faster and are a little more accurate than macros. The speed issue is so slight as to be nearly immeasurable. Macros are accurate to a minimum of 12 decimal digits, and scalars are accurate to roughly 16 decimal digits. Which you use makes little difference except in iterative calculations.

In addition, Stata has a serious matrix programming language called Mata, which is the subject of another manual. Mata can be used to write subroutines that are called by Stata programs. See the *Mata Reference Manual*, and in particular, see [M-1] **Using Mata with ado-files**.

18.6   Temporarily destroying the data in memory

It is sometimes necessary to modify the data in memory to accomplish a particular task. A well-behaved program, however, ensures that the user's data are always restored. The `preserve` command makes this easy:

```
code before the data need changing
preserve
code that changes data freely
```

When you give the `preserve` command, Stata makes a copy of the user's data on disk. When your program terminates—no matter how—Stata restores the data and erases the temporary file; see [P] **preserve**.

18.7   Temporary objects

If you write a substantial program, it will invariably require the use of temporary variables in the data, or temporary scalars, matrices, or files. Temporary objects are necessary while the program is making its calculations, and, once the program completes, they are discarded.

Stata provides three commands to create temporary objects: `tempvar` creates names for variables in the dataset, `tempname` creates names for scalars and matrices, and `tempfile` creates names for files. All are described in [P] **macro**, and all have the same syntax:

{ `tempvar` | `tempname` | `tempfile` } *macname* [*macname* ...]

The commands create local macros containing names you may use.

18.7.1   Temporary variables

Say that, in making a calculation, you need to add variables sum_y and sum_z to the data. You might be tempted to code

```
...
gen sum_y = ...
gen sum_z = ...
...
```

but that would be poor because (1) the dataset might already have variables named sum_y and sum_z in it, and (2) you will have to remember to drop the variables before your program concludes. Better is

```
...
tempvar sum_y
gen `sum_y' = ...
tempvar sum_z
gen `sum_z' = ...
...
```

or

```
...
tempvar sum_y sum_z
gen `sum_y' = ...
gen `sum_z' = ...
...
```

It is not necessary to explicitly drop `sum_y' and `sum_z' when you are finished, although you may if you wish. Stata will automatically drop any variables with names assigned by tempvar. After issuing the tempvar command, you must refer to the names with the enclosing quotes, which signifies macro expansion. Thus, after typing tempvar sum_y—the one case where you do not put single quotes around the name—refer thereafter to the variable `sum_y', with quotes. tempvar does not create temporary variables. Instead tempvar creates names that may subsequently be used to create new variables which will be temporary, and tempvar stores that name in the local macro whose name you provide.

A full description of tempvar can be found in [P] **macro**.

18.7.2   Temporary scalars and matrices

tempname works just like tempvar. For instance, a piece of your code might read

```
tempname YXX XXinv
matrix accum `YXX' = price weight mpg
matrix `XXinv' = invsym(`YXX'[2..., 2...])
tempname b
matrix `b' = `XXinv'*`YXX'[1..., 1]
```

The above code solves for the coefficients of a regression on price on weight and mpg; see [U] **14 Matrix expressions** and [P] **matrix** for more information on the matrix commands.

As with temporary variables, temporary scalars and matrices are automatically dropped at the conclusion of your program.

18.7.3 Temporary files

In cases where you ordinarily might think you need temporary files, you may not because of Stata's ability to preserve and automatically restore the data in memory; see [U] **18.6 Temporarily destroying the data in memory** above.

For more complicated programs, Stata does provide temporary files. A code fragment might read

```
preserve                      /* save original data */
tempfile males females
keep if sex==1
save "`males'"
restore, preserve             /* get back original data */
keep if sex==0
save "`females'"
```

As with temporary variables, scalars, and matrices, it is not necessary to delete the temporary files when you are through with them; Stata automatically erases them when your program ends.

18.8   Accessing results calculated by other programs

Stata commands that report results also save the results where they can be subsequently used by other commands or programs. This is documented in the *Saved Results* section of the particular command in the *Reference* manuals. Commands save results in one of three places:

1. r-class commands, such as `summarize`, save their results in `r()`; most commands are r-class.

2. e-class commands, such as `regress`, save their results in `e()`; e-class commands are Stata's model estimation commands.

3. s-class commands (there are no good examples) save their results in `s()`; this is a rarely used class that programmers sometimes find useful to help parse input.

Commands that do not save results are called n-class commands. More correctly, these commands require that you state where the result is to be saved, as in `generate` *newvar* `= ....`

▷ Example 1

You wish to write a program to calculate the standard error of the mean, which is given by the formula $\sqrt{s^2/n}$, where s^2 is the calculated variance. (You could obtain this statistic using the `ci` command, but we will pretend that is not true.) You look at [R] **summarize** and learn that the mean is stored in `r(mean)`, the variance in `r(Var)`, and the number of observations in `r(N)`. With that knowledge, you write the following program:

```
program meanse
        quietly summarize `1'
        display "     mean = " r(mean)
        display "SE of mean = " sqrt(r(Var)/r(N))
end
```

The result of executing this program is

```
. meanse mpg
      mean = 21.297297
SE of mean = .67255109
```

◁

If you run an r-class command and type `return list` or run an e-class command and type `ereturn list`, Stata will summarize what was saved:

```
. use http://www.stata-press.com/data/r9/auto
(1978 Automobile Data)

. regress mpg weight displ
(output omitted)

. ereturn list

scalars:
                   e(N) =  74
                e(df_m) =  2
                e(df_r) =  71
                   e(F) =  66.78504752026517
                  e(r2) =  .6529306984682528
                e(rmse) =  3.45606176570828
                 e(mss) =  1595.409691543724
                 e(rss) =  848.0497679157352
                e(r2_a) =  .643154098425105
                  e(ll) =  -195.2397979466294
                e(ll_0) =  -234.3943376482347

macros:
               e(title) : "Linear regression"
              e(depvar) : "mpg"
                 e(cmd) : "regress"
          e(properties) : "b V"
             e(predict) : "regres_p"
               e(model) : "ols"
           e(estat_cmd) : "regress_estat"

matrices:
                   e(b) :  1 x 3
                   e(V) :  3 x 3

functions:
              e(sample)

. summarize mpg if foreign
```

Variable	Obs	Mean	Std. Dev.	Min	Max
mpg	22	24.77273	6.611187	14	41

```
. return list

scalars:
                   r(N) =  22
               r(sum_w) =  22
                r(mean) =  24.77272727272727
                 r(Var) =  43.70779220779221
                  r(sd) =  6.611186898567625
                 r(min) =  14
                 r(max) =  41
                 r(sum) =  545
```

In the example above, we ran `regress` followed by `summarize`. As a result, `e(N)` records the number of observations used by `regress` (equal to 74), and `r(N)` records the number of observations used by `summarize` (equal to 22). `r(N)` and `e(N)` are not the same.

If we now ran another r-class command—say, `tabulate`—the contents of `r()` would change, but those in `e()` would remain unchanged. You might, therefore, think that if we then ran another e-class command, say, `probit`, the contents of `e()` would change, but `r()` would remain unchanged. While it is true that `e()` results remain in place until the next e-class command is executed, do not depend on `r()` remaining unchanged. If an e-class or n-class command were to use an r-class command as a subroutine, that would cause r() to change. Anyway, most commands are r-class, so the contents of r() change frequently.

❏ Technical Note

It is, therefore, of great importance that you access results stored in r() immediately after the command that sets them. If you need the mean and variance of the variable '1' for subsequent calculation, do *not* code

```
summarize '1'
...
... r(mean) ... r(Var) ...
```

Instead, code

```
summarize '1'
local mean = r(mean)
local var = r(Var)
...
... 'mean' ... 'var' ...
```

or

```
tempname mean var
summarize '1'
scalar 'mean' = r(mean)
scalar 'var' = r(Var)
...
... 'mean' ... 'var' ...
```

❏

Saved results, whether in r() or e(), come in three types: scalars, macros, and matrices. If you look back at the ereturn list and return list output, you will see that regress saves examples of all three, whereas summarize just saves scalars. (regress also saves the "function" e(sample), as do all the other e-class commands; see [U] **20.5 Specifying the estimation subsample**.)

Regardless of the flavor of e(*name*) or r(*name*), you can just refer to e(*name*) or r(*name*). That was the rule we gave in [U] **13.6 Accessing results from Stata commands**, and that rule is sufficient for most uses. There is, however, another way to refer to saved results. Rather than referring to r(*name*) and e(*name*), you can embed the reference in macro substitution characters ' ' to produce 'r(*name*)' and 'e(*name*)'. The result is the same as macro substitution; the saved result is evaluated, and then the evaluation is substituted:

```
. display "You can refer to " e(cmd) " or to 'e(cmd)'"
You can refer to regress or to regress
```

This means, for instance, that typing 'e(cmd)' is the same as typing regress because e(cmd) contains "regress":

```
. 'e(cmd)'
      Source |       SS       df       MS              Number of obs =      74
-------------+------------------------------           F(  2,    71) =   66.79
       Model | 1595.40969     2  797.704846            Prob > F      =  0.0000
 (remaining output omitted)
```

In the ereturn list, e(cmd) was listed as a macro, and when you place a macro's name in single quotes, the macro's contents are substituted, so this is hardly a surprise.

What is surprising is that you can do this with scalar and even matrix saved results. e(N) is a scalar equal to 74 and may be used as such in any expression such as "display e(mss)/e(N)" or "local meanss = e(mss)/e(N)". 'e(N)' substitutes to the string "74" and may be used in any context whatsoever, such as "local val'e(N)' = e(N)" (which would create a macro named val74). The rules for referring to saved results are

1. You may refer to r(*name*) or e(*name*) without single quotes in any expression, and only in an expression. (Referring to s-class s(*name*) without single quotes is not allowed.)

   1.1 If *name* does not exist, missing value (.) is returned; it is not an error to refer to a nonexisting saved result.

   1.2 If *name* is a scalar, the full double-precision value of *name* is returned.

   1.3 If *name* is a macro, it is examined to determine whether its contents can be interpreted as a number. If so, the number is returned; otherwise, the first 80 characters of *name* are returned.

   1.4 If *name* is a matrix, the full *matrix* is returned.

2. You may refer to 'r(*name*)', 'e(*name*)', or 's(*name*)'—note the presence of quotes indicating macro substitution—in any context whatsoever.

   2.1 If *name* does not exist, nothing is substituted; it is not an error to refer to a nonexisting saved result. The resulting line is the same as if you had never typed 'r(*name*)', 'e(*name*)', or 's(*name*)'.

   2.2 If *name* is a scalar, a string representation of the number accurate to no less than 12 digits of precision is substituted.

   2.3 If *name* is a macro, the full contents (up to 18,623 characters for Intercooled and 1,000 characters for Small Stata) are substituted.

   2.4 If *name* is a matrix, the word matrix is substituted.

In general, you should refer to scalar and matrix saved results without quotes—r(*name*) and e(*name*)—and to macro saved results with quotes—'r(*name*)', 'e(*name*)', and 's(*name*)'—but it is sometimes convenient to switch. For instance, say that returned result r(example) contains the number of time periods patients are observed, and assume that r(example) was saved as a macro and not as a scalar. You could still refer to r(example) without the quotes in an expression context and obtain the expected result. It would have made more sense for you to have stored r(example) as a scalar, but really it would not matter, and the user would not even have to know how the saved result was stored.

Switching the other way is sometimes useful, too. Say that returned result r(N) is a scalar that contains the number of observations used. You now want to use some other command that has an option n(#) that specifies the number of observations used. You could not type n(r(N)) because the syntax diagram says that option n() expects its argument to be a literal number. Instead, you could type n('r(N)').

18.9   Accessing results calculated by estimation commands

Estimation results are saved in e(), and you access them in the same way you access any saved result; see [U] **18.8 Accessing results calculated by other programs** above. In summary,

1. Estimation commands—regress, logistic, etc.—save results in e().

2. Estimation commands save their name in e(cmd). For instance, regress saves "regress" and poisson saves "poisson" in e(cmd).

3. Estimation commands save the number of observations used in e(N), and they identify the estimation subsample by setting e(sample). You could type, for instance, summarize if e(sample) to obtain summary statistics on the observations used by the estimator.

4. Estimation commands save the entire coefficient vector and variance–covariance matrix of the estimators in e(b) and e(V). These are matrices, and they may be manipulated like any other matrix:

```
. matrix list e(b)
e(b)[1,3]
          weight        displ         _cons
y1     -.00656711    .00528078     40.084522
. matrix y = e(b)*e(V)*e(b)'
. matrix list y
symmetric y[1,1]
            y1
y1    6556.982
```

5. Estimation commands set _b[*name*] and _se[*name*] as convenient ways to use coefficients and their standard errors in expressions; see [U] **13.5 Accessing coefficients and standard errors**.

6. Estimation commands may set other e() scalars, macros, or matrices containing additional information. This is documented in the *Saved Results* section of the particular command in the command reference.

▷ Example 2

If you are writing a command for use after regress, early in your code you should include the following:

```
if "`e(cmd)'" != "regress" {
        error 301
}
```

This is how you verify that the estimation results that are stored have been set by regress and not by some other estimation command. Error 301 is Stata's "last estimates not found" error.

◁

18.10   Saving results

If your program calculates something, it should save the results of the calculation so that other programs can access them. In this way your program not only can be used interactively, but also can be used as a subroutine for other commands.

Saving results is easy:

1. On the program line, specify the option rclass, eclass, or sclass according to whether you intend to return results in r(), e(), or s().

2. Code

return scalar *name* = *exp*	(same syntax as scalar without the return)
return local *name* ...	(same syntax as local without the return)
return matrix *name* *matname*	(moves *matname* to r(*name*))

to save results in r().

3. Code

ereturn *name = exp*   (same syntax as scalar without the ereturn)
ereturn local *name* ...   (same syntax as local without the ereturn)
ereturn matrix *name matname* (moves *matname* to e(*name*))

to save results in e(). You do not save the coefficient vector and variance matrix e(b) and e(V) in this way; instead you use ereturn post.

4. Code

sreturn local *name* ...   (same syntax as local without the sreturn)

to save results in s(). (The s-class has only macros.)

A program must be exclusively r-class, e-class, or s-class.

18.10.1   Saving results in r()

In [U] **18.8 Accessing results calculated by other programs**, we showed an example that reported the mean and standard error of the mean. A better version would save in r() the results of its calculations and would read

```
program meanse, rclass
        quietly summarize '1'
        local mean = r(mean)
        local sem  = sqrt(r(Var)/r(N))
        display "     mean = " 'mean'
        display "SE of mean = " 'sem'
        return scalar mean = 'mean'
        return scalar se = 'sem'
end
```

Running meanse now sets r(mean) and r(se):

```
. meanse mpg
     mean = 21.297297
SE of mean = .67255109

. return list

scalars:
       r(se)     =  .6725510870764975
       r(mean)   =  21.2972972972973
```

In this modification, we added option rclass to the program statement, and we added two return commands to the end of the program.

Although we placed the return statements at the end of the program, they may be placed at the point of calculation if that is more convenient. A more concise version of this program would read

```
program meanse, rclass
        quietly summarize '1'
        return scalar mean = r(mean)
        return scalar se = sqrt(r(Var)/r(N))
        display "     mean = " return(mean)
        display "SE of mean = " return(se)
end
```

The return() function is just like the r() function, except that return() refers to the results that this program *will* return rather than to the saved results that currently *are* returned (which in this case are due to summarize). That is, when you code the return command, the result is not immediately posted to r(). Rather, Stata holds onto the result in return() until your program concludes, and then it copies the contents of return() to r(). While your program is active, you may use the return() function to access results you have already "returned". (return() works just like r() works after your program returns, meaning that you may code 'return()' to perform macro substitution.)

18.10.2 Saving results in e()

Saving in e() is in most ways similar to saving in r(): you add the eclass option to the program statement, and then you use ereturn ... just as you used return ... to store results. There are, however, some significant differences:

1. Unlike r(), estimation results are saved in e() the instant you issue an ereturn scalar, ereturn local, or ereturn matrix command. This is because estimation results can consume considerable memory, and Stata does not want to have multiple copies of the results floating around. That means you must be more organized and post your results at the end of your program.

2. In your code when you have your estimates and are ready to begin posting, you will first clear the previous estimates, set the coefficient vector e(b) and corresponding variance matrix e(V), and set the estimation-sample function e(sample). How you do this depends on how you obtained your estimates:

   2.1 If you obtained your estimates using Stata's likelihood maximizer ml, this is automatically handled for you; skip to step 3.

   2.2 If you obtained estimates by "stealing" an already existing estimator, e(b), e(V), and e(sample) already exist, and you will not want to clear them; skip to step 3.

   2.3 If you write your own code from start to finish, you use the ereturn post command; see [P] **ereturn**. You will code something like "ereturn post 'b' 'V', esample('touse')", where 'b' is the name of the coefficient vector, 'V' is the name of the corresponding variance matrix, and 'touse' is the name of a variable containing 1 if the observation was used and 0 if it was ignored. ereturn post clears the previous estimates and moves the coefficient vector, variance matrix, and variable into e(b), e(V), and e(sample).

   2.4 A variation on (2.3) is when you use an already existing estimator to produce the estimates but do not want all the other e() results stored by the estimator. In that case, you code

   ```
   tempvar touse
   tempname b V
   matrix 'b' = e(b)
   matrix 'V' = e(V)
   qui gen byte 'touse' = e(sample)
   ereturn post 'b' 'V', esample('touse')
   ```

3. You now save anything else in e() that you wish by using the ereturn scalar, ereturn local, or ereturn matrix commands.

4. You code ereturn local cmd "*cmdname*". Stata does not consider estimation results complete until this command is posted, and Stata considers the results to be complete when this is posted, so you must remember to do this and to do this last. If you set e(cmd) too early and the user pressed *Break*, Stata would consider your estimates complete when they are not.

Say that you wish to write the estimation command with syntax

myest *depvar var*$_1$ *var*$_2$ [if *exp*] [in *range*], *optset1 optset2*

where *optset1* affects how results are displayed and *optset2* affects the estimation results themselves. One important characteristic of estimation commands is that, when typed without arguments, they redisplay the previous estimation results. The outline is

```
program myest, eclass
        local options "optset1"
        if replay() {
                if "'e(cmd)'"!="myest" {
                        error 301            /* last estimates not found      */
                }
                syntax [, 'options']
        }
        else {
                syntax varlist [if] [in] [, 'options' optset2]
                marksample touse
```
Code contains either this,
```
                tempnames b V
```
                commands for performing estimation
                assume produces 'b' *and* 'V'
```
                ereturn post 'b' 'V', esample('touse')
                ereturn local depvar "'depv'"
```
or this,
```
                ml model ... if 'touse' ...
```
and regardless, concludes,
                perhaps other ereturn *commands appear here*
```
                ereturn local cmd "myest"
        }
                                                /* (re)display results ...       */
```
code typically reads
        code to output header above coefficient table
```
        ereturn display                         /* displays coefficient table    */
```
or
```
        ml display                              /* displays header and coef. table */
end
```

Here is a list of the commonly saved e() results. Of course, you may create any e() results that you wish.

e(N) (scalar)
Number of observations.

e(df_m) (scalar)
Model degrees of freedom.

e(df_r) (scalar)
"Denominator" degrees of freedom if estimates are nonasymptotic.

e(r2_p) (scalar)
Value of the pseudo-R^2 if it is calculated. (If a "real" R^2 is calculated as it would be in linear regression, it is stored in (scalar) e(r2).)

e(F) (scalar)
Test of the model against the constant-only model, if relevant, and if results are nonasymptotic.

e(ll) (scalar)
Log-likelihood value, if relevant.

e(ll_0) (scalar)
Log-likelihood value for constant-only model, if relevant.

e(N_clust) (scalar)
  Number of clusters, if any.

e(chi2) (scalar)
  Test of the model against the constant-only model, if relevant, and if results are asymptotic.

e(cmd) (macro)
  Name of the estimation command.

e(depvar) (macro)
  Names of the dependent variables.

e(wtype) and e(wexp) (macros)
  If weighted estimation was performed, e(wtype) contains the weight type (fweight, pweight, etc.) and e(wexp) contains the weighting expression.

e(clustvar) (macro)
  Name of the cluster variable, if any.

e(vcetype) (macro)
  Text to appear above standard errors in estimation output; typically Robust, Bootstrap, Jack-knife, or "".

e(vce) (macro)
  vcetype specified in vce().

e(chi2type) (macro)
  LR or Wald or other depending on how e(chi2) was performed.

e(crittype) (macro)
  Type of optimization criterion used, such as log likelihood or deviance.

e(properties) (macro)
  Typically contains b V.

e(predict) (macro)
  Name of the command that predict is to use; if this is blank, predict uses the default _predict.

e(b) and e(V) (matrices)
  The coefficient vector and corresponding variance matrix. Saved when you coded ereturn post.

e(sample) (function)
  This function was defined by ereturn post's esample() option if you specified it. You specified a variable containing 1 if you used an observation and 0 otherwise. ereturn post stole the variable and created e(sample) from it.

18.10.3   Saving results in s()

S is a strange class because, whereas the other classes allow scalars, macros, and matrices, s allows only macros.

S is seldom used and is for subroutines that you might write to assist in parsing the user's input prior to evaluating any user-supplied expressions.

Here is the problem s solves: Say that you create a nonstandard syntax for some command so that you have to parse through it yourself. The syntax is so complicated that you want to create subroutines to take pieces of it and then return information to your main routine. Assume that your syntax contains expressions that the user might type. Now, say that one of the expressions the user types is, say, r(mean)/sqrt(r(Var))—perhaps the user is consuming results left behind by summarize.

If, in your parsing step, you call subroutines that return results in r(), you will wipe out r(mean) and r(Var) before you ever get around to seeing them, much less evaluating them. So, you must be careful to leave r() intact until your parsing is complete; you must use no r-class commands, and any subroutines you write must not touch r(). The way to do this is to use s-class subroutines because s-class routines return results in s() rather than r(). S-class provides macros only because that is all you need to solve parsing problems.

To create an s-class routine, specify the sclass option on the program line and then use sreturn local to return results.

S-class results are posted to s() at the instant you issue the sreturn() command, so you must organize your results. In addition, s() is never automatically cleared, so occasionally coding sreturn clear at appropriate points in your code is a good idea.

Very few programs need s-class subroutines.

18.11 Ado-files

Ado-files were introduced in [U] **17 Ado-files**.

When a user types 'gobbledegook', Stata first asks itself if gobbledegook is one of its built-in commands. If so, the command is executed. Otherwise, it asks itself if gobbledegook is a defined program. If so, the program is executed. Otherwise, Stata looks in various directories for gobbledegook.ado. If there is no such file, the process ends with the "unrecognized command" error.

If Stata finds the file, it quietly issues to itself the command 'run gobbledegook.ado' (specifying the path explicitly). If that runs without error, Stata asks itself again if gobbledegook is a defined program. If not, Stata issues the "unrecognized command" error. (In this case, somebody wrote a bad ado-file.) If the program is defined, as it should be, Stata executes it.

Thus, you can arrange for programs you write to be loaded automatically. For instance, if you were to create hello.ado containing

```
—————————————————————————————————— top of hello.ado ————————
program hello
        display "hi there"
end
—————————————————————————————————— end of hello.ado ————————
```

and store the file in your current directory or your personal directory (see [U] **17.5.2 Where is my personal ado directory?**), you could type hello and be greeted by a reassuring

```
. hello
hi there
```

You could, at that point, think of hello as just another part of Stata.

There are two places to put your personal ado-files. One is the current directory, and that is a good choice when the ado-file is unique to a project. You will want to use it only when you are in that directory. The other place is your *personal ado directory*, which is probably something like C:\ado\personal if you use Windows, ~/ado/personal if you use Unix, and ~/ado/personal if you use a Macintosh. We are guessing.

To find your personal ado directory, enter Stata and type

```
. personal
```

❏ Technical Note

Stata looks in various directories for ado-files, defined by the c-class value c(adopath), which contains

UPDATES;BASE;SITE;.;PERSONAL;PLUS;OLDPLACE

The words in capital letters are codenames for directories, and the mapping from codenames to directories can be obtained by typing the sysdir command. Here is what sysdir shows on one particular Windows computer:

```
. sysdir
    STATA:  C:\Program Files\Stata9\
  UPDATES:  C:\Program Files\Stata9\ado\updates\
     BASE:  C:\Program Files\Stata9\ado\base\
     SITE:  C:\Program Files\Stata9\ado\site\
     PLUS:  C:\ado\plus\
 PERSONAL:  C:\ado\personal\
 OLDPLACE:  C:\ado\
```

Even if you use Windows, your mapping might be different because it all depends on where you installed Stata. That is the point of the codenames. They make it possible to refer to directories according to their logical purposes rather than their physical location.

The c-class value c(adopath) is the search path, so in looking for an ado-file, Stata first looks in UPDATES, then in BASE, and so on, until it finds the file. Actually, Stata not only looks in UPDATES, it also takes the first letter of the ado-file it is looking for and looks in the lettered subdirectory. Say that Stata was looking for gobbledegook.ado. Stata would look up UPDATES (C:\Program Files\Stata9\ado\updates in our example) and, if the file were not found there, it would look in the g subdirectory of UPDATES (C:\Program Files\Stata9\ado\updates\g) before looking in BASE, whereupon it would follow the same rules.

Why the extra complication? We distribute literally hundreds of ado-files with Stata, and some operating systems have difficulty dealing with so many files in the same directory. All operating systems experience at least a performance degradation. To prevent this, the ado directory we ship is split 27 ways (letters *a–z* and underscore). Thus, the Stata command ci, which is implemented as an ado-file, can be found in the subdirectory c of BASE.

If you write ado-files, you can structure your personal ado directory this way, too, but there is no reason to do so until you have more than, say, 250 files in a single directory.

❏

❏ Technical Note

After finding and running *gobbledegook*.ado, Stata calculates the total size of all programs that it has automatically loaded. If this exceeds adosize (see [P] **sysdir**), Stata begins discarding the oldest automatically loaded programs until the total is less than adosize. Oldest here is measured by the time last used, not the time loaded. This discarding saves memory and does not affect you, since any program that was automatically loaded could be automatically loaded again if needed.

It does, however, affect performance. Loading the program takes time, and you will again have to wait if you use one of the previously loaded-and-discarded programs. Increasing adosize reduces this possibility, but at the cost of memory. The set adosize command allows you to change this parameter; see [P] **sysdir**. The default value of adosize is 500. A value of 500 for adosize means that up to 500K can be allocated to autoloaded programs. Experimentation has shown that this is a good number—increasing it does not improve performance much.

❏

18.11.1   Version

We recommend that the first line following `program` in your ado-file declare the Stata release under which you wrote the program; `hello.ado` would better read as

```
                                                     ─ top of hello.ado ─
program hello
        version 9
        display "hi there"
end
                                                     ─ end of hello.ado ─
```

We introduced the concept of version in [U] **16.1.1 Version**. In regular do-files, we recommend the `version` line appear as the first line of the do-file. For ado-files, the line appears after the `program` because loading the ado-file is one step and executing the program is another. It is when Stata executes the program defined in the ado-file that we want to stipulate the interpretation of the commands.

The inclusion of the `version` line is of more importance in ado-files than in do-files because (1) ado-files have longer lives than do-files, so it is more likely that you will use an ado-file with a later release, and (2) ado-files tend to use more of Stata's features, increasing the probability that any change to Stata will affect them.

18.11.2   Comments and long lines in ado-files

Comments in ado-files are handled the same way as in do-files: you enclose the text in `/*` comment `*/` brackets, or you begin the line with an asterisk (`*`), or you interrupt the line with `//`; see [U] **16.1.2 Comments and blank lines in do-files**.

Logical lines longer than physical lines are also handled as they are in do-files: either you change the delimiter to a semicolon (`;`) or you comment out the newline using `///` at the end of the previous physical line.

18.11.3   Debugging ado-files

Debugging ado-files is a little tricky because it is Stata and not you that is in control of when the ado-file is loaded.

Assume that you wanted to change `hello` to say "Hi, Mary". Assume that your editor is called `vi` and that you are in the habit of calling your editor from Stata, so you do this,

```
. !vi hello.ado
```

and make the obvious change to the program. Equivalently, you can pretend that you are using a windowed operating system and jump out of Stata—leaving it running—to modify the `hello.ado` file. Anyway, you change `hello.ado` to read

```
                                                     ─ top of hello.ado ─
program hello
        version 9
        display "hi, Mary"
end
                                                     ─ end of hello.ado ─
```

Back in Stata, you try it:

```
. hello
hi there
```

What just happened is that Stata ran the old copy of `hello`—the copy it still has in its memory. Stata wants to be fast about executing ado-files, so when it loads one, it keeps it around a while—waiting for memory to get short—before clearing it from its memory. Naturally, Stata can drop `hello` anytime because it can always reload it from disk.

You changed the copy on disk, but Stata still has the old copy loaded into memory. You type `discard` to tell Stata to forget these automatically loaded things and to force itself to get new copies of the ado-files from disk:

```
. discard
. hello
hi, Mary
```

Understand that you only had to type `discard` because you changed the ado-file while Stata was running. Had you exited Stata and returned later to use `hello`, the `discard` would not have been necessary because Stata forgets things between sessions, anyway.

18.11.4 Local subroutines

A single ado-file can contain more than one `program`, and if it does, the other programs defined in the ado-file are assumed to be subroutines of the main program. For example,

```
——————————————————————————————————— top of decoy.ado ——————————
program decoy
        ...
        duck ...
        ...
end
program  duck
        ...
end
——————————————————————————————————— end of decoy.ado ——————————
```

`duck` is considered a local subroutine of `decoy`. Even after `decoy.ado` was loaded, if you typed `duck`, you would be told "unrecognized command". To emphasize what *local* means, assume that you have also written an ado-file named `duck.ado`:

```
——————————————————————————————————— top of duck.ado ——————————
program duck
        ...
end
——————————————————————————————————— end of duck.ado ——————————
```

Even so, when `decoy` called `duck`, it would be the program `duck` defined in `decoy.ado` that was called. To further emphasize what *local* means, assume that `decoy.ado` contains

```
——————————————————————————————————— top of decoy.ado ——————————
program decoy
        ...
        manic ...
        ...
        duck ...
        ...
end
program duck
        ...
end
——————————————————————————————————— end of decoy.ado ——————————
```

and that `manic.ado` contained

```
——————————————————————————————— top of manic.ado ———————
    program manic
            ...
            duck ...
            ...
    end
——————————————————————————————— end of manic.ado ———————
```

Here is what would happen when you executed decoy:

1. `decoy` in `decoy.ado` would begin execution. `decoy` calls `manic`.

2. `manic` in `manic.ado` would begin execution. `manic` calls `duck`.

3. `duck` in `duck.ado` (yes) would begin execution. `duck` would do whatever and return.

4. `manic` regains control and eventually returns.

5. `decoy` is back in control. `decoy` calls `duck`.

6. `duck` in `decoy.ado` would execute, complete, and return.

7. `decoy` would regain control and return.

Note that, when `manic` called `duck`, it was the global ado-file `duck.ado` that was executed, yet when decoy called `duck`, it was the local program `duck` that was executed.

Stata does not find this confusing and neither should you.

18.11.5 Development of a sample ado-command

Below we demonstrate how to create a new Stata command. We will program an influence measure for use with linear regression. It is an interesting statistic in its own right, but even if you are not interested in linear regression and influence measures, the focus here is on programming, not on the particular statistic chosen.

Belsley, Kuh, and Welsch (1980, 24) present a measure of influence in linear regression defined as

$$\frac{\mathrm{Var}\left(\widehat{y}_i^{(i)}\right)}{\mathrm{Var}(\widehat{y}_i)}$$

which is the ratio of the variance of the ith fitted value based on regression estimates obtained by omitting the ith observation, to the variance of the ith fitted value estimated from the full dataset. This ratio is estimated using

$$\mathrm{FVARATIO}_i \equiv \frac{n-k}{n-(k+1)}\left\{1-\frac{d_i^2}{1-h_{ii}}\right\}(1-h_{ii})^{-1}$$

where n is the sample size; k is the number of estimated coefficients; $d_i^2 = e_i^2/\mathbf{e}'\mathbf{e}$ and e_i is the ith residual; and h_{ii} is the ith diagonal element of the hat matrix. The ingredients of this formula are all available through Stata, so, after estimating the regression parameters, we can easily calculate $\mathrm{FVARATIO}_i$. For instance, we might type

```
. regress mpg weight displ
. predict hii if e(sample), hat
. predict ei if e(sample), resid
. quietly count if e(sample)
. scalar nreg = r(N)
. gen eTe = sum(ei*ei)
. gen di2 = (ei*ei)/eTe[_N]
. gen FVi = (nreg - 3) / (nreg - 4) * (1 - di2/(1-hii)) / (1-hii)
```

The number 3 in the formula for `FVi` represents k, the number of estimated parameters (which is an intercept plus coefficients on `weight` and `displ`), and the number 4 represents $k + 1$.

❏ Technical Note

   Do you understand why this works? `predict` can create h_{ii} and e_i, but the trick is in getting $\mathbf{e'e}$—the sum of the squared e_is. Stata's `sum()` function creates a running sum. The first observation of `eTe` thus contains e_1^2; the second, $e_1^2 + e_2^2$; the third, $e_1^2 + e_2^2 + e_3^2$; and so on. The last observation, then, contains $\sum_{i=1}^{N} e_i^2$, which is $\mathbf{e'e}$. (We specified `if e(sample)` on our `predict` commands to restrict calculations to the estimation subsample, so `hii` and `eii` might have missing values, but that does not matter because `sum()` treats missing values as contributing zero to the sum.) We use Stata's explicit subscripting feature and then refer to `eTe[_N]`, the last observation. (See [U] **13.3 Functions** and [U] **13.7 Explicit subscripting**.) After that, we plug into the formula to obtain the result.

❏

   Assuming that we often wanted this influence measure, it would be easier and less prone to error if we canned this calculation in a program. Our first draft of the program reflects exactly what we would have typed interactively:

—————————————————————————————————— top of fvaratio.ado, version 1 ——————————

```
program fvaratio
        version 9
        predict hii if e(sample), hat
        predict ei if e(sample), resid
        quietly count if e(sample)
        scalar nreg = r(N)
        gen eTe = sum(ei*ei)
        gen di2 = (ei*ei)/eTe[_N]
        gen FVi = (nreg - 3) / (nreg - 4) * (1 - di2/(1-hii)) / (1-hii)
        drop hii ei eTe di2
end
```

——————————————————————————————————————— end of fvaratio.ado, version 1 ——————————

All we have done is to enter what we would have typed into a file, bracketing it with `program fvaratio` and `end`. Since our command is to be called `fvaratio`, the file must be named `fvaratio.ado` and must be stored in either the current directory or our personal ado directory (see [U] **17.5.2 Where is my personal ado directory?**).

   Now when we type `fvaratio`, Stata will be able to find it, load it, and execute it. In addition to copying the interactive lines into a program, we added the line '`drop hii ...`' to eliminate the working variables we had to create along the way.

   So, now we can interactively type

```
. regress mpg weight displ
. fvaratio
```

and add the new variable `FVi` to our data.

Our program is not general. It is suitable for use after fitting a regression model on two, and only two, independent variables because we coded a 3 in the formula for k. Stata statistical commands such as `regress` store information about the problem and answer in `e()`. Looking in *Saved Results* in [R] **regress**, we find that `e(df_m)` contains the model degrees of freedom, which is $k - 1$, assuming that the model has an intercept. Also, the sample size of the dataset used in the regression is stored in `e(N)`, eliminating our need to count the observations and define a scalar containing this count. Thus, the second draft of our program reads

───────────── top of fvaratio.ado, version 2 ─────────────
```
program fvaratio
        version 9
        predict hii if e(sample), hat
        predict ei if e(sample), resid
        gen eTe = sum(ei*ei)
        gen di2 = (ei*ei)/eTe[_N]
        gen FVi = (e(N)-(e(df_m)+1)) / (e(N)-(e(df_m)+2)) *      /// changed this
                  (1 - di2/(1-hii)) / (1-hii)                    // version
        drop hii ei eTe di2
end
```
───────────── end of fvaratio.ado, version 2 ─────────────

In the formula for `FVi`, we substituted `(e(df_m)+1)` for the literal number 3, `(e(df_m)+2)` for the literal number 4, and `e(N)` for the sample size.

Returning to the substance of our problem, `regress` also saves the residual sum of squares in `e(rss)`, so calculating `eTe` is not really necessary:

───────────── top of fvaratio.ado, version 3 ─────────────
```
program fvaratio
        version 9
        predict hii if e(sample), hat
        predict ei if e(sample), resid
        gen di2 = (ei*ei)/e(rss)                        // changed this version
        gen FVi = (e(N)-(e(df_m)+1)) / (e(N)-(e(df_m)+2)) *      ///
                  (1 - di2/(1-hii)) / (1-hii)
        drop hii ei di2
end
```
───────────── end of fvaratio.ado, version 3 ─────────────

Our program is now shorter and faster, and it is completely general. This program is probably good enough for most users; if you were implementing this solely for your own occasional use, you could stop right here. The program does, however, have the following deficiencies:

1. When we use it with data with missing values, the answer is correct, but we see messages about the number of missing values generated. (These messages appear when the program is generating the working variables.)

2. We cannot control the name of the variable being produced—it is always called `FVi`. Moreover, when `FVi` already exists (say from a previous regression), we get an error message that `FVi` already exists. We then have to `drop` the old `FVi` and type `fvaratio` again.

3. If we have created any variables named `hii`, `ei`, or `di2`, we also get an error that the variable already exists, and the program refuses to run.

Fixing these problems is not difficult. The fix for problem 1 is exceedingly easy; we embed the entire program in a `quietly` block:

```
———————————————————————————————— top of fvaratio.ado, version 4 ————————
program fvaratio
        version 9
        quietly {                                        // new this version
                predict hii if e(sample), hat
                predict ei if e(sample), resid
                gen di2 = (ei*ei)/e(rss)
                gen FVi = (e(N)-(e(df_m)+1)) / (e(N)-(e(df_m)+2)) *      ///
                        (1 - di2/(1-hii)) / (1-hii)
                drop hii ei di2
        }                                                // new this version
end
——————————————————————————————— end of fvaratio.ado, version 4 ————————
```

The output for the commands between the `quietly {` and `}` is now suppressed—the result is the same as if we had put `quietly` in front of each command.

Solving problem 2—that the resulting variable is always called `FVi`—requires use of the `syntax` command. Let's put that off and deal with problem 3—that the working variables have nice names like `hii`, `ei`, and `di2`, and so prevent users from using those names in their data.

One solution would be to change the nice names to unlikely names. We could change `hii` to `MyHiiVaR`, which would not guarantee the prevention of a conflict but would certainly make it unlikely. It would also make our program difficult to read, an important consideration should we want to change it in the future. There is a better solution. Stata's `tempvar` command (see [U] **18.7.1 Temporary variables**) places names into local macros that are guaranteed to be unique:

```
———————————————————————————————— top of fvaratio.ado, version 5 ————————
program fvaratio
        version 9
        tempvar hii ei di2                               // new this version
        quietly {
                predict `hii' if e(sample), hat    // changed, as are other lines
                predict `ei' if e(sample), resid
                gen `di2' = (`ei'*`ei')/e(rss)
                gen FVi = (e(N)-(e(df_m)+1)) / (e(N)-(e(df_m)+2)) *      ///
                        (1 - `di2'/(1-`hii')) / (1-`hii')
        }
end
——————————————————————————————— end of fvaratio.ado, version 5 ————————
```

At the top of our program, we declare the temporary variables. (We can do it outside or inside the `quietly`—it makes no difference—and we do not have to do it at the top or even all at once; we could declare them as we need them, but at the top is prettiest.) When we refer to a temporary variable, we do not refer directly to it (such as by typing `hii`), we refer to it indirectly by typing open and close single quotes around the name (`` `hii' ``). And at the end of our program, we no longer bother to `drop` the temporary variables—temporary variables are dropped automatically by Stata when a program concludes.

❏ Technical Note

Why do we type single quotes around the names? `tempvar` creates local macros containing the real temporary variable names. `hii` in our program is now a local macro, and `` `hii' `` refers to the contents of the local macro, which is the variable's actual name.

                                                     ❏

We now have an excellent program—its only fault is that we cannot specify the name of the new variable to be created. Here is the solution to that problem:

```
                                   ———————— top of fvaratio.ado, version 6 ————————
program fvaratio
        version 9
        syntax newvarname                              // new this version
        tempvar hii ei di2
        quietly {
                predict `hii' if e(sample), hat
                predict `ei' if e(sample), resid
                gen `di2' = (`ei'*`ei')/e(rss)
                gen `typlist' `varlist' = ///          changed this version
                        (e(N)-(e(df_m)+1)) / (e(N)-(e(df_m)+2)) *     ///
                        (1 - `di2'/(1-`hii')) / (1-`hii')
        }
end
                                   ———————— end of fvaratio.ado, version 6 ————————
```

It took a change to one line and the addition of another to obtain the solution. This magic all happens because of syntax (see [U] **18.4.4 Parsing standard Stata syntax** above).

'syntax newvarname' specifies that one new variable name must be specified (had we typed 'syntax [newvarname]', the new varname would have been optional; had we typed 'syntax newvarlist', the user would have been required to specify at least one new variable and allowed to specify more). In any case, syntax compares what the user types to what is allowed. If what the user types does not match what we have declared, syntax will issue the appropriate error message and stop our program. If it does match, our program will continue, and what the user typed will be broken out and stored in local macros for us. In the case of a newvarname, the new name typed by the user is placed in the local macro varlist, and the type of the variable (float, double, ...) is placed in typlist (even if the user did not specify a storage type, in which case, the type is the current default storage type).

This is now an excellent program. There are, however, two more improvements we could make. First, we have demonstrated that, by the use of 'syntax newvarname', we not only can allow the user to define the name of the created variable, but the storage type as well. However, when it comes to the creation of intermediate variables, such as 'hii' and 'di2', it is good programming practice to keep as much precision as possible. We want our final answer to be precise as possible, regardless of how we ultimately decide to store it. Any calculation that uses a previously generated variable would benefit if the previously generated variable were stored in double precision. Below we modify our program appropriately:

```
                                   ———————— top of fvaratio.ado, version 7 ————————
program fvaratio
        version 9
        syntax newvarname
        tempvar hii ei di2
        quietly {
                predict double `hii' if e(sample), hat     // changed, as are
                predict double `ei' if e(sample), resid    // other lines
                gen double `di2' = (`ei'*`ei')/e(rss)
                gen `typlist' `varlist' =  ///
                        (e(N)-(e(df_m)+1)) / (e(N)-(e(df_m)+2)) *     ///
                        (1 - `di2'/(1-`hii')) / (1-`hii')
        }
end
                                   ———————— end of fvaratio.ado, version 7 ————————
```

As for the second improvement we could make, `fvaratio` is intended to be used sometime after `regress`. How do we know the user is not misusing our program and executing it after, say, `logistic`? `e(cmd)` will tell us the name of the last estimation command; see [U] **18.9 Accessing results calculated by estimation commands** and [U] **18.10.2 Saving results in e()** above. We should change our program to read

```
─────────────────────────────────────── top of fvaratio.ado, version 9 ───────────
      program fvaratio
            version 9
            if "`e(cmd)'"!="regress" {                      // new this version
                  error 301
            }
            syntax newvarname
            tempvar hii ei di2
            quietly {
                  predict double `hii' if e(sample), hat
                  predict double `ei' if e(sample), resid
                  gen double `di2' = (`ei'*`ei')/e(rss)
                  gen `typlist' `varlist' = ///
                        (e(N)-(e(df_m)+1)) / (e(N)-(e(df_m)+2)) *       ///
                        (1 - `di2'/(1-`hii')) / (1-`hii')
            }
      end
─────────────────────────────────────── end of fvaratio.ado, version 8 ───────────
```

The `error` command issues one of Stata's prerecorded error messages and stops our program. Error 301 is "last estimates not found"; see [P] **error**. (Try typing `error 301` at the command line.)

In any case, this is a perfect program.

❏ Technical Note

You do not have to go to all the trouble we did to program the FVARATIO measure of influence or any other statistic that appeals to you. Whereas version 1 was not really an acceptable solution—it was too specialized—version 2 was acceptable. Version 3 was better, and version 4 better yet, but the improvements were of less and less importance.

Putting aside the details of Stata's language, you should understand that final versions of programs do not just happen—they are the results of drafts that have been refined. How much refinement depends on how often and who will be using the program. In this sense, the "official" ado-files that come with Stata are poor examples. They have been subject to substantial refinement because they will be used by strangers with no knowledge of how the code works. When writing programs for yourself, you may want to stop refining at an earlier draft.

❏

18.11.6   Writing online help

When you write an ado-file, you should also write a help file to go with it. This file is a standard ASCII text file, named *command*`.hlp`, that you place in the same directory as your ado-file *command*`.ado`. This way, when users type `help` followed by the name of your new command (or pulls down **Help**), they will see something better than "help for ... not found".

You can obtain examples of help files by examining the `.hlp` files in the official ado directory; type "`sysdir`" and look in the lettered subdirectories of the directory defined as `BASE`:

```
. sysdir
      STATA:  C:\Program Files\Stata9\
    UPDATES:  C:\Program Files\Stata9\ado\updates\
       BASE:  C:\Program Files\Stata9\ado\base\
       SITE:  C:\Program Files\Stata9\ado\site\
       PLUS:  C:\ado\plus\
   PERSONAL:  C:\ado\personal\
   OLDPLACE:  C:\ado\
```

In this case, you would find examples of `.hlp` files in the a, b, ... subdirectories of `C:\Program Files\Stata9\ado\base`.

Help files are physically written on the disk in ASCII text format, but their contents are SMCL—Stata Markup and Control Language. For the most part, you can ignore that. If the file contains a line that reads

```
Also see help for the finishup command
```

it will display in just that way. However, SMCL contains lots of special directives, so that if the line in the file were to read

```
Also see {hi:help} for the {help finishup} command
```

what would be displayed would be

Also see **help** for the **finishup** command

and moreover, **finishup** would appear as a hypertext link, meaning that if users clicked on it, they would see help on `finishup`.

You can read about the details of SMCL in [P] **smcl**. The following is a SMCL help file:

─── top of examplehelpfile.hlp ───────────

```
{smcl}
{* 04mar2005}{...}
{cmd:help whatever}
{hline}
{title:Title}

{p 4 8 2}
{bf:whatever -- Calculate whatever statistic}

{title:Syntax}

{p 8 17 2}
{cmdab:wh:atever}
[{it:varlist}]
[{it:weight}]
[{cmd:if} {it:exp}]
[{cmd:in} {it:range}]
[{cmd:,}
    {cmdab:d:etail}
    {cmdab:mean:only}
    {cmdab:median:only}
    {cmdab:gen:erate(}{it:newvar}{cmd:)}
]

{p 4 4 2}
{cmd:by} {it:...}{cmd::} may be used with {cmd:whatever}; see help {help by}.
{p 4 4 2}
{cmd:fweight}s are allowed; see help {help weights}.


{title:Description}
{p 4 4 2}
```

```
{cmd:whatever} calculates the whatever statistic for the variables in
{it:varlist} when the data are not stratified.


{title:Options}
{p 4 8 2}
{cmd:detail} presents detailed output of the calculation.
{p 4 8 2}
{cmd:meanonly} and {cmd:medianonly} restrict the calculation to be based on
only the means or medians.  The default is to use a trimmed mean.
{p 8 8 2}
{cmd:meanonly} is based on the idea of Rogers (1998) and should only
be specified when the data are known to be approximately symmetrically
distributed.
{p 8 8 2}
{cmd:medianonly} is a variation on meanonly that has not appeared in the
literature.  Use with caution.
{p 4 8 2}
{cmd:generate(}{it:newvar}{cmd:)} creates {it:newvar} containing the whatever
values.


{title:Remarks}
{p 4 4 2}
For detailed information on the whatever statistic, see Rogers (1998).


{title:Examples}
{p 4 8 2}{cmd:. whatever mpg weight}
{p 4 8 2}{cmd:. whatever mpg weight, meanonly}


{title:Author}
{p 4 4 2}
{browse "http://dbss.eul.edu/~hrogers":H. Rogers}, Department of Biostatistical
Social Science, Astronomy Building, Univ. of Equatorial London.  Email
{browse "mailto:hrogers@dbss.uel.edu":hrogers@dbss.eul.edu}
if you observe any problems.


{title:Also see}
{p 4 13 2}
Manual:  {hi:[U] 18.11.6 Writing online help}
{p 4 13 2}
Online:  help for {help help}, {help summarize}
```
———————————————————————————— end of examplehelpfile.hlp ————————

If you were to select **Help > Stata Command**, and type examplehelpfile and click **OK**, or if you were to type help examplehelpfile, this is what you would see:

help whatever
_____

Title

    whatever -- Calculate whatever statistic

Syntax

        whatever [varlist] [weight] [if exp] [in range] [, detail meanonly
            medianonly generate(newvar)]

    by ...: may be used with whatever; see help by.

fweights are allowed; see help **weights**.

Description

whatever calculates the whatever statistic for the variables in *varlist* when the data are not stratified.

Options

detail presents detailed output of the calculation.

meanonly and **medianonly** restrict the calculation to be based on only the means or medians. The default is to use a trimmed mean.

>**meanonly** is based on the idea of Rogers (1998) and should only be specified when the data are known to be approximately symmetrically distributed.

>**medianonly** is a variation on meanonly that has not appeared in the literature. Use with caution.

generate(*newvar*) creates *newvar* containing the whatever values.

Remarks

For detailed information on the whatever statistic, see Rogers (1998).

Examples

. **whatever mpg weight**

. **whatever mpg weight, meanonly**

Author

H. Rogers, Department of Biostatistical Social Science, Astronomy Building, Univ. of Equatorial London. Email **hrogers@dbss.eul.edu** if you observe any problems.

Also see

Manual: **[U] 18.11.6 Writing online help**
Online: help for **help**, **summarize**

Users will find it easier to understand your programs if you document them the same way that we document ours. We offer the following guidelines:

1. The first line must be

    {smcl}

    This notifies Stata that the help file is in SMCL format.

2. The second line should be

    {* *date*}{...}

    The * indicates a comment, and the {...} will suppress the blank line. Whenever you edit the help file, update the date found in the comment line.

3. The third line should contain

    {cmd:help *yourcmd*}
    {hline}

    assuming the help file is named *yourcmd*.hlp.

4. Include one blank line, and then place the title.

```
{title:Title}

{p 4 8 2}
{bf:yourcmd -- Your title}
```

5. Include two blank lines, and place the syntax title and diagram:

```
{title:Syntax}

{p 8 17 2}
syntax line

{p 8 17 2}
second syntax line, if necessary

{p 4 4 2}
clarifying text, if required
```

6. Include two blank lines, and place the Description title and text:

```
{title:Description}

{p 4 4 2}
description text
```

Provide a short description of what the command does. Do not burden the user with details yet. Assume that the user is at the point of asking whether this is what he or she is looking for.

7. If your command allows options, include two blank lines, and place the Options title and descriptions:

```
{title:Options}

{p 4 8 2}
{cmd:optionname} option description

{p 8 8 2}
continued option description, if necessary

{p 4 8 2}
{cmd:optionname} second option description
```

Options should be included in the order in which they appear in the syntax diagram. Option paragraphs are reverse indented, with the option name on the far left, where it is easily spotted. If an option requires more than one paragraph, subsequent paragraphs are set using {p 8 8 2}. One blank line separates one option from another.

8. Optionally include two blank lines, and place the Remarks title and text:

```
{title:Remarks}

{p 4 4 2}
text
```

Include whatever lengthy discussion you feel necessary. Stata's official online help files often omit this because the discussions appear in the manual. Stata's official help files for features added between releases (obtained from the *Stata Journal*, the Stata web site, etc.), however, include this because the appropriate SJ issue may not be as accessible as the manuals.

9. Optionally include two blank lines, and place the Examples title and text:

```
{title:Examples}

{p 4 8 2}
{cmd:. first example}

{p 4 8 2}
{cmd:. second example}
```

Nothing communicates better than providing something beyond theoretical discussion. Examples rarely need much explanation.

10. Optionally include two blank lines, and place the Author title and text:

```
{title:Author}

{p 4 4 2}
Name, affiliation, etc.
```

Exercise caution. If you include a telephone number, expect it to ring. An email address may be more appropriate.

11. Optionally include two blank lines, and place the References title and text:

```
{title:References}

{p 4 8 2}
Author. year.
Title. Location: Publisher.
```

12. Include two blank lines, and place the Also see title and text:

```
{title:Also see}

{p 4 13 2}
Manual: {hi:Manual reference, if any}

{p 4 13 2}
Online: help for {help name}, {help name}, ...
```

We also warn that it is easy to use too much {hi:highlighting}. Use it sparingly. In text, use {cmd:...} to show what would be shown in typewriter typeface it the documentation were printed in this manual.

❏ Technical Note

Sometimes it is more convenient to describe two or more related commands in the same .hlp file. Thus, xyz.hlp might document both the xyz and abc commands. To arrange that typing help abc displays xyz.hlp, create the file abc.hlp, containing

```
.h xyz
```

When a .hlp file contains a single line of the form '.h refname', Stata interprets that as an instruction to display help for refname.   ❏

❑ Technical Note

If you write a collection of programs, you need to somehow index the programs so that users (and you) can find the command they want. We do that with our `contents.hlp` entry. You should create a similar kind of entry. We suggest that you call your private entry `user.hlp` in your personal ado directory; see [U] **17.5.2 Where is my personal ado directory?**. This way, to review what you have added, you can type `help user`.

We suggest that Unix users at large sites also add `site.hlp` to the SITE directory (typically `/usr/local/ado`, but type `sysdir` to be sure). Then you can type `help site` for a list of the commands available site-wide.

❑

18.12   A compendium of useful commands for programmers

You can use literally any Stata command in your programs and ado-files. In addition, some commands are intended solely for use by Stata programmers. You should see the section under the *Programming* heading in the subject table of contents at the beginning of the *Quick Reference and Index*.

18.13   References

Belsley, D. A., E. Kuh, and R. E. Welsch. 1980. *Regression Diagnostics: Identifying Influential Data and Sources of Collinearity*. New York: Wiley.

Gould, W. 2001. pr0001: Statistical software certification. *The Stata Journal* 1(1): 29–50.

19 Immediate commands

Contents

19.1   Overview

An *immediate* command is a command that obtains data not from the data stored in memory, but from numbers typed as arguments. Immediate commands, in effect, turn Stata into a glorified hand calculator.

There are many instances when you may not have the data, but you do know something about the data, and what you know is adequate to perform statistical tests. For instance, you do not have to have individual-level data to obtain the standard error of the mean, and thereby a confidence interval, if you know the mean, standard deviation, and number of observations. In other instances, you may actually have the data, and you could enter the data and perform the test, but it would be easier if you could just ask for the statistic based on a summary. For instance, you flip a coin ten times, and it comes up heads twice. You could enter a ten-observation dataset with two ones (standing for heads) and eight zeros (meaning tails).

Immediate commands are meant to solve those problems. Immediate commands have the following properties:

1. They never disturb the data in memory. You can perform an immediate calculation as an aside without changing your data.

2. The syntax for these commands is the same, the command name followed by numbers, which are the summary statistics from which the statistic is calculated. The numbers are almost always summary statistics, and the order in which they are specified is in some sense "natural".

3. Immediate commands all end in the letter *i*, although the converse is not true. In most cases, if there is an immediate command, there is a nonimmediate form also, that is, a form that works on the data in memory. For every statistical command in Stata, we have included an immediate form if it is reasonable to assume that you might know the requisite summary statistics without having the underlying data, and if typing those statistics is not absurdly burdensome.

4. Immediate commands are documented along with their nonimmediate counterparts. Thus, if you want to obtain a confidence interval, whether it be from summary data with an immediate command or using the data in memory, use the table of contents or index to discover that [R] **ci** discusses confidence intervals. There, you learn that `ci` calculates confidence intervals using the data in memory, and that `cii` does the same with the data specified immediately following the command.

19.1.1   Examples

▷ Example 1

Let's take the example of confidence intervals. Professional papers often publish the mean, standard deviation, and number of observations for variables used in the analysis. Those statistics are sufficient for calculating a confidence interval. If we know that the mean mileage rating of cars in some sample is 24, the standard deviation is 6, and that there are 97 cars in the sample, we can calculate

```
. cii 97 24 6
```

Variable	Obs	Mean	Std. Err.	[95% Conf. Interval]
	97	24	.6092077	22.79073 25.20927

We learn that the mean's standard error is 0.61 and its 95% confidence interval is $[22.8, 25.2]$. To obtain this, we typed cii (the immediate form of the ci command) followed by the number of observations, the mean, and the standard deviation. We knew the order in which to specify the numbers because we had read [R] **ci**.

We could use the immediate form of the ttest command to test the hypothesis that the true mean is 22:

```
. ttesti 97 24 6 22
```
One-sample t test

	Obs	Mean	Std. Err.	Std. Dev.	[95% Conf. Interval]
x	97	24	.6092077	6	22.79073 25.20927

mean = mean(x)		t = 3.2830
Ho: mean = 22		degrees of freedom = 96
Ha: mean < 22	Ha: mean != 22	Ha: mean > 22
Pr(T < t) = 0.9993	Pr(|T| > |t|) = 0.0014	Pr(T > t) = 0.0007

The first three numbers were as we specified in the cii command. ttesti requires a fourth number, which is the constant against which the mean is being tested; see [R] **ttest**.

◁

▷ Example 2

We mentioned flipping a coin ten times and having it come up heads twice. The 99% confidence interval can also be obtained from ci:

```
. cii 10 2, level(99)
```

				— Binomial Exact —
Variable	Obs	Mean	Std. Err.	[99% Conf. Interval]
	10	.2	.1264911	.0108398 .6482422

In the previous example, we specified cii with three numbers following it; in this example, we specify 2. Immediate commands often determine what to do by the number of arguments following the command. With two arguments, ci assumes that we are specifying the number of trials and successes from a binomial experiment; see [R] **ci**.

The immediate form of the bitest command performs exact hypothesis testing:

```
. bitesti 10 2 .5
        N   Observed k   Expected k   Assumed p   Observed p

       10           2            5     0.50000      0.20000
Pr(k >= 2)              = 0.989258  (one-sided test)
Pr(k <= 2)              = 0.054688  (one-sided test)
Pr(k <= 2 or k >= 8) = 0.109375  (two-sided test)
```

For a full explanation of this output, see [R] **bitest**.

◁

▷ Example 3

Stata's `tabulate` command makes tables and calculates various measures of association. The immediate form, `tabi`, does the same, but we specify the contents of the table following the command:

```
. tabi 5 10 \ 2 14
            |        col
      row   |     1         2  |    Total

        1   |     5        10  |       15
        2   |     2        14  |       16

    Total   |     7        24  |       31
        Fisher's exact =                0.220
1-sided Fisher's exact =                0.170
```

The `tabi` command is slightly different from most immediate commands because it uses '\' to indicate where one row ends and another begins.

◁

19.1.2   A list of the immediate commands

Command	Reference	Description
bitesti	[R] **bitest**	Binomial probability test
cci csi iri mcci	[ST] **epitab**	Tables for epidemiologists
cii	[R] **ci**	Confidence intervals for means, proportions, and counts
prtesti	[R] **prtest**	One- and two-sample tests of proportions
sampsi	[R] **sampsi**	Sample size and power determination
sdtesti	[R] **sdtest**	Variance comparison tests
symmi	[R] **symmetry**	Symmetry and marginal homogeneity tests
tabi	[R] **tabulate twoway**	Two-way tables of frequencies
ttesti	[R] **ttest**	Mean comparison tests
twoway pci twoway pcarrowi twoway scatteri	[G] **graph twoway pci** [G] **graph twoway pcarrowi** [G] **graph twoway scatteri**	Paired-coordinate plot with spikes or lines Paired-coordinate plot with arrows Twoway scatterplot

19.2   The display command

display is not really an immediate command, but it can be used as a hand calculator.

```
. display 2+5
7
. display sqrt(2+sqrt(3^2-4*2*-2))/(2*3)
.44095855
```

See [R] **display**.

20 Estimation and postestimation commands

Contents

20.1 All estimation commands work the same way

All Stata commands that fit statistical models—commands such as `regress`, `logit`, `sureg`, and so on—work the same way. Most single-equation estimation commands have the syntax

$$command \; varlist \; \big[\,if\,\big] \; \big[\,in\,\big] \; \big[\,weight\,\big] \; \big[\,, \; options\,\big]$$

and most multiple-equation estimation commands have the syntax

$$command \; (varlist) \; (varlist) \; \dots \; (varlist) \; \big[\,if\,\big] \; \big[\,in\,\big] \; \big[\,weight\,\big] \; \big[\,, \; options\,\big]$$

Adopt a loose definition of single and multiple equation in interpreting this. For instance, `heckman` is a two-equation system, mathematically speaking, yet we categorize it, syntactically, with single-equation

251

commands because most researchers think of it as a linear regression with an adjustment for the censoring. The important thing is that most estimation commands have one or the other of these two syntaxes.

In single-equation commands, the first variable in the *varlist* is the dependent variable, and the remaining variables are the independent variables, with some exceptions. For instance, `anova` allows you to specify terms in addition to variables for the independent variables.

Prefix commands may be specified in front of an estimation command to modify what it does. The syntax is

$$prefix: command \ldots$$

where the prefix commands are

prefix command	description	manual entry
rolling	time-series rolling estimation	[TS] **rolling**
statsby	collect results across subsets of data	[D] **statsby**
*stepwise	stepwise estimation	[R] **stepwise**
*svy	estimation for complex survey data	[SVY] **svy**
xi	interaction expansion	[R] **xi**

* Available for some but not all estimation commands

Two other prefix commands—`bootstrap` and `jackknife`—also work with estimation commands—see [R] **bootstrap** and [R] **jackknife**—but in most cases it is easier to specify the estimation-command option `vce(bootstrap)` or `vce(jackknife)`.

In addition, all estimation commands—whether single or multiple equation—share the following features:

1. You can use the standard features of Stata's syntax—`if` *exp* and `in` *range*—to specify the estimation subsample; you do not have to make a special dataset.

2. You can retype the estimation command without arguments to redisplay the most recent estimation results. For instance, after fitting a model with `regress`, you can see the estimates again by typing `regress` by itself. You do not have to do this immediately—any number of commands can occur in between the estimation and the replaying, and, in fact, you can even replay the last estimates after the data have changed or you have dropped the data altogether. Stata never forgets (unless you type `discard`; see [P] **discard**).

3. You can specify option `level()` at the time of estimation, or when you redisplay results, to specify the width of the confidence intervals for the coefficients. The default is `level(95)`, meaning 95% confidence intervals. You can reset the default with `set level`; see [R] **level**.

4. You can use the postestimation command `mfx` to display model results in terms of marginal effects (dy/dx or even $df(y)/dx$), which can be displayed as either derivatives or elasticities; see [R] **mfx**.

5. You can use the postestimation command `adjust` to obtain tables of adjusted means; see [R] **adjust**.

6. You can use the postestimation command `estat` to obtain common statistics associated with the model. What statistics are available are documented in the postestimation section following the documentation of the estimation command, for instance, in [R] **regress postestimation** following [R] **regress**.

   In all cases, you can the use postestimation command `estat vce` to obtain the variance–covariance matrix of the estimators (VCE), presented as either a correlation matrix or a covariance matrix. (You can also obtain the estimated coefficients and covariance matrix as vectors and matrices and manipulate them with Stata's matrix capabilities; see [U] **14.5 Accessing matrices created by Stata commands**.)

7. You can use the postestimation command `predict` to obtain predictions, residuals, influence statistics, and the like, either for the data on which you just estimated or for some other data. You can use postestimation command `predictnl` to obtain point estimates, standard errors, etc., for customized predictions.

8. You can refer to the values of coefficients and standard errors in expressions (such as with `generate`) using standard notation; see [U] **13.5 Accessing coefficients and standard errors**. You can refer in expressions to the values of other estimation-related statistics using `e(`*resultname*`)`. For instance, all commands define `e(N)` recording the number of observations in the estimation subsample. After estimation, type command `ereturn list` to see a list of all that is available. See the *Saved results* section in the estimation command's documentation for their definition.

   An especially useful `e()` result is `e(sample)`: it returns 1 if an observation was used in the estimation and 0 otherwise, so you can add `if e(sample)` to the end of other commands to restrict them to the estimation subsample. You could type, for instance, `summarize if e(sample)`.

9. You can use the postestimation command `test` to perform tests on the estimated parameters (Wald tests of linear hypotheses), `testnl` to perform Wald tests of nonlinear hypotheses, and (`lrtest` to perform likelihood-ratio tests. You can use the postestimation command `lincom` to obtain point estimates and confidence intervals for linear combinations of the estimated parameters, and the postestimation command `nlcom` to obtain nonlinear combinations.

10. You can use the postestimation command `estimates` to store estimation results by name for later retrieval or for displaying/comparing multiple models using `estimates`; see [R] **estimates**.

11. You can use the postestimation command `_estimates` to hold estimates, perform other estimation commands, and then restore the prior estimates. This is of particular interest to programmers.

12. You can use the postestimation command `suest` to obtain the joint parameter vector and variance–covariance matrix for coefficients from two different models using seemingly unrelated estimation. This is especially useful for testing the equality, say, of coefficients across models; see [R] **suest**.

13. You can use the postestimation command `hausman` to perform Hausman model-specification tests using `hausman`; see [R] **hausman**.

14. With some exceptions, you can specify option `robust` at the time of estimation to obtain the Huber/White/robust alternate estimate of variance, and you can specify option `cluster()` to relax the assumption of independence of the observations.

   Also, many estimation commands also allow a `vce(`*vcetype*`)` option to specify other alternative variance estimators—which ones are allowed are documented with the estimator—and usually `vce(opg)`, `vce(bootstrap)` and `vce(jackknife)` are available.

20.2   Standard syntax

You can combine Stata's `if` *exp* and `in` *range* with any estimation command. Estimation commands also allow `by` *varlist:*, where it would be sensible.

▷ Example 1

We have data on 74 automobiles that record the mileage rating (`mpg`), weight (`weight`), and whether the car is domestic or foreign-produced (`foreign`). We can fit a linear regression model of `mpg` on `weight` and `weightsq`, using just the foreign-made automobiles, by typing

```
. use http://www.stata-press.com/data/r9/auto2
(1978 Automobile Data)
. regress mpg weight weightsq if foreign
```

Source	SS	df	MS		Number of obs =	22
					F(2, 19) =	8.31
Model	428.256889	2	214.128444		Prob > F =	0.0026
Residual	489.606747	19	25.7687762		R-squared =	0.4666
					Adj R-squared =	0.4104
Total	917.863636	21	43.7077922		Root MSE =	5.0763

mpg	Coef.	Std. Err.	t	P>|t|	[95% Conf.	Interval]
weight	-.0132182	.0275711	-0.48	0.637	-.0709252	.0444888
weightsq	5.50e-07	5.41e-06	0.10	0.920	-.0000108	.0000119
_cons	52.33775	34.1539	1.53	0.142	-19.14719	123.8227

We can run separate regressions for the domestic and foreign-produced automobiles with the `by` *varlist:* prefix:

```
. by foreign: regress mpg weight weightsq
```

-> foreign = Domestic

Source	SS	df	MS		
Model	905.395466	2	452.697733		
Residual	242.046842	49	4.93973146		
Total	1147.44231	51	22.4988688		

Number of obs = 52
F(2, 49) = 91.64
Prob > F = 0.0000
R-squared = 0.7891
Adj R-squared = 0.7804
Root MSE = 2.2226

mpg	Coef.	Std. Err.	t	P>|t|	[95% Conf. Interval]
weight	-.0131718	.0032307	-4.08	0.000	-.0196642 -.0066794
weightsq	1.11e-06	4.95e-07	2.25	0.029	1.19e-07 2.11e-06
_cons	50.74551	5.162014	9.83	0.000	40.37205 61.11896

-> foreign = Foreign

Source	SS	df	MS		
Model	428.256889	2	214.128444		
Residual	489.606747	19	25.7687762		
Total	917.863636	21	43.7077922		

Number of obs = 22
F(2, 19) = 8.31
Prob > F = 0.0026
R-squared = 0.4666
Adj R-squared = 0.4104
Root MSE = 5.0763

mpg	Coef.	Std. Err.	t	P>|t|	[95% Conf. Interval]
weight	-.0132182	.0275711	-0.48	0.637	-.0709252 .0444888
weightsq	5.50e-07	5.41e-06	0.10	0.920	-.0000108 .0000119
_cons	52.33775	34.1539	1.53	0.142	-19.14719 123.8227

Although all estimation commands allow if *exp* and in *range*, only some allow the by *varlist*: prefix. In the case of by(), the duration of Stata's memory is limited: it remembers the *last* set of estimates only. This means that, if we were to use any of the other features described below, they would use the last regression estimated, which right now is mpg on weight and weightsq for the Foreign subsample.

◁

20.3 Replaying prior results

When you type an estimation command without arguments, it redisplays prior results.

▷ Example 2

To perform a regression of mpg on the variables weight and displacement, we could type

(Continued on next page)

```
. regress mpg weight displacement

      Source |       SS       df       MS              Number of obs =      74
-------------+------------------------------          F(  2,     71) =   66.79
       Model | 1595.40969       2  797.704846          Prob > F      =  0.0000
    Residual | 848.049768      71  11.9443629          R-squared     =  0.6529
-------------+------------------------------          Adj R-squared =  0.6432
       Total | 2443.45946      73  33.4720474          Root MSE      =  3.4561

------------------------------------------------------------------------------
         mpg |      Coef.   Std. Err.      t    P>|t|     [95% Conf. Interval]
-------------+----------------------------------------------------------------
      weight |  -.0065671   .0011662    -5.63   0.000    -.0088925   -.0042417
displacement |   .0052808   .0098696     0.54   0.594    -.0143986    .0249602
       _cons |   40.08452    2.02011    19.84   0.000     36.05654    44.11251
------------------------------------------------------------------------------
```

We now go on to do other things, summarizing data, listing observations, performing hypothesis tests, or anything else. If we decide that we want to see the last set of estimates again, we type the estimation command without arguments.

```
. regress

      Source |       SS       df       MS              Number of obs =      74
-------------+------------------------------          F(  2,     71) =   66.79
       Model | 1595.40969       2  797.704846          Prob > F      =  0.0000
    Residual | 848.049768      71  11.9443629          R-squared     =  0.6529
-------------+------------------------------          Adj R-squared =  0.6432
       Total | 2443.45946      73  33.4720474          Root MSE      =  3.4561

------------------------------------------------------------------------------
         mpg |      Coef.   Std. Err.      t    P>|t|     [95% Conf. Interval]
-------------+----------------------------------------------------------------
      weight |  -.0065671   .0011662    -5.63   0.000    -.0088925   -.0042417
displacement |   .0052808   .0098696     0.54   0.594    -.0143986    .0249602
       _cons |   40.08452    2.02011    19.84   0.000     36.05654    44.11251
------------------------------------------------------------------------------
```

This feature works with *every* estimation command, so we could just as well have done it with, say, stcox or logit.

◁

20.4 Cataloging estimation results

Stata keeps only the results of the most-recently fitted model in active memory. You may, however, use Stata's estimates command to store estimation results by name for displaying, comparison, cross-model testing, etc., later. You may store up to twenty sets of estimation results.

▷ Example 3

Continuing with our automobile data, let's fit three models and store the results. We fit the models quietly so as to keep the output to a minimum.

```
. quietly regress mpg weight displ
. estimates store reg_model
. quietly logit foreign weight displ
. estimates store logit_model
. quietly ologit rep78 price weight gear_ratio
. estimates store ologit_model
```

We can now obtain a directory (listing) of our stored results, select a set to replay, and display a table comparing the estimated parameters and other model-summary statistics.

```
. estimates dir
```

model	command	depvar	npar	title
reg_model	regress	mpg	3	
logit_model	logit	foreign	3	
ologit_model	ologit	rep78	7	

```
. estimates replay logit_model
```

Model **logit_model**

Logistic regression	Number of obs	=	74
	LR chi2(2)	=	44.72
	Prob > chi2	=	0.0000
Log likelihood = -22.672104	Pseudo R2	=	0.4965

foreign	Coef.	Std. Err.	z	P>|z|	[95% Conf. Interval]	
weight	.0019774	.0019023	1.04	0.299	-.001751	.0057059
displacement	-.0651206	.0291647	-2.23	0.026	-.1222824	-.0079588
_cons	3.369612	2.104804	1.60	0.109	-.7557268	7.494952

```
. estimates table _all, stats(aic bic)
```

Variable	reg_model	logit_mo~l	ologit_m~l
weight	-.00656711	.00197745	
displacement	.00528078	-.06512057	
_cons	40.084522	3.3696124	
rep78			
price			.00020787
weight			-.00105961
gear_ratio			.82064832
cut1			
_cons			-3.3407627
cut2			
_cons			-1.5486078
cut3			
_cons			.88460924
cut4			
_cons			2.5450833
Statistics			
aic	396.4796	51.344208	183.22888
bic	403.39179	58.256404	198.86762

We can also select a set of results to be made active for use with Stata's postestimation commands.

```
. estimates restore logit_model
(results logit_model are active now)
```

```
. logit
Logistic regression                              Number of obs  =        74
                                                 LR chi2(2)     =     44.72
                                                 Prob > chi2    =    0.0000
Log likelihood = -22.672104                      Pseudo R2      =    0.4965
```

foreign	Coef.	Std. Err.	z	P>|z|	[95% Conf. Interval]	
weight	.0019774	.0019023	1.04	0.299	-.001751	.0057059
displacement	-.0651206	.0291647	-2.23	0.026	-.1222824	-.0079588
_cons	3.369612	2.104804	1.60	0.109	-.7557268	7.494952

◁

You can do a lot more with estimates; see [R] **estimates**. In particular, estimates makes it easy to perform cross-model tests, such as the Hausman specification test.

20.5 Specifying the estimation subsample

You specify the estimation subsample—the sample to be used in estimation—by specifying the if *exp* and/or in *range* modifiers with the estimation command.

Once an estimation command has been run or previous estimates restored, Stata remembers the estimation subsample, and you can use the modifier if e(sample) on the end of other Stata commands. The term *estimation subsample* refers to the set of observations used to produce the active estimation results. That might turn out to be all the observations (as it was in the above example) or a subset of the observations:

```
. generate excellent = rep78==5 if rep78 < .
(5 missing values generated)
. regress mpg weight excellent if foreign
```

Source	SS	df	MS		Number of obs =	21
					F(2, 18) =	10.21
Model	423.317154	2	211.658577		Prob > F =	0.0011
Residual	372.96856	18	20.7204756		R-squared =	0.5316
					Adj R-squared =	0.4796
Total	796.285714	20	39.8142857		Root MSE =	4.552

mpg	Coef.	Std. Err.	t	P>|t|	[95% Conf. Interval]	
weight	-.0131402	.0029684	-4.43	0.000	-.0193765	-.0069038
excellent	5.052676	2.13492	2.37	0.029	.5673764	9.537977
_cons	52.86088	6.540147	8.08	0.000	39.12054	66.60122

```
. summarize mpg weight excellent if e(sample)
```

Variable	Obs	Mean	Std. Dev.	Min	Max
mpg	21	25.28571	6.309856	17	41
weight	21	2263.333	364.7099	1760	3170
excellent	21	.4285714	.5070926	0	1

Note that 21 observations were used in the above regression and we subsequently obtained the means for those same 21 observations by typing summarize ... if e(sample). There are two reasons observations were dropped: we specified if foreign when we ran the regression, and there were observations for which excellent was missing. The reason does not matter; e(sample) is true if the observation was used and false otherwise.

You can use if e(sample) on the end of any Stata command that allows an if *exp*.

In this case, Stata has a shorthand command that produces the same results as summarize . . . if e(sample):

```
. estat summarize
Estimation sample regress                          Number of obs =      21
```

Variable	Mean	Std. Dev.	Min	Max	Label
mpg	25.28571	6.309856	17	41	Mileage (mpg)
weight	2263.333	364.7099	1760	3170	Weight (lbs.)
excellent	.4285714	.5070926	0	1	

See [R] **estat**.

20.6   Specifying the width of confidence intervals

You can specify the width of the confidence intervals for the coefficients using the level() option at estimation or when you play back the results.

▷ Example 4

To obtain narrower, 90% confidence intervals when we fit the model, we type

```
. regress mpg weight displ, level(90)
```

Source	SS	df	MS		Number of obs =	74
					F(2, 71) =	66.79
Model	1595.40969	2	797.704846		Prob > F =	0.0000
Residual	848.049768	71	11.9443629		R-squared =	0.6529
					Adj R-squared =	0.6432
Total	2443.45946	73	33.4720474		Root MSE =	3.4561

mpg	Coef.	Std. Err.	t	P>|t|	[90% Conf. Interval]	
weight	-.0065671	.0011662	-5.63	0.000	-.0085108	-.0046234
displacement	.0052808	.0098696	0.54	0.594	-.0111679	.0217294
_cons	40.08452	2.02011	19.84	0.000	36.71781	43.45124

If we subsequently typed regress, without arguments, 95% confidence intervals would be reported. If we initially fitted the model with 95% confidence intervals, we could later type regress, level(90) to redisplay results with 90% confidence intervals.

In addition, we could type set level 90 to make 90% intervals our default; see [R] **level**.

Stata allows noninteger confidence intervals between 10.00 and 99.99, with a maximum of two digits following the decimal point. For instance, we could type

```
. regress mpg weight displ, level(92.5)
```

Source	SS	df	MS		Number of obs =	74
					F(2, 71) =	66.79
Model	1595.40969	2	797.704846		Prob > F =	0.0000
Residual	848.049768	71	11.9443629		R-squared =	0.6529
					Adj R-squared =	0.6432
Total	2443.45946	73	33.4720474		Root MSE =	3.4561

mpg	Coef.	Std. Err.	t	P>|t|	[92.5% Conf. Interval]	
weight	-.0065671	.0011662	-5.63	0.000	-.0086745	-.0044597
displacement	.0052808	.0098696	0.54	0.594	-.0125535	.023115
_cons	40.08452	2.02011	19.84	0.000	36.43419	43.73485

◁

20.7   Obtaining the variance–covariance matrix

Typing estat vce displays the variance–covariance matrix of the estimators in active memory.

▷ Example 5

In example 2, we typed regress mpg weight displacement. The full variance–covariance matrix of the estimators can be displayed at any time after estimation:

```
. estat vce
Covariance matrix of coefficients of regress model
        e(V) |     weight  displace~t      _cons
-------------+-----------------------------------
      weight |  1.360e-06
displacement |  -.0000103   .00009741
       _cons |  -.00207455  .01188356  4.0808455
```

Typing estat vce with the corr option presents this matrix as a correlation matrix:

```
. estat vce, corr
Correlation matrix of coefficients of regress model
        e(V) |     weight   displa~t      _cons
-------------+----------------------------------
      weight |     1.0000
displacement |    -0.8949     1.0000
       _cons |    -0.8806     0.5960     1.0000
```

See [R] **estat**.

In addition, Stata's matrix commands understand that e(V) refers to the matrix:

```
. matrix list e(V)
symmetric e(V)[3,3]
                    weight  displacement       _cons
      weight     1.360e-06
displacement     -.0000103     .00009741
       _cons    -.00207455     .01188356   4.0808455
. mat Vinv = invsym(e(V))
. mat list Vinv
symmetric Vinv[3,3]
                    weight  displacement       _cons
      weight      60175851
displacement     4081161.2      292709.46
       _cons     18706.732      1222.3339   6.1953911
```

See [U] **14.5 Accessing matrices created by Stata commands**.

◁

20.8 Obtaining predicted values

Our discussion below, while cast in terms of predicted values, applies equally to the other statistics generated by predict; see [R] **predict**.

When Stata fits a model, whether it is regression or anything else, it internally saves the results, including the estimated coefficients and the variable names. The predict command allows you to use that information.

▷ Example 6

Let's perform a linear regression of mpg on weight and weightsq:

```
. regress mpg weight weightsq
```

Source	SS	df	MS		Number of obs =	74
					F(2, 71) =	72.80
Model	1642.52197	2	821.260986		Prob > F =	0.0000
Residual	800.937487	71	11.2808097		R-squared =	0.6722
					Adj R-squared =	0.6630
Total	2443.45946	73	33.4720474		Root MSE =	3.3587

mpg	Coef.	Std. Err.	t	P>|t|	[95% Conf. Interval]	
weight	-.0141581	.0038835	-3.65	0.001	-.0219016	-.0064145
weightsq	1.32e-06	6.26e-07	2.12	0.038	7.67e-08	2.57e-06
_cons	51.18308	5.767884	8.87	0.000	39.68225	62.68392

After the regression, predict is defined to be

$$-.0141581\texttt{weight} + 1.32 \cdot 10^{-6}\texttt{weightsq} + 51.18308$$

(Actually, it is more precise because the coefficients are internally stored at much higher precision than shown in the output.) Thus we can create a new variable—call it fitted—equal to the prediction by typing predict fitted and then use scatter to display the fitted and actual values separately for domestic and foreign automobiles:

```
. predict fitted
(option xb assumed; fitted values)
. scatter mpg fitted weight, by(foreign, total) c(. l) m(o i) sort
```

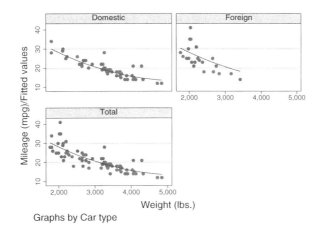

Graphs by Car type

predict can calculate much more than just predicted values. In the case of predict after linear regression, predict can calculate residuals, standardized residuals, studentized residuals, influence statistics, etc. In any case, we specify what is to be calculated via an option, so if we wanted the residuals stored in new variable r, we would type

```
. predict r, resid
```

The options that may be specified following predict vary according to the estimation command previously used; the predict options are documented along with the estimation command. For instance, to discover all the things predict can do following regress, see [R] **regress**.

◁

20.8.1   Using predict

The use of predict is not limited to linear regression. predict can be used after any estimation command.

▷ Example 7

You fit a logistic regression model of whether a car is manufactured outside the United States based on its weight and mileage rating using either the logistic or the logit command; see [R] **logistic** and [R] **logit**. We will use logit.

```
. use http://www.stata-press.com/data/r9/auto

. logit foreign weight mpg

Iteration 0:   log likelihood =  -45.03321
Iteration 1:   log likelihood = -29.898968
Iteration 2:   log likelihood = -27.495771
Iteration 3:   log likelihood = -27.184006
Iteration 4:   log likelihood = -27.175166
Iteration 5:   log likelihood = -27.175156

Logistic regression                            Number of obs   =         74
                                               LR chi2(2)      =      35.72
                                               Prob > chi2     =     0.0000
Log likelihood = -27.175156                    Pseudo R2       =     0.3966
```

foreign	Coef.	Std. Err.	z	P>|z|	[95% Conf. Interval]	
weight	-.0039067	.0010116	-3.86	0.000	-.0058894	-.001924
mpg	-.1685869	.0919174	-1.83	0.067	-.3487418	.011568
_cons	13.70837	4.518707	3.03	0.002	4.851864	22.56487

After logit, predict without options calculates the probability of a positive outcome (we learned that by looking at [R] **logit**). To obtain the predicted probabilities that each car is manufactured outside the U.S., we type

```
. predict probhat
(option p assumed; Pr(foreign))

. summarize probhat
```

Variable	Obs	Mean	Std. Dev.	Min	Max
probhat	74	.2972973	.3052979	.000729	.8980594

```
. list make mpg weight foreign probhat in 1/5
```

	make	mpg	weight	foreign	probhat
1.	AMC Concord	22	2,930	Domestic	.1904363
2.	AMC Pacer	17	3,350	Domestic	.0957767
3.	AMC Spirit	22	2,640	Domestic	.4220815
4.	Buick Century	20	3,250	Domestic	.0862625
5.	Buick Electra	15	4,080	Domestic	.0084948

◁

20.8.2   Making in-sample predictions

predict does not retrieve a vector of prerecorded values—it calculates the predictions based on the recorded coefficients and the data currently in memory. In the above examples, when we typed things like

```
. predict probhat
```

predict filled in the prediction everywhere that it could be calculated.

Sometimes we have more data in memory than were used by the estimation command, either because we explicitly ignored some of the observations by specifying an if *exp* with the estimation command or because there are missing values. In such cases, if we want to restrict the calculation to the estimation subsample, we would do that in the usual way by adding if e(sample) to the end of the command:

```
. predict probhat if e(sample)
```

20.8.3   Making out-of-sample predictions

Because predict makes its calculations based on the recorded coefficients and the data in memory, predict can do more than calculate predicted values for the data on which the estimation took place—it can make out-of-sample predictions, as well.

If you fit your model on a subset of the observations, you could then predict the outcome for all the observations:

```
. logit foreign weight mpg if rep78>3
. predict pall
```

If you do not specify if e(sample) at the end of the predict command, predict calculates the predictions for all observations possible.

In fact, because predict works from the active estimation results, you can use predict with *any* dataset that contains the necessary variables.

▷ Example 8

Continuing with our previous logit example, assume that we have a second dataset containing the mpg and weight of a different sample of cars. We have just fitted your model and now continue:

```
. use otherdat, clear
(Different cars)
```

```
. predict probhat                   Stata remembers the previous model
(option p assumed; Pr(foreign))
. summarize probhat foreign
```

Variable	Obs	Mean	Std. Dev.	Min	Max
probhat	12	.2505068	.3187104	.0084948	.8920776
foreign	12	.1666667	.3892495	0	1

◁

▷ Example 9

There are numerous ways to obtain out-of-sample predictions. Above, we estimated on one dataset and then used another. If our first dataset had contained both sets of cars, marked, say, by the variable difcars being 0 if from the first sample and 1 if from the second, we could type

```
. logit foreign weight mpg if difcars==0
same output as above appears

. predict probhat
(option p assumed; Pr(foreign))

. summarize probhat foreign if difcars==1
same output as directly above appears
```

If we just had a small number of additional cars, we could even input them after estimation. Assume that our data once again contain only the first sample of cars, and assume that we are interested in an additional sample of only 2 rather than 12 cars; we could type

```
. use http://www.stata-press.com/data/r9/auto

. keep make mpg weight foreign

. logit foreign weight mpg
same output as above appears

. input
                    make      mpg    weight    foreign
 75. "Merc. Zephyr" 20 2830 0            we type in our new data
 76. "VW Dasher" 23 2160 1
 77. end

. predict probhat                       obtain all the predictions
(option p assumed; Pr(foreign))

. list in -2/l
```

	make	mpg	weight	foreign	probhat
75.	Merc. Zephyr	20	2830	Domestic	.3275397
76.	VW Dasher	23	2160	Foreign	.8009743

◁

20.8.4   Obtaining standard errors, tests, and confidence intervals for predictions

When you use predict, you create, for each observation in the prediction sample, a statistic that is a function of the data and the estimated model parameters. You also could have generated your own customized predictions using generate. In either case, to get standard errors, Wald tests, and confidence intervals for your predictions, use predictnl. For example, if we wanted the standard errors for our predicted probabilities, we could type

```
. drop probhat
. predictnl probhat = predict(), se(phat_se)
. list in 1/5
```

	make	mpg	weight	foreign	probhat	phat_se
1.	AMC Concord	22	2,930	Domestic	.1904363	.0658386
2.	AMC Pacer	17	3,350	Domestic	.0957767	.0536296
3.	AMC Spirit	22	2,640	Domestic	.4220815	.0892845
4.	Buick Century	20	3,250	Domestic	.0862625	.0461927
5.	Buick Electra	15	4,080	Domestic	.0084948	.0093079

Comparing this output to our previous listing of the first five predicted probabilities, you will notice that the output is identical, except that we now have an additional variable, phat_se, which contains the estimated standard error for each predicted probability.

Note that we first had to drop probhat since predictnl will regenerate it. Note also the use of predict() within predictnl—it specified that we wanted to generate a point estimate (and standard error) for the default prediction after logit; see [R] **predictnl** for more details.

20.9   Accessing estimated coefficients

You can access coefficients and standard errors after estimation by referring to _b[*name*] and _se[*name*]; see [U] **13.5 Accessing coefficients and standard errors**.

▷ Example 10

Let us return to linear regression. We are doing a study of earnings of men and women at a particular company. In addition to each person's earnings, we have information on their educational attainment and tenure with the company. We type the following:

```
. generate femten = female*tenure
. generate femed = female*ed
. regress lnearn ed tenure female femed femten
  (output omitted )
```

We now wish to predict everyone's income as if they were male and then compare these as-if earnings with the actual earnings:

```
. generate asif = _b[_cons] + _b[ed]*ed + _b[tenure]*tenure
```

◁

▷ Example 11

We are analyzing the mileage of automobiles and are using a slightly more sophisticated model than any we have used so far. As we have previously, we assume that mpg is a function of weight and weightsq, but we also add the interaction of foreign multiplied by weight (called fweight), the car's gear ratio (gear_ratio), and foreign multiplied by gear_ratio (fgear_ratio).

```
. use http://www.stata-press.com/data/r9/auto2
(1978 Automobile Data)
. generate fweight = foreign*weight
```

```
. generate fgear_ratio = foreign*gear_ratio
. regress mpg weight weightsq fweight gear_ratio fgear_ratio
```

Source	SS	df	MS		
Model	1737.05293	5	347.410585	Number of obs =	74
Residual	706.406534	68	10.3883314	F(5, 68) =	33.44
				Prob > F =	0.0000
				R-squared =	0.7109
				Adj R-squared =	0.6896
Total	2443.45946	73	33.4720474	Root MSE =	3.2231

mpg	Coef.	Std. Err.	t	P>|t|	[95% Conf. Interval]	
weight	-.0118517	.0045136	-2.63	0.011	-.0208584	-.002845
weightsq	9.81e-07	7.04e-07	1.39	0.168	-4.25e-07	2.39e-06
fweight	-.0032241	.0015577	-2.07	0.042	-.0063326	-.0001157
gear_ratio	1.159741	1.553418	0.75	0.458	-1.940057	4.259539
fgear_ratio	1.597462	1.205313	1.33	0.189	-.8077036	4.002627
_cons	44.61644	8.387943	5.32	0.000	27.87856	61.35432

If you are not experienced in both regression technology and automobile technology, you may find it difficult to interpret this regression. Putting aside issues of statistical significance, we find that mileage decreases with a car's weight but increases with the square of weight; decreases even more rapidly with weight for foreign cars; increases with higher gear ratio; and increases even more rapidly with higher gear ratio in foreign cars.

Thus, do foreign cars yield better or worse gas mileage? Results are mixed. As the foreign cars' weight increases, they do more poorly in relation to domestic cars, but they do better at higher gear ratios. One way to compare the results is to predict what mileage foreign cars would have *if they were manufactured domestically*. The regression provides all the information necessary for making that calculation; mileage for domestic cars is estimated to be

$$-.012\,\text{weight} + 9.81 \cdot 10^{-7}\,\text{weightsq} + 1.160\,\text{gear\_ratio} + 44.6$$

We can use that equation to predict the mileage of foreign cars and then compare it with the true outcome. The _b[] function simplifies reference to the estimated coefficients. We can type

```
. gen asif=_b[weight]*weight + _b[weightsq]*weightsq +
        _b[gear_ratio]*gear_ratio + _b[_cons]
```

_b[weight] refers to the estimated coefficient on weight, _b[weightsq] to the estimated coefficient on weightsq, and so on.

We might now ask how the actual mileage of a Honda compares with the asif prediction:

```
. list make asif mpg if strpos(make,"Honda")
```

	make	asif	mpg
61.	Honda Accord	26.52597	25
62.	Honda Civic	30.62202	28

Notice the way we constructed our if clause to select Hondas. strpos() is the string function that returns the location in the first string where the second string is found or, if the second string does not occur in the first, zero. Thus, any recorded make that contains the string "Honda" anywhere in it would be listed; see [D] **functions**.

We find that both Honda models yield slightly lower gas mileage than the `asif` domestic-car-based prediction. (Please note that we do not endorse this model as a complete model of the determinants of mileage, nor do we single out the Honda for any special scorn. In fact, please note that the observed values are within the root mean squared error of the average prediction.)

We might wish to compare the overall average `mpg` and the `asif` prediction over all foreign cars in the data:

```
. summarize mpg asif if foreign

    Variable │      Obs        Mean    Std. Dev.        Min         Max
─────────────┼─────────────────────────────────────────────────────────
         mpg │       22    24.77273    6.611187         14          41
        asif │       22    26.67124    3.142912   19.70466    30.62202
```

We find that, on average, foreign cars yield slightly lower mileage than our `asif` prediction. This might lead us to ask if any foreign cars do better than the `asif` prediction:

```
. list make asif mpg if foreign & mpg>asif, sep(0)

        ┌──────────────────────────────────┐
        │ make                 asif     mpg │
        ├──────────────────────────────────┤
    55. │ BMW 320i          24.31697     25 │
    57. │ Datsun 210        28.96818     35 │
    63. │ Mazda GLC         29.32015     30 │
    66. │ Subaru            28.85993     35 │
    68. │ Toyota Corolla    27.01144     31 │
    71. │ VW Diesel         28.90355     41 │
        └──────────────────────────────────┘
```

We find six such automobiles.

◁

20.10 Performing hypothesis tests on the coefficients

20.10.1 Linear tests

After estimation, `test` is used to perform tests of linear hypotheses based on the variance–covariance matrix of the estimators (Wald tests).

▷ Example 12

(`test` has numerous syntaxes and features, so do not use this example as an excuse for not reading [R] **test**.) Using the automobile data, we perform the following regression:

```
. use http://www.stata-press.com/data/r9/auto
(1978 Automobile Data)
. generate weightsq=weight^2
```

(Continued on next page)

```
. regress mpg weight weightsq foreign
```

Source	SS	df	MS
Model	1689.15372	3	563.05124
Residual	754.30574	70	10.7757963
Total	2443.45946	73	33.4720474

```
Number of obs =      74
F(  3,    70) =   52.25
Prob > F      =  0.0000
R-squared     =  0.6913
Adj R-squared =  0.6781
Root MSE      =  3.2827
```

mpg	Coef.	Std. Err.	t	P>|t|	[95% Conf. Interval]	
weight	-.0165729	.0039692	-4.18	0.000	-.0244892	-.0086567
weightsq	1.59e-06	6.25e-07	2.55	0.013	3.45e-07	2.84e-06
foreign	-2.2035	1.059246	-2.08	0.041	-4.3161	-.0909002
_cons	56.53884	6.197383	9.12	0.000	44.17855	68.89913

We can use the `test` command to calculate the joint significance of `weight` and `weightsq`:

```
. test weight weightsq
 ( 1)  weight = 0
 ( 2)  weightsq = 0
       F(  2,    70) =   60.83
             Prob > F =    0.0000
```

We are not limited to testing whether the coefficients are zero. We can test whether the coefficient on `foreign` is -2 by typing

```
. test foreign = -2
 ( 1)  foreign = -2
       F(  1,    70) =    0.04
             Prob > F =    0.8482
```

We can even test more complicated hypotheses since `test` can perform basic algebra. Here is an absurd hypothesis:

```
. test 2*(weight+weightsq)=-3*(foreign-(weight-weightsq))
 ( 1) - weight + 5.0 weightsq + 3.0 foreign = 0
       F(  1,    70) =    4.31
             Prob > F =    0.0416
```

`test` simplified the algebra of our hypothesis and then presented the test results. We discover that the hypothesis may be absurd but we cannot reject it at the 1% or even 4% level. We can also use `test`'s `accumulate` option to combine this test with another test:

```
. test foreign+weight=0, accum
 ( 1) - weight + 5.0 weightsq + 3.0 foreign = 0
 ( 2)  weight + foreign = 0
       F(  2,    70) =    9.12
             Prob > F =    0.0003
```

There are limitations. `test` can test only linear hypotheses. If we attempt to test a nonlinear hypothesis, `test` will tell us that it is not possible:

```
. test weight/foreign=0
not possible with test
r(131);
```

Testing nonlinear hypotheses is discussed in [U] **20.10.4 Nonlinear Wald tests** below.

20.10.2   Using test

test bases its results on the estimated variance–covariance matrix of the estimators (i.e., performs a Wald test), so it can be used after any estimation command. In the case of maximum likelihood estimation, you will have to decide whether you want to perform tests based on the information matrix instead of constraining the equation, re-estimating it, and then calculating the likelihood-ratio test (see [U] **20.10.3 Likelihood-ratio tests**). Since test bases its results on the information matrix, its results have exactly the same standing as the asymptotic Z statistic presented in the coefficient table.

▷ Example 13

Let's examine the repair records of the cars in our automobile data as rated by *Consumer Reports*:

```
. tabulate rep78 foreign
```

Repair Record 1978	Car type Domestic	Foreign	Total
1	2	0	2
2	8	0	8
3	27	3	30
4	9	9	18
5	2	9	11
Total	48	21	69

The values are coded 1–5, corresponding to well below average to well above average. We will fit this variable using a maximum-likelihood ordered logit model (the nolog option suppresses the iteration log, saving us some paper):

```
. ologit rep78 price foreign weight weightsq displ, nolog
```

Ordered logistic regression

Number of obs   =   69
LR chi2(5)   =   33.12
Prob > chi2   =   0.0000
Pseudo R2   =   0.1767

Log likelihood = −77.133082

rep78	Coef.	Std. Err.	z	P>|z|	[95% Conf. Interval]	
price	−.000034	.0001188	−0.29	0.775	−.0002669	.000199
foreign	2.685648	.9320398	2.88	0.004	.8588833	4.512412
weight	−.0037447	.0025609	−1.46	0.144	−.0087639	.0012745
weightsq	7.87e-07	4.50e-07	1.75	0.080	−9.43e-08	1.67e-06
displacement	−.0108919	.0076805	−1.42	0.156	−.0259455	.0041617
/cut1	−9.417196	4.298201			−17.84151	−.9928766
/cut2	−7.581864	4.23409			−15.88053	.7168002
/cut3	−4.82209	4.147679			−12.95139	3.307212
/cut4	−2.79344	4.156219			−10.93948	5.352599

We now wonder whether all our variables other than foreign are jointly significant. We test the hypothesis just as we would after linear regression:

```
. test weight weightsq displ price
 ( 1)   [rep78]weight = 0
 ( 2)   [rep78]weightsq = 0
 ( 3)   [rep78]displacement = 0
 ( 4)   [rep78]price = 0
              chi2(  4) =     3.63
          Prob > chi2 =     0.4590
```

To compare this with the results performed by a likelihood-ratio test, see [U] **20.10.3 Likelihood-ratio tests**. In this case, results differ little.

◁

20.10.3 Likelihood-ratio tests

After maximum likelihood estimation, you can obtain likelihood-ratio tests by fitting both the unconstrained and constrained models, storing the results using estimates store, and then running lrtest. See [R] **lrtest** for the full details.

▷ Example 14

In [U] **20.10.2 Using test** above, we fitted an ordered logit on rep78 and then tested the significance of all the explanatory variables except foreign.

To obtain the likelihood-ratio test, sometime after fitting the full model, we type estimates store *full_model_name*, where *full_model_name* is just a label that we assign to these results.

```
. ologit rep78 price foreign weight weightsq displ
(output omitted )
. estimates store myfullmodel
```

This command saves the current model results with the name myfullmodel.

Next we fit the constrained model. After that, typing 'lrtest myfullmodel .' compares the current model with the model we saved:

```
. ologit rep78 foreign
Iteration 0:   log likelihood = -93.692061
Iteration 1:   log likelihood = -79.696089
Iteration 2:   log likelihood = -79.044933
Iteration 3:   log likelihood = -79.029267
Iteration 4:   log likelihood = -79.029243
Ordered logistic regression                     Number of obs   =         69
                                                LR chi2(1)      =      29.33
                                                Prob > chi2     =     0.0000
Log likelihood = -79.029243                     Pseudo R2       =     0.1565
```

rep78	Coef.	Std. Err.	z	P>|z|	[95% Conf. Interval]	
foreign	2.98155	.6203637	4.81	0.000	1.76566	4.197441
/cut1	-3.158382	.7224269			-4.574313	-1.742452
/cut2	-1.362642	.3557343			-2.059868	-.6654154
/cut3	1.232161	.3431227			.5596533	1.90467
/cut4	3.246209	.5556646			2.157127	4.335292

```
. lrtest myfullmodel .
```

Likelihood-ratio test   LR chi2(4)   =   3.79
(Assumption: . nested in myfullmodel)   Prob > chi2 =   0.4348

When we tested the same constraint with `test` (which performed a Wald test), we obtained a χ^2 of 3.63 and a significance level of 0.4590. Note that we used . (the dot) to specify the results in active memory, although we also could have stored them with `estimates store` and referred to them by name instead. Also, the order in which you specify the two models to `lrtest` doesn't matter; `lrtest` is smart enough to know the full model from the constrained model.

◁

Two other postestimation commands work in the same way as `lrtest`, meaning that they accept names of stored estimation results as their input: `hausman` for performing Hausman specification tests and `suest` for seemingly unrelated estimation. We do not cover these commands here; see [R] **hausman** and [R] **suest** for more details.

20.10.4 Nonlinear Wald tests

`testnl` can be used to test nonlinear hypotheses about the parameters of the active estimation results. `testnl`, like `test`, bases its results on the variance–covariance matrix of the estimators (i.e., performs a Wald test), so it can be used after any estimation command; see [R] **testnl**.

▷ Example 15

We fit the model

```
. regress price mpg weight foreign
  (output omitted )
```

and then type

```
. testnl (38*_b[mpg]^2 = _b[foreign]) (_b[mpg]/_b[weight]=4)
  (1)   38*_b[mpg]^2 = _b[foreign]
  (2)   _b[mpg]/_b[weight]=4
              F(2, 70) =        0.02
              Prob > F =       0.9806
```

We performed this test on linear regression estimates, but tests of this type could be performed after any estimation command.

◁

(Continued on next page)

20.11   Obtaining linear combinations of coefficients

lincom computes point estimates, standard errors, t or z statistics, p-values, and confidence intervals for a linear combination of coefficients after any estimation command. Results can optionally be displayed as odds ratios, incidence-rate ratios, or relative-risk ratios.

▷ Example 16

We fit a linear regression:

```
. use http://www.stata-press.com/data/r9/regress

. regress y x1 x2 x3
```

Source	SS	df	MS
Model	3259.3561	3	1086.45203
Residual	1627.56282	144	11.3025196
Total	4886.91892	147	33.2443464

	Number of obs = 148
	F(3, 144) = 96.12
	Prob > F = 0.0000
	R-squared = 0.6670
	Adj R-squared = 0.6600
	Root MSE = 3.3619

y	Coef.	Std. Err.	t	P>|t|	[95% Conf. Interval]
x1	1.457113	1.07461	1.36	0.177	-.6669339 3.581161
x2	2.221682	.8610358	2.58	0.011	.5197797 3.923583
x3	-.006139	.0005543	-11.08	0.000	-.0072345 -.0050435
_cons	36.10135	4.382693	8.24	0.000	27.43863 44.76407

Suppose that we want to see the difference of the coefficients of x2 and x1. We type

```
. lincom x2 - x1

 ( 1)  - x1 + x2 = 0
```

y	Coef.	Std. Err.	t	P>|t|	[95% Conf. Interval]
(1)	.7645682	.9950282	0.77	0.444	-1.20218 2.731316

◁

lincom is very handy for computing the odds ratio of one covariate group relative to another.

▷ Example 17

We estimate the parameters of a logistic model of low birthweight:

```
. use http://www.stata-press.com/data/r9/lbw3
(Hosmer & Lemeshow data)
```

```
. logit low age lwd black other smoke ptd ht ui

Iteration 0:   log likelihood =    -117.336
Iteration 1:   log likelihood = -99.431174
Iteration 2:   log likelihood = -98.785718
Iteration 3:   log likelihood =    -98.778
Iteration 4:   log likelihood = -98.777998
```

Logistic regression				Number of obs	=	189
				LR chi2(8)	=	37.12
				Prob > chi2	=	0.0000
Log likelihood = -98.777998				Pseudo R2	=	0.1582

low	Coef.	Std. Err.	z	P>|z|	[95% Conf. Interval]	
age	-.0464796	.0373888	-1.24	0.214	-.1197603	.0268011
lwd	.8420615	.4055338	2.08	0.038	.0472299	1.636893
black	1.073456	.5150752	2.08	0.037	.0639273	2.082985
other	.815367	.4452979	1.83	0.067	-.0574008	1.688135
smoke	.8071996	.404446	2.00	0.046	.0145001	1.599899
ptd	1.281678	.4621157	2.77	0.006	.3759478	2.187408
ht	1.435227	.6482699	2.21	0.027	.1646415	2.705813
ui	.6576256	.4666192	1.41	0.159	-.2569313	1.572182
_cons	-1.216781	.9556797	-1.27	0.203	-3.089878	.656317

If we want to obtain the odds ratio for black smokers relative to white nonsmokers (the reference group), we type

```
. lincom black + smoke, or

( 1)   black + smoke = 0
```

low	Odds Ratio	Std. Err.	z	P>|z|	[95% Conf. Interval]	
(1)	6.557805	4.744692	2.60	0.009	1.588176	27.07811

`lincom` computed $\exp(\beta_{\text{black}} + \beta_{\text{smoke}}) = 6.56$.

◁

20.12   Obtaining nonlinear combinations of coefficients

`lincom` is limited to estimating linear combinations of coefficients, e.g., `black + smoke`, or exponentiated linear combinations, as in the above. For general nonlinear combinations, use `nlcom`.

▷ Example 18

Continuing our previous example, suppose that we wanted the ratio of the coefficients (and standard errors, Wald test, confidence interval, etc.) of `black` and `other`:

```
. nlcom _b[black]/_b[other]

      _nl_1:  _b[black]/_b[other]
```

low	Coef.	Std. Err.	z	P>|z|	[95% Conf. Interval]	
_nl_1	1.316531	.7359262	1.79	0.074	-.1258574	2.75892

The Wald test given is that of the null hypothesis that the nonlinear combination is zero versus the two-sided alternative—this is probably not very informative in the case of a ratio. If we would instead like to test whether this ratio is one, we can rerun `nlcom`, this time subtracting one from our ratio estimate.

```
. nlcom _b[black]/_b[other] - 1
    _nl_1:  _b[black]/_b[other] - 1
```

low	Coef.	Std. Err.	z	P>|z|	[95% Conf. Interval]	
_nl_1	.3165314	.7359262	0.43	0.667	-1.125857	1.75892

We can interpret this as not very much evidence that the "ratio minus 1" is different from zero, meaning that we cannot reject the null hypothesis that the ratio equals one.

Note that when using `nlcom`, we needed to refer to the model coefficients by their "proper" names, e.g. `_b[black]`, and not by the shorthand `black`, such as when using `lincom`. If we had typed

```
. nlcom black/other
```

Stata would have reported an error. Consider this a limitation of Stata.

◁

20.13   Obtaining marginal effects

Stata's `mfx` command computes the marginal effects of the independent variables on predicted values.

▷ Example 19

Consider the logistic regression model that we previously fitted on the automobile data:

```
. use http://www.stata-press.com/data/r9/auto

. logit foreign weight mpg
Iteration 0:   log likelihood =  -45.03321
Iteration 1:   log likelihood = -29.898968
Iteration 2:   log likelihood = -27.495771
Iteration 3:   log likelihood = -27.184006
Iteration 4:   log likelihood = -27.175166
Iteration 5:   log likelihood = -27.175156
```

```
Logistic regression                              Number of obs  =        74
                                                 LR chi2(2)     =     35.72
                                                 Prob > chi2    =    0.0000
Log likelihood = -27.175156                      Pseudo R2      =    0.3966
```

foreign	Coef.	Std. Err.	z	P>|z|	[95% Conf. Interval]	
weight	-.0039067	.0010116	-3.86	0.000	-.0058894	-.001924
mpg	-.1685869	.0919174	-1.83	0.067	-.3487418	.011568
_cons	13.70837	4.518707	3.03	0.002	4.851864	22.56487

Typing `mfx compute` gives the marginal effects for the default prediction, which, in the case of `logit`, is the predicted probability that the automobile is manufactured outside the U.S.

```
. mfx compute
Marginal effects after logit
      y  = Pr(foreign) (predict)
         = .15733364
```

variable	dy/dx	Std. Err.	z	P>|z|	[95% C.I.]	X
weight	-.0005179	.00014	-3.73	0.000	-.00079 -.000246	3019.46
mpg	-.0223512	.0127	-1.76	0.079	-.04725 .002548	21.2973

Given the above output, we see that both `weight` and `mpg` have a negative effect on the predicted probability. For example, increased weight (from the mean weight of 3019.46 lbs.) decreases the likelihood that the automobile is `foreign` when controlling for gas mileage.

◁

`mfx` can also calculate elasticities, calculate at covariate values other than the covariate means (the default), and calculate marginal effects for predictions other than the default prediction; see [R] **mfx** for details.

20.14 Obtaining robust variance estimates

Estimates of variance refer to estimated standard errors or, more completely, the estimated variance–covariance matrix of the estimators of which the standard errors are a subset, being the square root of the diagonal elements. Call this matrix the variance. All estimation commands produce an estimate of variance and, using that, produce confidence intervals and significance tests.

In addition to the conventional estimator of variance, there is another estimator that has been called by various names because it has been derived independently in different ways by different authors. Two popular names associated with the calculation are Huber and White, but it is also known as the sandwich estimator of variance (because of how the calculation formula physically appears) and the robust estimator of variance (because of claims made about it). In addition, this estimator also has an independent and long tradition in the survey literature.

The conventional estimator of variance is derived by starting with a model. Let's start with the regression model

$$y_i = \mathbf{x}_i \boldsymbol{\beta} + \epsilon_i, \qquad \epsilon_i \sim N(0, \sigma^2)$$

although it is not important for the discussion that we are using regression. Under the model-based approach, we assume that the model is true and thereby derive an estimator for $\boldsymbol{\beta}$ and its variance.

The estimator of the standard error of $\widehat{\boldsymbol{\beta}}$ we develop is based on the assumption that the model is true in every detail. The reason that y_i is not exactly equal to $\mathbf{x}_i \boldsymbol{\beta}$ (so that we would only need to solve an equation to obtain precisely that value of $\boldsymbol{\beta}$) is that the observed y_i has noise ϵ_i added to it, the noise is Gaussian, and it has constant variance. That noise leads to the uncertainty about $\boldsymbol{\beta}$, and it is from the characteristics of that noise that we are able to calculate a sampling distribution for $\widehat{\boldsymbol{\beta}}$.

The key thought here is that the standard error of $\widehat{\boldsymbol{\beta}}$ arises because of ϵ and is valid only because the model is absolutely, without question, true; we just do not happen to know the particular values of $\boldsymbol{\beta}$ and σ^2 that make the model true. The implication is that, in an infinite-sized sample, the estimator $\widehat{\boldsymbol{\beta}}$ for $\boldsymbol{\beta}$ would converge to the true value of $\boldsymbol{\beta}$ and that its variance would go to 0.

Now, here is another interpretation of the estimation problem: We are going to fit the model

$$y_i = \mathbf{x}_i \mathbf{b} + e_i$$

and, to obtain estimates of $\mathbf{b}$, we are going to use the calculation formula

$$\widehat{\mathbf{b}} = (\mathbf{X}'\mathbf{X})^{-1}\mathbf{X}'\mathbf{y}$$

Please note that we have made no claims that the model is true nor any claims about e_i or its distribution. We shifted our notation from β and ϵ_i to $\mathbf{b}$ and e_i to emphasize this. All we have stated are the physical actions we intend to carry out on the data. Interestingly, it is possible to calculate a standard error for $\widehat{\mathbf{b}}$ in this case! At least, it is possible if you will agree with us on what the standard error measures are.

We are going to define the standard error as measuring the standard error of the calculated $\widehat{\mathbf{b}}$ if we were to repeat the data collection followed by estimation over and over again.

Also note that this is a different concept of the standard error from the conventional, model-based ideas, but it is not unrelated. Both measure uncertainty about $\mathbf{b}$ (or β). The regression model-based derivation states from where the variation arises and so can make grander statements about the applicability of the measured standard error. The weaker second interpretation makes fewer assumptions and so produces a standard error suitable for one purpose.

There is a subtle difference in interpretation of these identically calculated point estimates. $\widehat{\beta}$ is the estimate of β under the assumption that the model is true. $\widehat{\mathbf{b}}$ is the estimate of $\mathbf{b}$, which is merely what the estimator would converge to if we collected more and more data.

Is the estimate of $\mathbf{b}$ unbiased? If we mean, does $\mathbf{b} = \beta$, that depends on whether the model is true. $\widehat{\mathbf{b}}$ is, however, an unbiased estimate of $\mathbf{b}$, which, admittedly, is not saying much.

What if $\mathbf{x}$ and e are correlated? Don't we have a problem in that case? We may have an interpretation problem—$\mathbf{b}$ may not measure what we want to measure, namely, β—but we measure $\widehat{\mathbf{b}}$ to be such and such and expect, if the experiment and estimation were repeated, that we would observe results in the range we have reported.

So, we have two very different understandings of what the parameters mean and how the variance in their estimators arises. However, both interpretations must confront the issue of how to make valid statistical inference about the coefficient estimates when the data do not come from either a simple random sample or the distribution of $(\mathbf{x}_i, \epsilon_i)$ is not independent and identically distributed (i.i.d.). In essence, we need an estimator of the standard errors that is robust to this deviation from the standard case.

Hence, the name *the robust estimate of variance*; its associated authors are Huber (1967) and White (1980, 1982) (who developed it independently), although many others have extended its development, including Gail, Tan, and Piantadosi (1988), Kent (1982), Royall (1986), and Lin and Wei (1989). In the survey literature, this same estimator has been developed; see, for example, Kish and Frankel (1974), Fuller (1975), and Binder (1983).

Many of Stata's estimation commands can produce this alternative estimate of variance, and, if they can, they have a `robust` option. Without `robust`, we get one measure of variance:

```
. use http://www.stata-press.com/data/r9/auto7
(1978 Automobile Data)

. regress mpg weight foreign
```

Source	SS	df	MS
Model	1619.2877	2	809.643849
Residual	824.171761	71	11.608053
Total	2443.45946	73	33.4720474

```
                                                   Number of obs =      74
                                                   F(  2,    71) =   69.75
                                                   Prob > F      =  0.0000
                                                   R-squared     =  0.6627
                                                   Adj R-squared =  0.6532
                                                   Root MSE      =  3.4071
```

mpg	Coef.	Std. Err.	t	P>|t|	[95% Conf. Interval]
weight	-.0065879	.0006371	-10.34	0.000	-.0078583 -.0053175
foreign	-1.650029	1.075994	-1.53	0.130	-3.7955 .4954422
_cons	41.6797	2.165547	19.25	0.000	37.36172 45.99768

With `robust`, we get another:

```
. regress mpg weight foreign, robust

Linear regression                                  Number of obs =      74
                                                   F(  2,    71) =   73.81
                                                   Prob > F      =  0.0000
                                                   R-squared     =  0.6627
                                                   Root MSE      =  3.4071
```

mpg	Coef.	Robust Std. Err.	t	P>|t|	[95% Conf. Interval]
weight	-.0065879	.0005462	-12.06	0.000	-.007677 -.0054988
foreign	-1.650029	1.132566	-1.46	0.150	-3.908301 .6082424
_cons	41.6797	1.797553	23.19	0.000	38.09548 45.26392

Either way, the point estimates are the same. (See [R] **regress** for an example where specifying `robust` produces strikingly different standard errors.)

How do we interpret these results? Let's consider the model-based interpretation. Suppose that

$$y_i = \mathbf{x}_i \boldsymbol{\beta} + \epsilon_i,$$

where $(\mathbf{x}_i, \epsilon_i)$ are independently and identically distributed (i.i.d.) with variance σ^2. For the model-based interpretation, we also must assume that $\mathbf{x}_i$ and ϵ_i are uncorrelated. With these assumptions and a few technical regularity conditions, our first regression gives us consistent parameter estimates and standard errors that we can use for valid statistical inference about the coefficients. Now suppose that we weaken our assumptions so that $(\mathbf{x}_i, \epsilon_i)$ are independently and—but not necessarily—identically distributed. Our parameter estimates are still consistent, but the standard errors from the first regression can no longer be used to make valid inference. We need estimates of the standard errors that are robust to the fact that the error term is not identically distributed. The standard errors in our second regression are just what we need. We can use them to make valid statistical inference about our coefficients, even though our data are not identically distributed.

Now consider a nonmodel-based interpretation. If our data come from a survey design that ensures that $(\mathbf{x}_i, e_i)$ are i.i.d., then we can use the nonrobust standard errors for valid statistical inference about the population parameters $\mathbf{b}$. Note that, for this interpretation, we do not need to assume that $\mathbf{x}_i$ and e_i are uncorrelated. If they are uncorrelated, the population parameters $\mathbf{b}$ and the model parameters $\boldsymbol{\beta}$ are the same. However, if they are correlated, then the population parameters $\mathbf{b}$ that we are estimating are not the same as the model-based β. So, what we are estimating is different, but we still need standard errors that allow us to make valid statistical inference. So, if the process that we used to collect the data caused $(\mathbf{x}_i, e_i)$ to be independently but not identically distributed, then we need to use the robust standard errors to make valid statistical inference about the population parameters $\mathbf{b}$.

The robust estimator of variance has one feature that the conventional estimator does not have: the ability to relax the assumption of independence of the observations. That is, if you specify the cluster() option, it can produce "correct" standard errors (in the measurement sense), even if the observations are correlated.

In the case of the automobile data, it is difficult to believe that the models of the various manufacturers are truly independent. Manufacturers, after all, use common technology, engines, and drive trains across their model lines. The VW Dasher in the above regression has a measured residual of -2.80. Having been told that, do you really believe that the residual for the VW Rabbit is as likely to be above 0 as below? (The residual is -2.32.) Similarly, the measured residual for the Chevrolet Malibu is 1.27. Does that provide information about the expected value of the residual of the Chevrolet Monte Carlo (which turns out to be 1.53)?

We need to be careful about picking examples out of data; we have not told you about the Datsun 210 and 510 (residuals $+8.28$ and -1.01) or the Cadillac Eldorado and Seville (residuals -1.99 and $+7.58$), but you should, at least, question the assumption of independence. It may be believable that the measured mpg given the weight of one manufacturer's vehicles is independent of other manufacturers' vehicles, but it is at least questionable whether a manufacturer's vehicles are independent of one another.

In commands with the robust option, another option—cluster()—relaxes the independence assumption and requires only that the observations be independent across the clusters:

```
. regress mpg weight foreign, robust cluster(manufacturer)
Linear regression                                Number of obs =        74
                                                 F(  2,    22) =     90.93
                                                 Prob > F      =    0.0000
                                                 R-squared     =    0.6627
Number of clusters (manufacturer) = 23           Root MSE      =    3.4071
```

mpg	Coef.	Robust Std. Err.	t	P>|t|	[95% Conf. Interval]
weight	-.0065879	.0005339	-12.34	0.000	-.0076952 -.0054806
foreign	-1.650029	1.039033	-1.59	0.127	-3.804852 .5047939
_cons	41.6797	1.844559	22.60	0.000	37.85432 45.50508

It turns out that, in these data, whether or not we specify cluster() makes little difference. The VW and Chevrolet examples above were not representative; had they been, the confidence intervals would have widened. (In the above, manuf is a variable that takes on values such as "Chev.", "VW", etc., recording the manufacturer of the vehicle. This variable was created from variable make, which contains values such as "Chev. Malibu", "VW Rabbit", etc., by extracting the first word.)

As a demonstration of how well clustering can work, in [R] **regress** we fitted a random-effects model with regress, robust cluster() and then compare the results with ordinary least squares and the GLS random-effects estimator. Here we will simply summarize the results.

We start with a dataset on 4,782 women aged 16 to 46. Subjects appear an average of 7.14 times in this dataset, so there are a total of 34,139 observations. The model we use is log wage on age, age-squared, and grade of schooling completed. The focus of the example is the estimated coefficient on schooling. We obtain the following results:

Estimator	point estimate	confidence interval
(inappropriate) least squares	.081	[.079, .083]
robust, cluster	.081	[.077, .085]
GLS random effects	.080	[.076, .083]

Notice how well `robust` with the `cluster()` option does compared with the GLS random-effects model. We then run a Hausman specification test, obtaining $\chi^2(2) = 62$, which casts grave doubt on the assumptions justifying the use of the GLS estimator, and hence on the GLS results. At this point, we will simply quote our comments:

> Meanwhile, our robust regression results still stand as long as we are careful about the interpretation. The correct interpretation is that, if the data collection were repeated (on women sampled the same way as in the original sample), and if we re-estimated the model parameters, 95% of the time we would expect our obtained range to contain the true coefficient on grade.

> Even with robust regression, we must be careful about going beyond that statement. In this case, the Hausman test is probably picking up something that differs within and between persons, casting doubt on our robust regression model in terms of interpreting $[.077, .085]$ to contain the rate of return to additional schooling, across the economy, for all women, without exception.

The formula for the robust estimator of variance is

$$\widehat{\mathcal{V}} = \widehat{\mathbf{V}} \left(\sum_{j=1}^{N} \mathbf{u}_j' \mathbf{u}_j \right) \widehat{\mathbf{V}}$$

where $\widehat{\mathbf{V}} = (-\partial^2 \ln L / \partial \boldsymbol{\beta}^2)^{-1}$ (the conventional estimator of variance) and $\mathbf{u}_j$ (a row vector) is the contribution from the jth observation to $\partial \ln L / \partial \boldsymbol{\beta}$.

In the example above, observations are assumed to be independent. Assume, for a moment, that the observations denoted by j are not independent but that they can be divided into M groups G_1, G_2, ..., G_M that are independent. The robust estimator of variance is

$$\widehat{\mathcal{V}} = \widehat{\mathbf{V}} \left(\sum_{k=1}^{M} \mathbf{u}_k^{(G)\prime} \mathbf{u}_k^{(G)} \right) \widehat{\mathbf{V}}$$

where $\mathbf{u}_k^{(G)}$ is the contribution of the kth group to $\partial \ln L / \partial \boldsymbol{\beta}$. That is, application of the robust variance formula merely involves using a different decomposition of $\partial \ln L / \partial \boldsymbol{\beta}$, namely, $\mathbf{u}_k^{(G)}$, $k = 1, \ldots, M$ rather than $\mathbf{u}_j$, $j = 1, \ldots, N$. Moreover, if the log-likelihood function is additive in the observations denoted by j

$$\ln L = \sum_{j=1}^{N} \ln L_j$$

then $\mathbf{u}_j = \partial \ln L_j / \partial \boldsymbol{\beta}$, so

$$\mathbf{u}_k^{(G)} = \sum_{j \in G_k} \mathbf{u}_j$$

That is what the `cluster()` option does. (This point was first made in writing by Rogers [1993], although he considered the point an obvious generalization of Huber [1967] and the calculation—implemented by Rogers—had appeared in Stata a year earlier.)

❑ Technical Note

What is written above is asymptotically correct but ignores a finite-sample adjustment to $\widehat{\mathcal{V}}$. For maximum likelihood estimators, when you specify robust but not cluster(), a better estimate of variance is $\widehat{\mathcal{V}}^* = \{N/(N-1)\}\widehat{\mathcal{V}}$. When you also specify the cluster() option, this becomes $\widehat{\mathcal{V}}^* = \{M/(M-1)\}\widehat{\mathcal{V}}$.

In the case of linear regression, the finite-sample adjustment is $N/(N-k)$ without cluster()—where k is the number of regressors—and $\{M/(M-1)\}\{(N-1)/(N-k)\}$ with cluster(). In addition, two data-dependent modifications to the calculation for $\widehat{\mathcal{V}}^*$, suggested by MacKinnon and White (1985), are also provided by regress; see [R] **regress**.

❑

20.15 Obtaining scores

Many of the estimation commands that provide the robust option also provide the ability to generate equation-level score variables via the predict command. With the score option, predict returns an important ingredient into the robust variance calculation that is sometimes useful in its own right. As explained in [U] **20.14 Obtaining robust variance estimates** above, ignoring the finite-sample corrections, the robust estimate of variance is

$$\widehat{\mathcal{V}} = \widehat{\mathbf{V}}\left(\sum_{j=1}^{N} \mathbf{u}_j' \mathbf{u}_j\right)\widehat{\mathbf{V}}$$

where $\widehat{\mathbf{V}} = (-\partial^2 \ln L/\partial \boldsymbol{\beta}^2)^{-1}$ is the conventional estimator of variance. Let us consider likelihood functions that are additive in the observations

$$\ln L = \sum_{j=1}^{N} \ln L_j$$

then $\mathbf{u}_j = \partial \ln L_j/\partial \boldsymbol{\beta}$. In general, function L_j is a function of $\mathbf{x}_j$ and $\boldsymbol{\beta}$, $L_j(\boldsymbol{\beta}; \mathbf{x}_j)$. For many likelihood functions, however, it is only the linear form $\mathbf{x}_j\boldsymbol{\beta}$ that enters the function. In those cases,

$$\frac{\partial \ln L_j(\mathbf{x}_j\boldsymbol{\beta})}{\partial \boldsymbol{\beta}} = \frac{\partial \ln L_j(\mathbf{x}_j\boldsymbol{\beta})}{\partial(\mathbf{x}_j\boldsymbol{\beta})}\frac{\partial(\mathbf{x}_j\boldsymbol{\beta})}{\partial \boldsymbol{\beta}} = \frac{\partial \ln L_j(\mathbf{x}_j\boldsymbol{\beta})}{\partial(\mathbf{x}_j\boldsymbol{\beta})}\mathbf{x}_j$$

Writing $u_j = \partial \ln L_j(\mathbf{x}_j\boldsymbol{\beta})/\partial(\mathbf{x}_j\boldsymbol{\beta})$, this becomes simply $u_j\mathbf{x}_j$. Thus, the formula for the robust estimate of variance can be rewritten as

$$\widehat{\mathcal{V}} = \widehat{\mathbf{V}}\left(\sum_{j=1}^{N} u_j^2 \mathbf{x}_j' \mathbf{x}_j\right)\widehat{\mathbf{V}}$$

We refer to u_j as the equation-level score (in the singular), and it is u_j that is returned when you use predict with the score option. u_j is like a residual in that

1. $\sum_j u_j = 0$, and

2. correlation of u_j and $\mathbf{x}_j$, calculated over $j = 1, \ldots, N$, is 0.

In fact, in the case of linear regression, u_j is the residual, normalized,

$$\frac{\partial \ln L_j}{\partial (\mathbf{x}_j \boldsymbol{\beta})} = \frac{\partial}{\partial (\mathbf{x}_j \boldsymbol{\beta})} \ln f \left\{ (y_j - \mathbf{x}_j \boldsymbol{\beta}) / \sigma \right\}$$
$$= (y_j - \mathbf{x}_j \boldsymbol{\beta}) / \sigma$$

where $f()$ is the standard normal density.

▷ Example 20

probit provides both the robust option and predict, score. Equation-level scores play an important role in calculating the robust estimate of variance, but we can use predict, score regardless of whether we specify robust:

```
. use http://www.stata-press.com/data/r9/auto

. probit foreign mpg weight

Iteration 0:   log likelihood =  -45.03321
Iteration 1:   log likelihood = -29.244141
Iteration 2:   log likelihood = -27.041557
Iteration 3:   log likelihood =  -26.84658
Iteration 4:   log likelihood = -26.844189
Iteration 5:   log likelihood = -26.844189
```

```
Probit regression                               Number of obs   =         74
                                                LR chi2(2)      =      36.38
                                                Prob > chi2     =     0.0000
Log likelihood = -26.844189                     Pseudo R2       =     0.4039
```

foreign	Coef.	Std. Err.	z	P>|z|	[95% Conf. Interval]
mpg	-.1039503	.0515689	-2.02	0.044	-.2050235 -.0028772
weight	-.0023355	.0005661	-4.13	0.000	-.003445 -.0012261
_cons	8.275464	2.554142	3.24	0.001	3.269438 13.28149

```
. predict double u, scores

. summarize u
```

Variable	Obs	Mean	Std. Dev.	Min	Max
u	74	-3.87e-16	.5988325	-1.655439	1.660787

```
. correlate u mpg weight
(obs=74)
```

	u	mpg	weight
u	1.0000		
mpg	-0.0000	1.0000	
weight	-0.0000	-0.8072	1.0000

```
. list make foreign mpg weight u if abs(u)>1.65
```

	make	foreign	mpg	weight	u
24.	Ford Fiesta	Domestic	28	1,800	-1.6554395
64.	Peugeot 604	Foreign	14	3,420	1.6607871

The light, high-mileage Ford Fiesta is surprisingly domestic, while the heavy, low-mileage Peugeot 604 is surprisingly foreign.

◁

❑ Technical Note

For some estimation commands, one score is not enough. Consider a likelihood that can be written as $L_j(\mathbf{x}_j\boldsymbol{\beta}_1, \mathbf{z}_j\boldsymbol{\beta}_2)$, a function of two linear forms (or linear equations). Then $\partial\ln L_j/\partial\boldsymbol{\beta}$ can be written $(\partial\ln L_j/\partial\boldsymbol{\beta}_1, \partial\ln L_j/\partial\boldsymbol{\beta}_2)$. Each of the components can in turn be written as $[\partial\ln L_j/\partial(\beta_1\mathbf{x})]\mathbf{x} = u_1\mathbf{x}$ and $[\partial\ln L_j/\partial(\beta_2\mathbf{z})]\mathbf{z} = u_2\mathbf{z}$. There are then two equation-level scores, u_1 and u_2, and, in general, there could be more.

Stata's `streg, distribution(weibull)` command is an example of this: it estimates $\boldsymbol{\beta}$ and a shape parameter, $\ln p$, the latter of which can be thought of as a degenerate linear form $(\ln p)\mathbf{z}$ with $\mathbf{z} = 1$. `predict, scores` after this command requires that you specify two new variable names, or you can specify *stub**, which will generate two variables, *stub*1 and *stub*2; the first will be defined containing u_1—the score associated with $\boldsymbol{\beta}$—and the second will be defined containing u_2—the score associated with $\ln p$.

❑

❑ Technical Note

Using Stata's matrix commands—see [P] **matrix**—we can make the robust variance calculation for ourselves and then compare it with that made by Stata.

```
. quietly probit foreign mpg weight, score(u)

. matrix accum S =  mpg weight [iweight=u^2*74/73]
(obs=26.53642547)

. matrix rV = e(V)*S*e(V)

. matrix list rV

symmetric rV[3,3]
                  mpg        weight        _cons
   mpg      .00352299
weight      .00002216    2.434e-07
 _cons    -.14090346    -.00117031    6.4474172

. quietly probit foreign mpg weight, robust

. matrix list e(V)

symmetric e(V)[3,3]
                  mpg        weight        _cons
   mpg      .00352299
weight      .00002216    2.434e-07
 _cons    -.14090346    -.00117031    6.4474172
```

The results are the same.

There is an important lesson here for programmers. Given the scores, conventional variance estimates can be easily transformed to robust estimates. If we were writing a new estimation command, it would not be difficult to include a `robust` option.

It is, in fact, easy if we ignore clustering. With clustering, it is more work since the calculation involves forming sums within clusters. For programmers interested in implementing robust variance calculations, Stata provides an `_robust` command to ease the task. This is documented in [P] **_robust**.

To use `_robust`, you first produce conventional results (a vector of coefficients and covariance matrix) along with a variable containing the scores u_j (or variables if the likelihood function has more than one stub). You then call `_robust`, and it will transform your conventional variance estimate into the robust estimate. `_robust` will handle the work associated with clustering and the details of the finite-sample adjustment, and it will even label your output so that the word *Robust* appears above the standard error when the results are displayed.

Of course, this is all even easier if you write your commands using Stata's `ml` maximum likelihood optimization, in which case, you merely pass the `robust` and `cluster()` options on to `ml`. `ml` will then call `_robust` itself and do all the work for you.

❏

20.16   Weighted estimation

[U] **11.1.6 weight** introduced the syntax for weights. Stata provides four kinds of weights: `fweights`, or frequency weights; `pweights`, or sampling weights; `aweights`, or analytic weights; and `iweights`, or importance weights. The syntax for using each is the same. Type

    . regress y x1 x2

and you obtain unweighted estimates; type

    . regress y x1 x2 [pweight=pop]

and you obtain (in this example) `pweight`ed estimation.

The sections below explain in detail how each type of weight is used in estimation.

20.16.1   Frequency weights

Frequency weights—fweights—are integers and are nothing more than replication counts. The weight is statistically uninteresting, but, from a data processing perspective, it is of great importance. Consider the following data

y	x1	x2
22	1	0
22	1	0
22	1	1
23	0	1
23	0	1
23	0	1

and the estimation command

    . regress y x1 x2

Exactly equivalent is the following, more compressed data

y	x1	x2	pop
22	1	0	2
22	1	1	1
23	0	1	3

and the corresponding estimation command

    . regress y x1 x2 [fweight=pop]

When you specify frequency weights, you are treating each observation as one or more real observations.

❏ Technical Note

You might occasionally run across a command that does not allow weights at all, especially among user-written commands. expand (see [D] **expand**) can be used with such commands to obtain frequency-weighted results. The expand command duplicates observations so that the data become self-weighting. For example, suppose that you want to run the command usercmd, which does something or other, and you would very much like to type usercmd y x1 x2 [fw=pop]. Unfortunately, usercmd does not allow weights. Instead, you type

```
. expand pop
. usercmd y x1 x2
```

to obtain your result. Moreover, there is an important principle here: The results of running any command with frequency weights should be exactly the same as running the command on the unweighted, expanded data. Unweighted, duplicated data and frequency-weighted data are merely two ways of recording identical information.

❏

20.16.2 Analytic weights

Analytic weights—*analytic* is a term we made up—statistically arise in one particular problem: linear regression on data that are themselves observed means. That is, think of the model

$$y_i = \mathbf{x}_i\boldsymbol{\beta} + \epsilon_i, \qquad \epsilon_i \sim N(0, \sigma^2)$$

and now think about estimating this model on data $(\overline{y}_j, \overline{\mathbf{x}}_j)$ that are themselves observed averages. For instance, a piece of the underlying data for $(y_i, \mathbf{x}_i)$ might be $(3, 1)$, $(4, 2)$, and $(2, 2)$, but you do not know that. Instead, you have a single observation $\{(3+4+2)/3, (1+2+2)/3\} = (3, 1.67)$ and know only that the $(3, 1.67)$ arose as the average of 3 underlying observations. All your data are like that.

regress with aweights is the solution to that problem:

```
. regress y x [aweight=pop]
```

There is a history of misusing such weights. A researcher does not have cell-mean data but instead has a probability-weighted random sample. Long before Stata existed, some researchers were using aweights to produce estimates from such samples. We will come back to this point in [U] **20.16.3 Sampling weights** below.

Anyway, the statistical problem that aweights resolve can be written as

$$y_i = \mathbf{x}_i\boldsymbol{\beta} + \epsilon_i, \qquad \epsilon_i \sim N(0, \sigma^2/w_i)$$

where the w_i are the analytic weights. The details of the solution, it turns out, are to make linear regression calculations using the weights as if they were fweights, but to normalize them to sum to N before doing that.

Most commands that allow aweights handle them in this manner. That is, if you specify aweights, they are

1. normalized to sum to N, and then

2. inserted in the calculation formulas in the same way as fweights.

20.16.3 Sampling weights

Sampling weights—probability weights or pweights—refer to probability-weighted random samples. Actually, what you specify in [pweight=...] is a variable recording the number of subjects in the full population that the sampled observation in your data represents. That is, an observation that had probability $1/3$ of being included in your sample has pweight 3.

As noted above, some researchers have used aweights with this kind of data. If they do, they are probably making a mistake. Consider the regression model

$$y_i = \mathbf{x}_i \boldsymbol{\beta} + \epsilon_i, \qquad \epsilon_i \sim N(0, \sigma^2)$$

Begin by considering the exact nature of the problem of fitting this model on cell-mean data—for which aweights are the solution: heteroskedasticity arising from the grouping. Note that the error term ϵ_i is homoskedastic (meaning that it has constant variance σ^2). Say that the first observation in the data is the mean of three underlying observations. Then

$$y_1 = \mathbf{x}_1 \boldsymbol{\beta} + \epsilon_1, \qquad \epsilon_i \sim N(0, \sigma^2)$$
$$y_2 = \mathbf{x}_2 \boldsymbol{\beta} + \epsilon_2, \qquad \epsilon_i \sim N(0, \sigma^2)$$
$$y_3 = \mathbf{x}_3 \boldsymbol{\beta} + \epsilon_3, \qquad \epsilon_i \sim N(0, \sigma^2)$$

and taking the mean

$$(y_1 + y_2 + y_3)/3 = \{(\mathbf{x}_1 + \mathbf{x}_2 + \mathbf{x}_3)/3\}\boldsymbol{\beta} + (\epsilon_1 + \epsilon_2 + \epsilon_3)/3$$

For another observation in the data—which may be the result of summing of a different number of observations—the variance will be different. Hence, the model for the data is

$$\bar{y}_j = \bar{x}_j \boldsymbol{\beta} + \bar{\epsilon}_j, \qquad \bar{\epsilon}_j \sim N(0, \sigma^2/N_j)$$

This makes intuitive sense. Consider two observations, one recording means over two subjects and the other means over 100,000 subjects. You would expect the variance of the residual to be less in the 100,000-subject observation, or, said differently, there is more information in the 100,000-subject observation than in the two-subject observation.

Now instead say that you are fitting the same model, $y_i = \mathbf{x}_i \boldsymbol{\beta} + \epsilon_i, \epsilon_i \sim N(0, \sigma^2)$, on probability-weighted data. Each observation in your data is a single subject, but the different subjects have differing chances of being included in your sample. Therefore, for each subject in your data

$$y_i = \mathbf{x}_i \boldsymbol{\beta} + \epsilon_i, \qquad \epsilon_i \sim N(0, \sigma^2)$$

That is, there is no heteroskedasticity problem. The use of the aweighted estimator cannot be justified on these grounds.

As a matter of fact, based on the argument just given, you do not need to adjust for the weights at all, although the argument does not justify not making an adjustment. If you do not adjust, you are holding tightly to the assumed truth of your model. Two issues arise when considering adjustment for sampling weights:

1. the efficiency of the point estimate $\widehat{\boldsymbol{\beta}}$ of $\boldsymbol{\beta}$ and

2. the reported standard errors (and, more generally, the variance matrix of $\widehat{\boldsymbol{\beta}}$).

Efficiency argues in favor of adjustment, and that, by the way, is why many researchers have used aweights with pweighted data. The adjustment implied by pweights to the point estimates is the same as the adjustment implied by aweights.

With regard to the second issue, the use of aweights produces incorrect results because it interprets larger weights as designating more accurately measured points. In the case of pweights, however, the point is no more accurately measured—it is still just one observation with a single residual ϵ_j and variance σ^2. In [U] **20.14 Obtaining robust variance estimates** above, we introduced another estimator of variance that measures the variation that would be observed if the data collection followed by the estimation were repeated. Those same formulas provide the solution to pweights, and they have the added advantage that they are not conditioned on the model's being true. If we have any hopes of measuring the variation that would be observed were the data collection followed by estimation repeated, we must include the probability of the observations being sampled in the calculation.

In Stata, when you type

```
. regress y x1 x2 [pw=pop]
```

the results are the same as if you had typed

```
. regress y x1 x2 [pw=pop], robust
```

That is, specifying pweights implies the `robust` option and, hence, the robust variance calculation (but weighted). In this example, we use `regress` simply for illustration. The same is true of `probit` and all of Stata's estimation commands. Estimation commands that do not have a `robust` option (there are a few) do not allow pweights.

pweights are adequate for handling random samples where the probability of being sampled varies. pweights may be all you need. If, however, the observations are not sampled independently but are sampled in groups—called clusters in the jargon—you should specify the estimator's `cluster()` option as well:

```
. regress y x1 x2 [pw=pop], cluster(block)
```

There are two ways of thinking about this:

1. The robust estimator answers the question of which variation would be observed were the data collection followed by the estimation repeated; if that question is to be answered, the estimator must account for the clustered nature of how observations are selected. If observations 1 and 2 are in the same cluster, then you cannot select observation 1 without selecting observation 2 (and, by extension, you cannot select observations like 1 without selecting observations like 2).

2. If you prefer, you can think about potential correlations. Observations in the same cluster may not really be independent—that is an empirical question to be answered by the data. For instance, if the clusters are neighborhoods, it would not be surprising that the individual neighbors are similar in their income, their tastes, and their attitudes, and even more similar than two randomly drawn persons from the area at large with similar characteristics, such as age and sex.

Either way of thinking leads to the same (robust) estimator of variance.

Sampling weights usually arise from complex sampling designs, which often involve not only unequal probability sampling and cluster sampling, but also stratified sampling. There is a family of commands in Stata designed to work with the features of complex survey data, and those are the commands that begin with `svy`. To fit a linear regression model with stratification, for example, you would use the `svy:regress` command.

Non-svy commands that allow pweights and clustering give essentially identical results to the svy commands. If the sampling design is simple enough that it can be accommodated by the non-svy command, that is a fine way to perform the analysis. The svy commands differ in that they have additional features, and they do all the little details correctly for real survey data. See [SVY] **survey** for a brief discussion of some of the issues involved in the analysis of survey data and a list of all the differences between the svy and non-svy commands.

Not all model estimation commands in Stata allow pweights. In many of these cases, this is because they are computationally or statistically difficult to implement.

20.16.4   Importance weights

Stata's iweights—importance weights—are the emergency exit. These weights are for those who want to take control and create special effects. For example, programmers have used regress with iweights to compute iteratively reweighted least-squares solutions for various problems.

iweights are treated much like aweights, except that they are not normalized. Stata's iweight rule is that

1. the weights are not normalized and

2. they are generally inserted into calculation formulas in the same way as fweights. There are exceptions; see the *Methods and Formulas* for the particular command.

iweights are used mostly by programmers who are often on the way to implementing one of the other kinds of weights.

20.17   A list of postestimation commands

The following commands can be used after estimation:

[R] **adjust**	adjusted predictions of $\mathbf{x}\beta$, probabilities, or $\exp(\mathbf{x}\beta)$
[R] **estat**	AIC, BIC, VCE, and estimation sample summary
[R] **estimates**	cataloging estimation results
[R] **hausman**	Hausman specification test
[R] **lincom**	point estimates, standard errors, testing, and inference for linear combinations of coefficients
[R] **linktest**	specification link test for single-equation models
[R] **lrtest**	likelihood-ratio test
[R] **mfx**	marginal effects or elasticities
[R] **nlcom**	point estimates, standard errors, testing, and inference for generalized predictions
[R] **predict**	predictions, residuals, influence statistics, and other diagnostic measures
[R] **predictnl**	point estimates, standard errors, testing, and inference for generalized predictions
[R] **suest**	seemingly unrelated estimation
[R] **test**	Wald tests for simple and composite linear hypotheses
[R] **testnl**	Wald tests of nonlinear hypotheses

Also see [U] **13.5 Accessing coefficients and standard errors** for accessing coefficients and standard errors.

20.18 References

Binder, D. A. 1983. On the variances of asymptotically normal estimators from complex surveys. *International Statistical Review* 51: 279–292.

Deaton, A. 1997. *The Analysis of Household Surveys: A Microeconometric Approach to Development Policy.* Baltimore, MD: Johns Hopkins University Press.

Fuller, W. A. 1975. Regression analysis for sample survey. *Sankhyā, Series C* 37: 117-132.

Gail, M. H., W. Y. Tan, and S. Piantadosi. 1988. Tests for no treatment effect in randomized clinical trials. *Biometrika* 75: 57–64.

Huber, P. J. 1967. The behavior of maximum likelihood estimates under non-standard conditions. In *Proceedings of the Fifth Berkeley Symposium on Mathematical Statistics and Probability.* Berkeley: University of California Press, 1, 221–233.

Kent, J. T. 1982. Robust properties of likelihood ratio tests. *Biometrika* 69: 19–27.

Kish, L. and M. R. Frankel. 1974. Inference from complex samples. *Journal of the Royal Statistical Society, Series B* 36: 1–37.

Lin, D. Y. and L. J. Wei. 1989. The robust inference for the Cox proportional hazards model. *Journal of the American Statistical Association* 84: 1074–1078.

Long, J. S. and J. Freese. 2000a. sg145: Scalar measures of fit for regression models. *Stata Technical Bulletin* 56: 34–40. Reprinted in *Stata Technical Bulletin Reprints*, vol. 10, pp. 197–205.

——. 2000b. sg152: Listing and interpreting transformed coefficients from certain regression models. *Stata Technical Bulletin* 57: 27–34. Reprinted in *Stata Technical Bulletin Reprints*, vol. 10, pp. 231–240.

MacKinnon, J. G. and H. White. 1985. Some heteroskedasticity consistent covariance matrix estimators with improved finite sample properties. *Journal of Econometrics* 29: 305–325.

Rogers, W. H. 1993. sg17: Regression standard errors in clustered samples. *Stata Technical Bulletin* 13: 19–23. Reprinted in *Stata Technical Bulletin Reprints*, vol. 3, 88–94.

Royall, R. M. 1986. Model robust confidence intervals using maximum likelihood estimators. *International Statistical Review* 54: 221–226.

Weesie, J. 2000. sg127: Summary statistics for estimation sample. *Stata Technical Bulletin* 53: 32–35. Reprinted in *Stata Technical Bulletin Reprints*, vol. 9, pp. 275–277.

White, H. 1980. A heteroskedasticity-consistent covariance matrix estimator and a direct test for heteroskedasticity. *Econometrica* 48: 817–830.

——. 1982. Maximum likelihood estimation of misspecified models. *Econometrica* 50: 1–25.

Advice

Chapters

21 Inputting data

21.1 Overview

To input data into Stata, you can use

[D] **edit** and [D] **input**	to enter data from the keyboard
[D] **insheet**	to read tab- or comma-separated data
[D] **infile (free format)**	to read unformatted data
[D] **infile (fixed format)** or [D] **infix (fixed format)**	to read formatted data
[D] **xmlsave** (where xmluse is documented)	to use datasets in XML and Excel format
[D] **odbc**	to read from an ODBC source
[D] **fdasave** (where fdause is documented)	to read datasets in SAS XPORT format
[TS] **haver**	to read data in Haver Analytics' format
[U] **21.4 Transfer programs**	to transfer data

Since dataset formats differ, you should familiarize yourself with each method.

Note that [D] **infile (fixed format)** and [D] **infix (fixed format)** are two different commands that do the same thing. Read about both, and then use whichever appeals to you.

Alternatively, edit and input both allow you to enter data from the keyboard. edit opens a Viewer, and input allows you to type as the command line.

After you have read this chapter, also see [D] **infile** for additional examples of the different commands to input data.

21.2 Determining which input method to use

Below are several rules that, when applied sequentially, will direct you to the appropriate method for entering your data. Following the rules is a description of each command, as well as a reference to the corresponding entry in the *Reference* manuals.

1. If you have a small amount of data and simply wish to type the data directly into Stata at the keyboard, see [D] **input**—there are many examples, and you should have little difficulty. Also see [D] **edit**.

2. If your dataset is in binary format or the internal format of some software package, you have several options:

    a. If the data are in a spreadsheet, copy and paste the data into Stata's Data Editor; see [D] **edit** for details.

    b. If the data are in an Excel spreadsheet, use Excel to export the data as XML, and then use Stata's xmluse command to read them; see [D] **xmlsave**.

    c. If the data are in SAS XPORT format, use fdause to read the data; see [D] **fdasave**.

    d. If the data in Haver Analytics' .dat format (Haver Analytics provides economics and financial databases), and you are using Stata for Windows, use haver to read the data; see [TS] **haver**.

    e. Translate the data into ASCII (also known as character) format using the other software. For instance, in Excel you can save spreadsheets as tab-delimited or comma-separated text. Then, see [D] **insheet**.

    f. Other software packages are available that will convert non-Stata-format data files into Stata-format files; see [U] **21.4 Transfer programs**.

    g. If the data are located in an ODBC source, which typically includes databases and spreadsheets, you can use the odbc load command to import the data; see [D] **odbc**. Currently odbc is available for Windows, Macintosh, and Linux versions of Stata.

3. If the dataset has one observation per line and the data are tab or comma separated, use insheet; see [D] **insheet**. This is the easiest way to read data.

4. If the dataset is formatted and that formatting information is required to interpret the data, you can use infile with a dictionary or infix; see [D] **infile (fixed format)** or [D] **infix (fixed format)**.

5. If there are no string variables, you can use infile without a dictionary: see [D] **infile (free format)**.

6. If all the string variables in the data are enclosed in (single or double) quotes, you can use infile without a dictionary; see [D] **infile (free format)**.

7. If the undelimited string variables have no blanks, you can use infile without a dictionary; see [D] **infile (free format)**.

8. If you make it to here, see [D] **infile (fixed format)** or [D] **infix (fixed format)**.

21.2.1 Entering data interactively

If you have a small amount of data, you can type the data directly into Stata; see [D] **input** or [D] **edit**. Otherwise, we assume that your data are stored on disk.

21.2.2   If the dataset is in binary format

Stata can read ASCII datasets, which is technical jargon for datasets composed of characters—datasets that can be typed on your screen or printed on your printer. The alternative, binary datasets, cannot be read by Stata. Binary datasets are quite popular, and almost every software package has its own binary format. Stata .dta datasets are an example of a binary format Stata can read. The Excel format is a binary format that Stata cannot read.

If your dataset is in binary format or in the internal format of another software package, you must either translate it into an XML format that Stata can understand (see [D] **xmlsave**), translate it into ASCII, or use some other program for conversion to Stata format. If this dataset is located in a database or an ODBC source, see [U] **21.5 ODBC sources**. If the dataset is in SAS XPORT format, you can read it using Stata's fdause command; see [D] **fdasave**. If the dataset is in Haver Analytics' .dat format, you can read it using Stata's haver command; see [TS] **haver**.

Detecting whether data are stored in binary format can be tricky. For instance, many Windows users wish to read data that have been entered into a word processor—let's assume Word. Unwittingly, they have stored the dataset as a Word document. The dataset looks like ASCII to them: When they look at it in Word, they see readable characters. The dataset seems to even pass the printing test in that Word can print it. Nevertheless, the dataset is not ASCII; it is stored in an internal Word format, and the data cannot really pass the printing test since only Word can print it. To read the dataset, Windows users must use it in Word and then store it as an MS-DOS text file, MS-DOS text being the term Word decided to use to mean ASCII.

So, how do you know whether your dataset is binary? Here's a simple test: Regardless of the operating system you use, start Stata and type type followed by the name of the file:

```
. type myfile.raw
output will appear
```

You do not have to print the entire file; press *Break* when you have seen enough.

Do you see things that look like hieroglyphics? If so, the dataset is binary. See [U] **21.4 Transfer programs** below.

If it looks like data, however, the file is (probably) ASCII.

Let us assume that you have an ASCII dataset that you wish to read. The data's format will determine the command you need to use. The different formats are discussed in the following sections.

21.2.3   If the data are simple

The easiest way to read data is with insheet; see [D] **insheet**.

insheet is smart: it looks at the dataset, determines what it contains, and then reads it. That is, insheet is smart given certain restrictions, such as that the dataset has one observation per line and that the values are tab or comma separated. insheet can read this

———————————————————————————————— top of data1.raw ————

```
M,Joe Smith,288,14
M,K Marx,238,12
F,Farber,211,7
```

———————————————————————————————— end of data1.raw ————

or this (which has variable names on the first line)

```
——————————————————————————————————— top of data2.raw ———————————
sex, name, dept, division
M,Joe Smith,288,14
M,K Marx,238,12
F,Farber,211,7
——————————————————————————————————— end of data2.raw ———————————
```

or this (which has one tab character separating the values):

```
——————————————————————————————————— top of data3.raw ———————————
M       Joe Smith    288      14
M       K Marx  238  12
F       Farber  211  7
——————————————————————————————————— end of data3.raw ———————————
```

This looks odd because of how tabs work; `data3.raw` could similarly have a variable header, but `insheet` cannot read

```
——————————————————————————————————— top of data4.raw ———————————
M       Joe Smith    288      14
M       K Marx       238      12
F       Farber       211      7
——————————————————————————————————— end of data4.raw ———————————
```

which has spaces rather than tabs!

There is a way to tell `data3.raw` from `data4.raw`: Ask Stata to type the data and show the tabs by typing

```
. type data3.raw, showtabs
M<T>Joe Smith<T>288<T>14
M<T>K Marx<T>238<T>12
F<T>Farber<T>211<T>7
. type data4.raw, showtabs
M       Joe Smith    288      14
M       K Marx       238      12
F       Farber       211      7
```

21.2.4   If the dataset is formatted and the formatting is significant

If the dataset is formatted and formatting information is required to interpret the data, see [D] **infile (fixed format)** or [D] **infix (fixed format)**.

Using `infix` or `infile` with a data dictionary is something new users want to avoid if at all possible.

The purpose of this section is only to take you to the most complicated of all cases if there is no alternative. Otherwise, you should wait and see if it is necessary. Do not misinterpret this section and say, "Ah, my dataset is formatted, so at last I have a solution."

Just because a dataset is formatted does not mean that you have to exploit the formatting information. The following dataset is formatted,

```
————————————————————————————————————————— top of data5.raw ———————————
   1    27.39      12
   2     1.00       4
   3   100.10     100
————————————————————————————————————————— end of data5.raw ———————————
```

in that the numbers line up in neat columns, but you do not need to know the information to read it. Alternatively, consider the same data run together:

```
——————————————————————————————————————————— top of data6.raw ———————————
   1 27.39 12
   2  1.00  4
   3100.10100
——————————————————————————————————————————— end of data6.raw ———————————
```

This dataset is formatted, too, and you must know the formatting information in order to make sense of "3100.10100". You must know that variable 2 starts in column 4 and is six characters long to extract the 100.10. It is datasets like data6.raw that you should be looking for at this stage—datasets that only make sense if you know the starting and ending columns of data elements. To read data such as data6.raw, you must use either infix or infile with a data dictionary.

It should be obvious why reading unformatted data is easier. If you need the formatting information to interpret the data, then you must communicate that information to Stata, which means that you will have to type it. This is the hardest kind of data to read, but Stata can do it. See [D] **infile (fixed format)** or [D] **infix (fixed format)**.

Looking back at data4.raw,

```
——————————————————————————————————————————— top of data4.raw ———————————
   M        Joe Smith      288      14
   M        K Marx         238      12
   F        Farber         211       7
——————————————————————————————————————————— end of data4.raw ———————————
```

you may be uncertain whether you have to read it with a data dictionary. If you are uncertain, do not jump yet.

Finally, here is an obvious example of unformatted data:

```
——————————————————————————————————————————— top of data7.raw ———————————
   1 27.39             12
   2 1 4
   3 100.1 100
——————————————————————————————————————————— end of data7.raw ———————————
```

In this case, blanks separate one data element from the next and, in one case, lots of blanks, although there is no special meaning attached to more than one blank.

The following sections discuss datasets that are unformatted or formatted in a way that do not require a data dictionary.

21.2.5   If there are no string variables

If there are no string variables, see [D] **infile (free format)**.

Although the dataset data7.raw is unformatted, it can still be read using infile without a dictionary. This is not the case with data4.raw because this dataset contains undelimited string variables with embedded blanks.

❏ Technical Note

Some Stata users prefer to read data with a data dictionary, even when we suggest differently, as above. They like the convenience of the data dictionary—they can sit in front of an editor and carefully compose the list of variables and attach variable labels rather than having to type the variable list (correctly) on the Stata command line. What they should understand, however, is that they can create a do-file containing the `infile` statement and thus have all the advantages of a data dictionary without some of the (extremely technical) disadvantages of data dictionaries.

Nevertheless, we do tend to agree with such users—we, too, prefer data dictionaries. Our recommendations, however, are designed to work in all cases. If the dataset is unformatted and contains no string variables, it can always be read without a data dictionary, whereas only in some cases can it be read with a data dictionary.

The distinction is that `infile` without a data dictionary performs stream I/O, whereas with a data dictionary it performs record I/O. The difference is intentional—it guarantees that you will be able to read your data into Stata somehow. Some datasets require stream I/O, others require record I/O, and still others can be read either way. Recommendations 1–5 identify datasets that either require stream I/O or can be read either way.

❏

We are now left with datasets that contain at least one string variable.

21.2.6   If all the string variables are enclosed in quotes

If all the string variables in the data are enclosed in (single or double) quotes, see [D] **infile (free format)**.

See [U] **23 Dealing with strings** for a formal definition of strings, but as a quick guide, a string variable is a variable that takes on values like "bob", "joe", etc., as opposed to numeric variables that take on values like 1, 27.5, and –17.393. Undelimited strings—strings not enclosed in quotes—can be difficult to read.

Here is an example including delimited string variables:

```
──────────────────────────────────────── top of data8.raw ────────────
"M" "Joe Smith" 288 14
"M" "K Marx" 238 12
"F" "Farber" 211 7
──────────────────────────────────────── end of data8.raw ────────────
```

or

```
──────────────────────── top of data8.raw, alternative format ─────────
"M" "Joe Smith" 288   14
"M" "K Marx"     238   12
"F" "Farber"     211    7
──────────────────────── end of data8.raw, alternative format ─────────
```

Both of these are merely variations on `data4.raw` except that the strings are enclosed in quotes. In this case, `infile` without a dictionary can be used to read the data.

Here is another version of `data4.raw` without delimiters or even formatting:

```
──────────────────────────────────────── top of data9.raw ────────────
M Joe Smith 288 14
M K Marx 238 12
F Farber 211 7
──────────────────────────────────────── end of data9.raw ────────────
```

What makes these data difficult? Blanks sometimes separate values and sometimes are nothing more than a blank within a string. For instance, you cannot tell whether Farber has first initial F with missing sex or is instead female with a missing first initial.

Fortunately, such data rarely happen. Either the strings are delimited, as we showed in `data8.raw`, or the data are in columns, as in `data4.raw`.

21.2.7   If the undelimited strings have no blanks

There is a case in which uncolumnized, undelimited strings cause no confusion — when they contain no blanks. For instance, if our data contained only last names,

```
───────────────────────────────────────────── top of data10.raw ───────────
   Smith 288 14
   Marx 238 12
   Farber 211 7
───────────────────────────────────────────── end of data10.raw ───────────
```

Stata could read it without a data dictionary. Caution: the last names must contain no blanks — no Van Owen's or von Beethoven's.

If the undelimited string variables have no blanks, see [D] **infile (free format)**.

21.2.8   If you make it to here

If you make it to here, see [D] **infile (fixed format)** or [D] **infix (fixed format)**.

Remember `data4.raw`?

```
───────────────────────────────────────────── top of data4.raw ───────────
   M        Joe Smith       288       14
   M        K Marx          238       12
   F        Farber          211        7
───────────────────────────────────────────── end of data4.raw ───────────
```

It can be read using either `infile` with a dictionary or `infix`.

21.3   If you run out of memory

You can increase the amount of memory allocated to Stata; see [U] **6 Setting the size of memory**.

You can also try to conserve memory.

When you read the data, did you specify variable types? Stata can store integers more compactly than floats, and small integers more compactly than large integers; see [U] **12 Data**.

If that is not sufficient, you will have to resort to reading the data in pieces. Both `infile` and `infix` allow you to specify an in *range* modifier, and, in this case, the range is interpreted as the observation range to read. Thus, `infile ... in 1/100` would read observations 1 through 100 of your data and stop.

`infile ... in 101/200` would read observations 101 through 200. The end of the range may be specified as larger than the actual number of observations in the data. If the dataset contained only 150 observations, `infile ... in 101/200` would read observations 101 through 150.

Another way of reading the data in pieces is to specify the if *exp* modifier. Say that your data contained an equal number of males and females, coded as the variable sex (which you will read) being 0 or 1, respectively. You could type infile ... if sex==0 to read the males. infile will read an observation, determine if sex is zero, and if not, throw the observation away. Obviously, you could read just the females by typing infile ... if sex==1.

If the dataset is really big, perhaps you only need a random sample of the data—you never intended to analyze the entire dataset. Since infile and infix allow if *exp*, you could type infile ... if uniform()<.1. uniform() is the uniformly distributed random-number generator; see [D] **functions**. This method would read an approximate 10% sample of the data. If you are serious about using random samples, do not forget to set the seed before using uniform(); see [D] **generate**.

The final approach is to read all the observations but only a subset of the variables. When reading data without a data dictionary, you can specify _skip for variables, indicating that the variable is to be skipped over. When reading with a data dictionary or using infix, you can specify the actual columns to read, skipping any columns you wish to ignore.

21.4   Transfer programs

To import data from, say, Excel, you can save the data as a text file and then read it in to Stata according to the rules above, read it via an ODBC source, or purchase a program to translate the dataset from Excel's format to Stata's format.

One such program is Stat/Transfer, which is available for Windows, Macintosh OS X, and Unix. It reads and writes data in a variety of formats, including Microsoft Access, dBASE, Epi Info, Excel, FoxPro GAUSS, JMP, Lotus 1-2-3, MATLAB, ODBC, Paradox, Quattro Pro, S-Plus, SAS, SPSS, Statistica, SYSTAT, and, of course, Stata.

Stat/Transfer, available from...	is manufactured by...
StataCorp	Circle Systems
4905 Lakeway Drive	1001 Fourth Avenue Plaza, Suite 3200
College Station, Texas 77845	Seattle, Washington 98154
Telephone: 979-696-4600	Telephone: 206-682-3783
Fax: 979-696-4601	Fax: 206-328-4788
Email: *stata@stata.com*	*sales@circlesys.com*

There are other transfer programs available, too. Our web site, *http://www.stata.com*, lists programs available from other sources.

21.5   ODBC sources

If your dataset is located in a network database or shared spreadsheet, you may be able to import your data via ODBC. ODBC (Open Database Connectivity) is a standard for exchanging data between programs. Stata supports the ODBC standard for importing data via the odbc command and can read from any ODBC source on your computer.

This process requires a data source, such as a database located on a network. To use the odbc command to import data from a database requires that the database first be set up as an ODBC source on the same machine that is running Stata. The database itself does not have to be on the same machine, just the definition of that database as the ODBC source. On a Windows machine, an ODBC source is added via a Control Panel called "Data Sources." Additionally, typing odbc list from Stata displays all the ODBC sources that are provided by the computer.

Assuming that the database is functioning and that the appropriate data source has been set up on the same machine as Stata, a single call using `odbc load` is all that is needed to import data. For a more thorough description of this process, see [D] **odbc**.

21.6   Reference

Swagel, P. 1994. os14: A program to format raw data files. *Stata Technical Bulletin* 20: 10–12. Reprinted in *Stata Technical Bulletin Reprints*, vol. 4, pp. 80–82.

22 Combining datasets

You have two datasets that you wish to combine. Below, we will draw a dataset as a box where, in the box, the variables go across and the observations go down.

See [D] **append** if you want to combine datasets vertically:

append adds observations to the existing variables. That is an oversimplification because append does not require that the datasets have exactly the same variables. append is appropriate, for instance, when you have data on hospital patients and then receive data on more patients.

See [D] **merge** if you want to combine datasets horizontally:

merge adds variables to the existing observations. That is an oversimplification because merge does not require that the datasets have exactly the same observations. merge is appropriate, for instance, when you have data on survey respondents and then receive data on part 2 of the questionnaire.

See [D] **joinby** when you want to combine datasets horizontally but form all pairwise combinations within group:

joinby is similar to merge, but forms all combinations of the observations where it makes sense. joinby would be appropriate, for instance, where A contained data on parents and B contained data on their children. joinby *familyid* would form a dataset of each parent joined with each of his or her children.

Also see [D] **cross** for a less frequently used command that forms every pairwise combination of two datasets.

23 Dealing with strings

Contents

Please read [U] **12 Data** before reading this entry.

23.1 Description

The word *string* is shorthand for a string of characters. "Male" and "Female", "yes" and "no", and "R. Smith" and "P. Jones" are examples of strings. The alternative to strings is numbers—0, 1, 2, 5.7, and so on. Variables containing strings—called *string variables* —occur in data for a variety of reasons. Four of these reasons are listed below.

A variable might contain strings because it is an identifying variable. Employee names in a payroll file, patient names in a hospital file, and city names in a city data file are all examples of this. This is a proper use of string variables.

A variable might contain strings because it records categorical information. "Male" and "Female" and "yes" and "no" are examples of such use, but this is not an appropriate use of string variables. It is not appropriate because the same information could be coded numerically, and, if it were, (1) it would take less memory to store the data, and (2) the data would be more useful. We will explain how to convert categorical strings to categorical numbers below.

In addition, a variable might contain strings because of a mistake. For example, the variable contains things like 1, 5, 8.2, but due to an error in reading the data, the data were mistakenly put into a string variable. We will explain how to fix such mistakes.

Finally, a variable might contain strings because the data simply could not be coerced into being stored numerically. "15 Jan 1992", "1/15/92", and "1A73" are examples of such use. We will explain how to deal with such complexities.

23.2 Categorical string variables

A variable might contain strings because it records categorical information.

Suppose that you have read in a dataset that contains a variable called `sex`, recorded as "male" and "female", yet when you attempt to run an ANOVA, the following message is displayed:

```
. use http://www.stata-press.com/data/r9/hbp2
. anova hbp sex
no observations
r(2000);
```

There are no observations because `anova`, along with most of Stata's "analytic" commands, cannot deal with string variables. Commands want to see numbers, and when they do not, they treat the variable as if it contained numeric missing values. Despite this limitation, it is possible to obtain tables:

```
. encode sex, gen(gender)
. anova hbp gender
```

	Number of obs =	1128	R-squared	=	0.0123
	Root MSE = .214223		Adj R-squared =		0.0114

Source	Partial SS	df	MS	F	Prob > F
Model	.644485682	1	.644485682	14.04	0.0002
gender	.644485682	1	.644485682	14.04	0.0002
Residual	51.6737767	1126	.045891454		
Total	52.3182624	1127	.046422593		

The magic here is to convert the string variable `sex` into a numeric variable called `gender` with an associated value label, a trick accomplished by `encode`; see [U] **12.6.3 Value labels** and [D] **encode**.

23.3   Mistaken string variables

A variable might contain strings because of a mistake.

Suppose that you have numeric data in a variable called `x`, but due to a mistake, `x` was made a string variable when you read the data. When you `list` the variable, it looks fine:

```
. list x
```

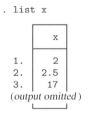

```
        ┌──────┐
        │   x  │
        ├──────┤
   1.   │   2  │
   2.   │  2.5 │
   3.   │  17  │
        └──────┘
 (output omitted)
```

Yet, when you attempt to obtain summary statistics on `x`,

```
. summarize x
```

Variable	Obs	Mean	Std. Dev.	Min	Max
x	0				

If this happens to you, type `describe` to confirm that `x` is stored as a string:

```
. describe
Contains data
  obs:           10
  vars:           3
  size:         160 (99.9% of memory free)
```

variable name	storage type	display format	value label	variable label
x	str4	%9s		
y	float	%9.0g		
z	float	%9.0g		

```
Sorted by:
```

`x` is stored as an `str4`.

The problem is that `summarize` does not know how to calculate the mean of string variables—how to calculate the mean of "Joe" plus "Bill" plus "Roger"—even when the string variable contains what could be numbers. By using the `destring` command ([D] **destring**), the variable mistakenly stored as a `str4` can be converted to a numeric variable.

```
. destring x, replace
. summarize x
    Variable |      Obs        Mean    Std. Dev.       Min        Max
-------------+--------------------------------------------------------
        newx |       10        1.76    .8071899         .7          3
```

An alternative to using the `destring` command is to use `generate` with the `real()` function; see [D] **functions**.

23.4   Complex strings

A variable might contain strings because the data simply could not be coerced into being stored numerically.

A complex string is a string that contains more than one piece of information. The most common example of a complex string is a date: "15 Jan 1992" contains three pieces of information—a day, a month, and a year. If your complex strings are dates, see [U] **24 Dealing with dates**.

Although Stata has functions for dealing with dates, you will have to deal with other complex strings yourself. For instance, assume that you have data that include part numbers:

```
. list partno
```

```
     | partno |
     |--------|
  1. | 5A2713 |
  2. | 2B1311 |
  3. | 8D2712 |
     (output omitted )
```

The first digit of the part number is a division number, and the character that follows identifies the plant at which the part was manufactured. The next three digits represent the major part number and the last digit is a modifier indicating the color. This complex variable can be decomposed using the `substr()` and `real()` functions described in [D] **functions**:

```
. gen byte div = real(substr(partno,1,1))
. gen str1 plant = substr(partno,2,1)
. gen int part = real(substr(partno,3,3))
. gen byte color = real(substr(partno,6,1))
```

We use the `substr()` function to extract pieces of the string and use the `real()` function, when appropriate, to translate the piece into a number.

For an extended discussion of numeric and string data types and how to convert from one kind to another, see Cox (2002).

23.5   Reference

Cox, N. J. 2002. Speaking Stata: On numbers and strings. *Stata Journal* 2: 314–329.

24 Dealing with dates

24.1 Overview

You can record dates however you want, but there is one format that Stata understands, called elapsed dates or %d dates. A %d date is the number of days from January 1, 1960. In this format,

0	corresponds to	01jan1960			
1	corresponds to	02jan1960	−1	corresponds to	31dec1959
2	corresponds to	03jan1960	−2	corresponds to	30dec1959
⋮			⋮		
31	corresponds to	01feb1960	−31	corresponds to	01dec1959
⋮			⋮		
15,000	corresponds to	25jan2001	−15,000	corresponds to	01dec1918
⋮			⋮		
2,936,549	corresponds to	31dec9999	−679,350	corresponds to	01jan0100

This format can be used with dates 01jan0100–31dec9999, although caution should be exercised in dealing with dates before Friday, 15oct1582, because that is when the Gregorian calendar went into effect.

Stata provides functions to convert dates into %d dates, formats to print %d dates in understandable forms, and other functions to manipulate %d dates.

Use of %d dates is described in [U] **24.2 Dates** below.

In addition to %d dates, Stata has five other date formats, called %t dates, that it understands:

Format	Description		Coding	
%td	daily (same as %d)	−1 = 31dec1959,	0 = 01jan1960,	1 = 02jan1960
%tw	weekly	−1 = 1959w52,	0 = 1960w1,	1 = 1960w2
%tm	monthly	−1 = 1959m12,	0 = 1960m1,	1 = 1960m2
%tq	quarterly	−1 = 1959q4,	0 = 1960q1,	1 = 1960q2
%th	half-yearly	−1 = 1959h2,	0 = 1960h1,	1 = 1960h2
%ty	yearly	1959 = 1959,	1960 = 1960,	1961 = 1961

Use of %t dates is described in [U] **24.3 Time-series dates** below.

24.2 Dates

In this section, we discuss %d dates, also called elapsed dates.

24.2.1 Inputting dates

The trick to inputting dates in Stata is to forget they are dates. Input them as strings and then later convert them into Stata's elapsed dates. You might have

——————————————————————————————————— top of bdays.raw ———————

```
Bill  21 Jan 1952  22
May   11 Jul 1948  18
Sam   12 Nov 1960  25
Kay    9 Aug 1975  16
```

——————————————————————————————————— end of bdays.raw ———————

and if you did, you could read these data by typing

```
. infix str name 1-5  str bday 7-17  x 20-21 using bdays
(4 observations read)
```

If you now listed the data, the data would look fine,

```
. list
```

	name	bday	x
1.	Bill	21 Jan 1952	22
2.	May	11 Jul 1948	18
3.	Sam	12 Nov 1960	25
4.	Kay	9 Aug 1975	16

but you would find there is not much you could do with bday because it is just a string variable. Turning it into a date Stata understands is easy,

```
. gen birthday = date(bday,"dmy")
. list
```

	name	bday	x	birthday
1.	Bill	21 Jan 1952	22	−2902
2.	May	11 Jul 1948	18	−4191
3.	Sam	12 Nov 1960	25	316
4.	Kay	9 Aug 1975	16	5699

and making the numeric birthday variable look like a date is equally easy:

```
. format birthday %d
. list
```

	name	bday	x	birthday
1.	Bill	21 Jan 1952	22	21jan1952
2.	May	11 Jul 1948	18	11jul1948
3.	Sam	12 Nov 1960	25	12nov1960
4.	Kay	9 Aug 1975	16	09aug1975

The convenient thing about the variable `birthday` is that it is numeric, which means you can make calculations on it. How old was each of these people on January 1, 2000? It is easy to add such a variable:

```
. gen age2000 = (mdy(1,1,2000)-birthday)/365.25
. list
```

	name	bday	x	birthday	age2000
1.	Bill	21 Jan 1952	22	21jan1952	47.94524
2.	May	11 Jul 1948	18	11jul1948	51.47433
3.	Sam	12 Nov 1960	25	12nov1960	39.13484
4.	Kay	9 Aug 1975	16	09aug1975	24.39699

24.2.2   Conversion into elapsed dates

Two functions are provided—`mdy()` and `date()`—for converting variables into elapsed dates.

24.2.2.1   The mdy() function

`mdy()` takes three numeric arguments—a month, day, and year—and returns the corresponding elapsed date. For instance,

```
. list
```

	month	day	year
1.	7	11	1948
2.	1	21	1952
3.	11	2	1994
4.	8	12	93

```
. gen edate = mdy(month,day,year)
(1 missing value generated)
. list
```

	month	day	year	edate
1.	7	11	1948	-4191
2.	1	21	1952	-2902
3.	11	2	1994	12724
4.	8	12	93	.

Note that in the last observation, `mdy()` produced missing. It did this because the year was 93, and `mdy()` does not assume 93 means 1993.

24.2.2.2 The date() function

The second way to convert to elapsed dates is with the date() function. date() takes two string arguments. There is a variation on this—two strings arguments and a numeric argument—but let's postpone that. The first argument is the date to be converted. The second argument tells date() the order of the month, day, and year in the first argument. Typing date(*strvar*,"mdy") means that *strvar* contains the month, day, and year in that order. Typing date(*strvar*,"dmy") means *strvar* contains the day, month, and year. Knowing the order, date() allows *strvar* to be in almost any format. For instance,

```
. list
```

	mydate
1.	7-11-1948
2.	1/21/52
3.	11.2.1994
4.	Aug 12,1993
5.	Sept 11,2002
6.	November 13, 2005

```
. gen edate = date(mydate, "mdy")
(1 missing value generated)
. list
```

	mydate	edate
1.	7-11-1948	-4191
2.	1/21/52	.
3.	11.2.1994	12724
4.	Aug 12,1993	12277
5.	Sept 11,2002	15594
6.	November 13, 2005	16753

or, if you prefer,

```
. list
```

	mydate
1.	11-7-1948
2.	21/1/52
3.	2.11.1994
4.	12 Aug 1993
5.	11Sept2002
6.	13 November 2005

```
. gen edate = date(mydate, "mdy")
(1 missing value generated)
```

```
. list
```

	mydate	edate
1.	11-7-1948	-4191
2.	21/1/52	.
3.	2.11.1994	12724
4.	12 Aug 1993	12277
5.	11Sept2002	15594
6.	13 November 2005	16753

date() can deal with virtually any date format: all it needs to know is the order of the month, day, and year, and that you indicate by the second argument using the letters m, d, and y. Second argument "mdy" means month–day–year order, "dmy" means day–month–year order, and so on.

Note, however, that like mdy(), date() refused to translate two-digit years: 1/21/52 and 21/1/52 both translated to missing. Unlike mdy(), date() would be willing to assume that 52 means 1952 or 2052 if you will tell it which. There are two ways to do this.

The first way involves specifying a default century, and you do that using date()'s second argument. Specify "md19y" or "dm20y", and date() will assume that two-digit years should be interpreted as being prefixed by 19 or 20; four-digit years will still be correctly interpreted no matter which default you specify.

The second way involves specifying date()'s third argument. Specify date(...,...,2040), and date() will assume that two-digit years should be interpreted as the maximum year not greater than 2040. 52 would be interpreted as 1952, but 39 would be interpreted as 2039. 40 would be interpreted as 2040. You can specify whatever third argument works best for your data.

If you do neither, then two-digit years cannot be translated, and that is why we saw the missing values in the examples above; date() could not translate 1/21/52 (21/1/52 when we varied the order). We could have translated 1/21/52 had we typed

```
. gen edate = date(mydate, "md19y")
```

or

```
. gen edate = date(mydate, "mdy",2040)
```

Either method would translate 1/21/52 as 21jan1952, but they would differ on how they would treat dates with two-digit years 00, 01, ..., 40. The first method would treat them as 1900, 1901, ..., 1940, whereas the second would treat them as 2000, 2001, ..., 2040.

To summarize: To get dates into Stata, either create three numeric variables containing the month, day, and year and use mdy() to convert them, or create a string variable containing the date in whatever format and use date() to translate it. If you are reading date data into Stata, the easiest way is to read the data into a string and then use date().

❏ Technical Note

How date() *works.* The date to be converted has three pieces of information, the month, day, and year, and the second argument specifies the order, of which there are six possibilities, "dmy", "mdy", "ymd", "ydm", "dym", and "myd", although the last three rarely occur. Knowing the order, date() examines the contents of the first argument and looks for transitions, meaning any separating character such as blanks, commas, dashes, and slashes, or changes from numeric to alpha or alpha to numeric. This allows dividing the source into its three components for translation. If the source divides into other than three components, or if it divides into three but they do not make sense, date() returns a missing value.

If you have two-digit years, date() will return missing unless you specify a default century on the second argument—e.g., "md19y"—or you specify a third argument.

date() can translate virtually any format except formats where all three elements run together and the months are indicated numerically, such as 012152 or 520121. It is the lack of blanks or other separating characters that confuses date(); date could translate 01 21 52 or 52 01 21. date() could also translate 21Jan52—note the absence of blanks—because the string Jan makes clear the separation.

Let us assume you have a date of the form 520121—the order is year, month, and day—stored in a numeric variable called ymd. Here is how you might translate it:

```
. gen year = int(ymd/10000)
. gen month = int((ymd-year*10000)/100)
. gen day = ymd - year*10000  - month*100
. gen edate = mdy(month, day, 1900 + year)
```

❏

24.2.3   Displaying dates

%d elapsed dates are convenient for computers and sometimes even for humans—you can, for instance, subtract them to obtain the number of days between dates. Nevertheless, they are unreadable. For instance, here are some birth dates:

```
. list
```

	bdate
1.	-2902
2.	-4191
3.	316
4.	5699

All you need do to make such dates readable is assign Stata's %d format to the variable:

```
. format bdate %d
. list bdate
```

	bdate
1.	21jan1952
2.	11jul1948
3.	12nov1960
4.	09aug1975

If you now saved the data, the date would forevermore be displayed in this format.

You may find the format 21jan1952 unappealing, and, if so, you can modify it. The %d format is equivalent to %dD1CY, meaning first display the day (D), then the month abbreviation in lowercase (1), then the century (C), and finally the year without the century (Y). This default was selected because, by using an abbreviated month, it makes clear the order of the day and month, and, by omitting the blanks, it is short.

Here is a variation on the format:

```
. format bdate %dD_m_CY
. list
```

	bdate
1.	21 Jan 1952
2.	11 Jul 1948
3.	12 Nov 1960
4.	09 Aug 1975

And here are two more variations:

```
. format bdate %dN/D/Y
. list
```

	bdate
1.	01/21/52
2.	07/11/48
3.	11/12/60
4.	08/09/75

```
. format bdate %dM_D,_CY
. list
```

	bdate
1.	January 21, 1952
2.	July 11, 1948
3.	November 12, 1960
4.	August 09, 1975

You can specify simply %d or you can follow the %d with up to 11 characters that tell Stata what to display. Here is what the characters mean:

(Continued on next page)

C	Display the century of the year with a leading 0; year 500 is 05, year 1994 is 19, year 2002 is 20.
c	Display the century of the year without a leading 0; year 500 is 5, year 1994 is 19.
Y	Display the year within century with a leading 0; 1908 is 08, 1994 is 94, 2002 is 02.
y	Display the year within century without a leading 0; 1908 is 8, 1994 is 94, 2002 is 2.
M	Display the month spelled out; January is January, February is February, ..., December is December.
m	Display the 3-letter abbreviation of month; January is Jan, February is Feb, ..., December is Dec.
L	Same as M except the month is presented in all lowercase; January is january, February is february, ..., December is december.
l	Same as m except the 3-letter abbreviation is in all lowercase; January is jan, February is feb, ..., December is dec.
N	Display the numeric month with a leading 0; January is 01, February is 02, ..., December is 12.
n	Display the numeric month without a leading 0; January is 1, February is 2, ..., December is 12.
D	Display the day with a leading 0; 1 is 01, 2 is 02, ..., 31 is 31.
d	Display the day without a leading 0; 1 is 1, 2 is 2, ..., 31 is 31.
J	display day-within-year with leading 0s (001 to 366)
j	display day-within-year without leading 0s (1 to 366)
h	display half of year (1 to 2)
q	display quarter of year (1, 2, 3, or 4)
W	display week of year with leading 0 (01 to 52)
w	display week of year without leading 0 (1 to 52)
_	Display a blank.
.	Display a period.
,	Display a comma.
:	Display a colon.
-	Display a dash.
/	Display a slash.
'	Display a close single quote.
!c	display character c (code !! to display exclamation point)

When using the detail characters, you need not specify all the components of the date. In our birth dates, if we wanted to see just the month and year, we might

```
. format bdate %dm,_CY
. list
```

	bdate
1.	Jan, 1952
2.	Jul, 1948
3.	Nov, 1960
4.	Aug, 1975

24.2.4 Other date functions

How you display the date does not matter; the variable itself always contains the number of days from January 1, 1960. Given a date variable, the following functions extract information from it:

year(*date*)	returns four-digit year; e.g., 1980, 2002
month(*date*)	returns month; 1, 2, ..., 12
day(*date*)	returns day within month; 1, 2, ..., 31
halfyear(*date*)	returns the half of year; 1 or 2
quarter(*date*)	returns quarter of year; 1, 2, 3, or 4
week(*date*)	returns week of year; 1, 2, ..., 52
dow(*date*)	returns day of week; 0, 1, ..., 6; 0 = Sunday
doy(*date*)	returns day of year; 1, 2, ..., 366

For example,

```
. gen m = month(bdate)
. gen d = day(bdate)
. gen y = year(bdate)
. gen week_d = dow(bdate)

. list
```

	bdate	m	d	y	week_d
1.	Jan, 1952	1	21	1952	1
2.	Jul, 1948	7	11	1948	0
3.	Nov, 1960	11	12	1960	6
4.	Aug, 1975	8	9	1975	6

dow() returns 0–6, 0 meaning Sunday, 1 Monday, ..., 6 Saturday. Thus, the person born on January 21, 1952 was born on a Monday.

24.2.5 Specifying particular dates (date literals)

If you work with dates, you will want to type dates in expressions. For instance, in a previous example when we needed to calculate the age of persons as of 1jan2000, we typed

```
. gen age2000 = (mdy(1,1,2000)-birthday)/365.25
```

We used mdy() to obtain 1jan2000 as an elapsed date. Alternatively, we could have used Stata's d(*constant*) function

```
. gen age2000 = (d(1jan2000)-birthday)/365.25
```

d() is unusual in that you cannot type any expression inside the parentheses; instead, you must type a day followed by a month followed by a four-digit year. Do that, however, and d() returns the corresponding %d date value.

You may type the day, month, and year however you wish, but you must specify them in the order day, month, and year. Stata will understand d(1/1/2000) or d(1-1-2000) or d(1.1.2000) or d(1 jan 2000) or d(1 January 2000), etc., but it would not understand d(jan.1,2000):

```
. gen age2000 = (d(jan.1,2000)-birthday)/365.25
d(jan.1,2000) invalid
r(198);
```

In addition, if you type the date in a style where two numbers appear one after the other, you must put some form of punctuation other than a space between the two numbers: You would think Stata would understand d(1 1 2000), but it does not because spaces disappear in expressions:

```
. gen age2000  = (d(1 1 2000)-birthday)/365.25
d(112000) invalid
r(198);
```

When you spell out the month, you can include spaces or not:

```
. gen age2000 = (d(1 jan 2000)-birthday)/365.25
```

Finally, d() is allowed only in expressions. There may be occasions when you need to specify the numeric value of a date in the option of some command, such as

```
. lowess value time, xline(d(15apr1998))
xline() invalid
r(198);
```

Unfortunately, Stata does not understand xline(d(15apr1998)). In such cases, you can use display to obtain the numeric equivalent:

```
. display d(15apr1998)
13984

. lowess value time, xline(13984)
  (output omitted )
```

24.3   Time-series dates

In addition to %d date formats, Stata has five other date formats called %t or time-series dates. These dates work like %d in that 1jan1960 is mapped to 0, but the meaning of 1 varies:

Format	Description		Coding	
%td	daily (same as %d)	−1 = 31dec1959,	0 = 01jan1960,	1 = 02jan1960
%tw	weekly	−1 = 1959w52,	0 = 1960w1,	1 = 1960w2
%tm	monthly	−1 = 1959m12,	0 = 1960m1,	1 = 1960m2
%tq	quarterly	−1 = 1959q4,	0 = 1960q1,	1 = 1960q2
%th	half-yearly	−1 = 1959h2,	0 = 1960h1,	1 = 1960h2
%ty	yearly	1959 = 1959,	1960 = 1960,	1961 = 1961

To best understand these formats, think of what happens when variable d contains a date and you add 1 to it. If d is in %td format, you obtain the next day. If d is in %tw format, you obtain the next week. If d is in %tm format, you obtain the next month. If d is in %tq format, you obtain the next quarter. If d is in %th format, you obtain the next half-year. If d is in %ty format, you obtain the next year.

Or, think of it like this: subtract two dates and you obtain the number of days, weeks, months, quarters, half-years, or years between them.

%td daily format.
   This format is equivalent to %d. In any context, whether a variable is %td or %d makes no difference. Examples of dates in this format are 12jun1998, 22dec2002, etc.

%tw weekly format.
   The year is divided into 52 weeks: week 1 is the first 7 days of year, week 2 the second 7 days, and so on. Since years have just over 52 weeks in them, the 52nd week is defined as having 8 or 9 days. Examples of dates in this format are 1998w24, 2002w52, etc.

%tm monthly format.
> The year is divided into the 12 calendar months. Examples of dates in this format are 1998m6, 2002m12, etc.

%tq quarterly format.
> The year is divided into 4 quarters based on months; quarter 1 is January through March; quarter 2 is April through June; quarter 3 is July through September; and quarter 4 is October through December. Examples of dates in this format are 1998q2, 2002q4, etc.

%th half-yearly format.
> The year is divided into 2 halves based on months; half 1 is January through June and half 2 is July through December. Examples of dates in this format are 1998h1 and 2002h2.

%ty yearly format.
> The year is not divided at all. Examples of dates in this format are 1998 and 2002.

24.3.1 Inputting time variables

Our advice for inputting %d variables, summarized in [U] **24.2.1 Inputting dates** and [U] **24.2.2 Conversion into elapsed dates**, was

1. use the mdy() function if you have three integers recording the month, day, and year; or

2. input the date as a string and then use the date() function to convert it.

We offer the same advice for %t variables; just the names of the functions change:

format	integer conversion	string conversion
%td	mdy(*month*, *day*, *year*)	date(*string*, "md[##]y" or "dm[##]Y" ... , [*topyear*])
%tw	yw(*year*, *week*)	weekly(*string*, "w[##]y" or "[##]yw", [*topyear*])
%tm	ym(*year*, *month*)	monthly(*string*, "m[##]y" or "[##]ym", [*topyear*])
%tq	yq(*year*, *quarter*)	quarterly(*string*, "q[##]y" or "[##]yq", [*topyear*])
%th	yh(*year*, *halfyear*)	halfyearly(*string*, "q[##]y" or "[##]yq", [*topyear*])
%ty	*year*	yearly(*string*, "[##]y", [*topyear*])

For instance, just as mdy(5,30,1998) = 14,029 (30may1998),

    yw(1998,22) = 1,997 (1998w22)
    ym(1998,5) = 460 (1998m5)
    yq(1998,2) = 153 (1998q2)
    yh(1998,1) = 76 (1998h1)
    1998 = 1998

A dataset containing numeric variables year and quarter could be translated into a Stata %tq variable by typing

```
. gen date = yq(year,quarter)
(1 missing value generated)

. format date %tq
```

```
. list
```

	year	quarter	date
1.	1998	1	1998q1
2.	1999	5	.
3.	2005	3	2005q3

Note that the mistaken year 1999 quarter 5 translated to a missing value. Had our years all been in the 20th century and recorded in two digits (e.g., 95, 98, etc.), we would have typed 'generate date = yq(1900+year,quarter)'.

The string-conversion functions work just like the date() function: the second argument specifies the order in which the components of the date are expected to occur in the string; the y of the second argument may be prefixed with 19 or 20 as one way of handling two-digit years; a third argument specifying the maximum year may be specified as another way of handling two-digit years; and if you do not prefix y and do not specify a third argument, years in the string must be four digits.

For example, monthly(s,"my") could translate s containing "jan 1999" or "January, 1999" or "jan1999" or "1/1999" or "1-1999", but would return missing for "jan99". "jan99" could be decoded by specifying monthly(s,"m19y") (in which case it would be interpreted as 1jan1999) or by specifying, say, monthly(s,"my",2040) (in which case it would also be interpreted as 1jan1999 because 1999 is the maximum year not greater than 2040).

Below we translate a string variable sdate containing quarterly dates to a Stata date:

```
. use a different dataset
. gen date = quarterly(sdate,"yq",2040)
(1 missing value generated)
. format date %tq
. list
```

	sdate	date
1.	1995q2	1995q2
2.	1996 3	1996q3
3.	1996 5	.
4.	1997 quarter 1	1997q1
5.	98q.4	1998q4
6.	2001-3	2001q3
7.	2002q2	2002q2

24.3.2 Specifying particular dates (date literals)

Just as you can use the d(*constant*) function to type a date in an expression, such as

```
. gen age2000 = (d(1jan2000)-birthday)/365.25
```

Stata provides one-letter functions for typing weekly, monthly, quarterly, half-yearly, and yearly dates:

format	function	argument	examples
%td	d()	type day, month, year	d(15feb1998), d(15-5-2002)
%tw	w()	type year, week	w(1998w7), w(2002-25)
%tm	m()	type year, month	m(1998m2), m(2002-5)
%tq	q()	type year, quarter	q(1998q1), q(2002-2)
%th	h()	type year, half	h(1998h1), h(2002-1)
%ty	y()	type year	y(1998), y(2002)

For instance, if variable qtr contained a %tq date, you could type

```
. list if qtr>=q(1998q1)
```
(*output omitted*)

The y() function is included largely for completeness. For the %ty format, the year maps to the year, so 1960 means 1960, and there is little reason to type y(1960). Programmers, however, sometimes find y() useful in terms of syntax checking. y()—just as all the single-letter functions—produces an error when given an invalid date, which, in this case, means year < 100 or year > 9999.

24.3.3 Time-series formats

Just as with the %d format, the %td, %tw, %tm, %tq, %th, and %ty formats may be modified so that you can display the date in the form you wish. This is done using the same letter codes used with the %d format; see [U] **24.2.3 Displaying dates**. The default formats for each of the types are

format	means
%td	%tdD1CY
%tw	%twCY!ww
%tm	%tmCY!mn
%tq	%tqCY!qq
%th	%thCY!hh
%ty	%tyCY

Think of the %t format as

    %t⟨*character stating how data encoded*⟩⟨*optional characters saying how displayed*⟩

If you had variable qtr containing %tq dates and you wanted the dates displayed as, for instance, 1980 Q.1, you could type

```
. format qtr %tqCY_!Q.q
```

24.3.4   Translating between time units

A time unit, such as %tq quarterly, can be translated to any other time unit, such as %th half-yearly. Stata provides functions to translate any time unit to and from %td daily units. The trick is to combine these functions:

Input	Output %td daily	%tw weekly	%tm monthly	%tq quarterly	%th half-yearly	%ty yearly
%td		wofd(d)	mofd(d)	qofd(d)	hofd(d)	yofd(d)
%tw	dofw(w)		mofd(dofw(w))	qofd(dofw(w))	hofd(dofw(w))	yofd(dofw(w))
%tm	dofm(m)	wofd(dofm(m))		qofd(dofm(m))	hofd(dofm(m))	yofd(dofm(m))
%tq	dofq(q)	wofd(dofq(q))	mofd(dofq(q))		hofd(dofq(q))	yofd(dofq(q))
%th	dofh(h)	wofd(dofh(h))	mofd(dofh(h))	qofd(dofh(h))		yofd(dofh(h))
%ty	dofy(y)	wofd(dofy(y))	mofd(dofy(y))	qofd(dofy(y))	hofd(dofy(y))	

The functions that translate *from* %td dates—wofd(d), mofd(d), qofd(d), hofd(d), and yofd(d)—return the date containing d. Thus, qofd() of 6apr1998 is 1998q2.

The functions that translate *to* %td dates—dofw(), dofm(), dofq(), dofh(), and dofy()—return the %td date of the beginning of the period. Thus, dofq() of 1998q2 is 01apr1998.

24.3.5   Extracting components of time

With %d dates, functions month(), day(), and year() extract the month, day, and year. In fact, if you look back at [U] **24.2.4 Other date functions**, you will find that there are functions to extract the week (week()), quarter (quarter()), and so on.

To extract values from a %t date, combine these extraction functions with the dof*() functions. For instance, if variable qtr contains a %tq date and you want new variable q to contain the quarter, type

```
. gen q = quarter(dofq(qtr))
```

If you wanted to create a variable equal to 1 for the first two quarters of each year and 0 otherwise, you could type

```
. gen first = quarter(dofq(qtr))<=2
```

or

```
. gen first = halfyear(dofq(qtr))==1
```

24.3.6   Creating time variables

If you have data for which you know that the first observation is for the first quarter of 1990, the second for the second quarter of 1990, and so on, but the dataset contains no variable recording that fact, you can generate one by typing

```
. generate time = q(1990q1)+_n-1
. format time %tq
```

Remember that _n is Stata's built-in observation counter variable; _n = 1 in the first observation, 2 in the second, and so on. The single-letter functions make it easy to type a date.

24.3.7   Setting the time variable

If you find the %t format useful, it is likely you are performing time-series analysis and will find Stata's other time-series features, such as time-series varlists, useful as well. If so, you need to set the time variable to turn on those other features:

        . tsset time

The time variable you set does not need to be a %t variable, but results will be more readable if it is; see [TS] **tsset**.

24.3.8   Selecting periods of time

Once you have `tsset` a time variable, two new functions become available to you: `tin()` and `twithin()`. These functions are useful for selecting contiguous subsets of data:

        . regress y x l.x if tin(01jan1998,31dec2000)

        . list if tin(01jan1998,)

$tin(t_0,t_1)$ and $twithin(t_0,t_1)$ select observations in the range t_0 to t_1 and differ only in whether observations for which $t = t_0$ and $t = t_1$ are included (`tin()` includes them and `twithin()` excludes them).

The time variable t is assumed to be the time variable you previously `tsset`.

The arguments you specify should be typed in the same way you would type them using the single-letter d(), w(), m(), q(), h(), or y() functions; see [U] **24.3.2 Specifying particular dates (date literals)**. That means that what you type depends on the time variable's %t format.

If the time variable t does not have a %t variable, you type numbers.

        . list if tin(5,20)

means the same as

        . list if t>=5 & t<=20

where t is the name of the time variable you previously `tsset`.

If the time variable t has a %td or %d format, then you can type things like

        . list if tin(5jun1995,20jun1995)

and that means the same as

        . list if t>=d(5jun1995) & t<=d(20jun1995)

If the time variable t has a %tq format, then you can type things like

        . list if tin(1998q1,1998q4)

which means the same as

        . list if t>=q(1998q1) & t<=d(1998q4)

`tin()` and `twithin()` work by accessing the time variable previously `tsset` and then examining its display format to determine how the arguments you type should appear.

In typing `tin()` and `twithin()`, you may omit either or both arguments, but you must type the comma unless you omit both arguments. The following are all valid:

        . list if tin(1998q1,)   (meaning 1998q1 and thereafter)

. list if tin(,1998q4)   (meaning up to and including 1998q4)

. list if tin(,)   (meaning all the data)

. list if tin()   (also meaning all the data)

There is no reason to type if tin(,) or if tin(), but you can.

24.3.9   The %tg format

In addition to the %t formats documented above, there is one more: %tg. The g stands for generic. The %tg format is like not putting a %t format on your variable at all—it is equivalent to %9.0g. It is included for those who have a time variable, and wish to emphasize that it is a time variable, but the variable is in units that Stata does not understand.

When you have such a variable, it does not matter whether you place a %tg format on it.

25 Dealing with categorical variables

Contents

25.1   Continuous, categorical, and indicator variables

Although to Stata a variable is a variable, it is helpful to distinguish among three conceptual types:

- A *continuous variable* measures something. Such a variable might measure, for instance, a person's age, height, or weight; a city's population or land area; or a company's revenues or costs.

- A *categorical variable* identifies a group to which the thing belongs. For example, you could categorize persons according to their race or ethnicity, cities according to their geographic location, or companies according to their industry. Often, but not always, categorical variables are stored as strings.

- An *indicator variable* denotes whether something is true. For example, is a person a veteran, does a city have a mass transit system, or is a company profitable?

Indicator variables are a special case of categorical variables. Consider a variable that records a person's sex. Examined one way, it is a categorical variable. A categorical variable identifies the group to which a thing belongs, and in this case, the thing is a person and the basis for categorization is anatomy. Looked at another way, however, it is an indicator variable. It indicates whether a person is, say, female.

In fact, we can use the same logic on any categorical variable that divides the data into two groups. It is a categorical variable since it identifies whether an observation is a member of this or that group; it is an indicator variable since it denotes the truth value of the statement "the observation is in this group".

All indicator variables are categorical variables, but the opposite is not true. A categorical variable might divide the data into more than two groups. For clarity, let's reserve the term *categorical variable* for variables that divide the data into more than two groups, and let's use the term *indicator variable* for categorical variables that divide the data into exactly two groups.

Stata can convert continuous variables to categorical and indicator variables and categorical variables to indicator variables.

25.1.1   Converting continuous variables to indicator variables

Stata treats logical expressions as taking on the values *true* or *false*, which it identifies with the numbers 1 and 0; see [U] **13 Functions and expressions**. For instance, if you have a continuous variable measuring a person's age and you wish to create an indicator variable denoting persons aged 21 and over, you could type

```
. generate age21p = age>=21
```

The variable `age21p` takes on the value 1 for persons aged 21 and over and 0 for persons under 21.

Since `age21p` can take on only 0 or 1, it would be more economical to store the variable as a `byte`. Thus, it would be better to type

```
. generate byte age21p = age>=21
```

This solution has a problem. The value of `age21` is set to 1 for all persons whose `age` is missing because Stata defines missing to be larger than all other numbers. In our data, we have no such missing ages, but it still would have been safer to type

```
. generate byte age21p = age>=21 if age<.
```

That way, persons whose age is missing would also have a missing `age21p`.

❑ Technical Note

Put aside missing values and consider the following alternative to `generate age21p = age>=21` that may have occurred to you:

```
. generate age21p = 1 if age>=21
```

That does not produce the desired result. This statement makes `age21p` 1 (*true*) for all persons aged 21 and above but makes `age21p` missing for everyone else.

If you followed this second approach, you would have to combine it with

```
. replace age21p = 0 if age<21
```

to make the result identical to that produced by the single statement `gen age21p = age>=21`.

❑

25.1.2   Converting continuous variables to categorical variables

Suppose that you wish to categorize persons into four groups based on their age. You want a variable to denote whether a person is 21 or under, between 22 and 38, between 39 and 64, or 65 and above. Although most people would label these categories 1, 2, 3, and 4, there is really no reason to restrict ourselves to such a meaningless numbering scheme. Let's call this new variable `agecat` and make it so that it takes on the topmost value for each group. Thus, persons in the first group will be identified with an `agecat` of 21, persons in the second with 38, persons in the third with 64, and persons in the last (drawing a number out of the air) with 75. Here is one way that will work, but it is not the best method for doing so:

```
. generate byte agecat=21 if age<=21
(176 missing values generated)

. replace agecat=38 if age>21 & age<=38
(148 real changes made)
```

```
. replace agecat=64 if age>38 & age<=64
(24 real changes made)
. replace agecat=75 if age>64 & age<.
(4 real changes made)
```

We mechanically created the categorical variable according to the definition by using the `generate` and `replace` commands. The only thing that deserves comment is the opening `generate`. We (wisely) told Stata to `generate` the new variable `agecat` as a `byte`, thus conserving memory.

We can create the same result with one command using the `recode()` function:

```
. generate byte agecat=recode(age,21,38,64,75)
```

`recode()` takes three or more arguments. It examines the first argument (in this case, `age`) against the remaining arguments in the list. It returns the first element in the list that is greater than or equal to the first argument or, failing that, the last argument in the list. Thus, for each observation, `recode()` asked if `age` was less than or equal to 21. If so, the value is 21. If not, is it less than or equal to 38? If so, the value is 38. If not, is it less than or equal to 64? If so, the value is 64. If not, the value is 75.

Most researchers typically make tables of categorical variables, so we will `tabulate` the result:

```
. tabulate agecat
```

agecat	Freq.	Percent	Cum.
21	28	13.73	13.73
38	148	72.55	86.27
64	24	11.76	98.04
75	4	1.96	100.00
Total	204	100.00	

There is another way to convert continuous variables into categorical variables, and it is even more automated: `autocode()` works like `recode()`, except that all you tell the function is the range and the total number of cells that you want that range broken into:

```
. generate agecat=autocode(age,4,18,65)
. tabulate agecat
```

agecat	Freq.	Percent	Cum.
29.75	96	47.06	47.06
41.5	92	45.10	92.16
53.25	8	3.92	96.08
65	8	3.92	100.00
Total	204	100.00	

In one instruction, we told Stata to break `age` into four evenly spaced categories from 18 to 65. When we `tabulate agecat`, we see the result. In particular, we see that the break points of the four categories are 29.75, 41.5, 53.25, and 65. The first category contains everyone aged 29.75 years or less; the second category contains persons over 29.75 who are 41.5 years old or less; the third category contains persons over 41.5 who are 53.25 years old or less; and the last category contains all persons over 53.25.

❑ Technical Note

We chose the range 18 to 65 arbitrarily. Although you cannot tell from the table above, there are persons in this dataset who are under 18, and there are persons over 65. Those persons are counted in the first and last cells, but we have not divided the age range in the data evenly. We could split the full age range into four categories by obtaining the overall minimum and maximum ages (by typing `summarize`) and substituting the overall minimum and maximum for the 18 and 65 in the `autocode()` function:

```
. summarize age
    Variable |      Obs       Mean    Std. Dev.       Min        Max
─────────────┼───────────────────────────────────────────────────────
         age |      204   29.64706    9.805645         2         66
. generate agecat2=autocode(age,4,2,66)
```

Alternatively, we could `sort` the data into ascending order of `age` and tell Stata to construct four categories over the range `age[1]` (the minimum) to `age[_N]` (the maximum):

```
. sort age
. generate agecat2=autocode(age,4,age[1],age[_N])
. tabulate agecat2
     agecat2 |      Freq.     Percent        Cum.
─────────────┼───────────────────────────────────
          18 |         20        9.80        9.80
          34 |        148       72.55       82.35
          50 |         28       13.73       96.08
          66 |          8        3.92      100.00
─────────────┼───────────────────────────────────
       Total |        204      100.00
```

❑

25.1.3 Converting categorical variables to indicator variables

The easiest way to convert categorical variables to indicator variables is to use the `xi` command, which will construct indicator variables on the fly. Here we use `xi`, an extremely useful command, with the `logistic` command; `grp` is a variable taking on values 1, 2, and 3:

```
. xi: logistic outcome i.grp age
i.grp             _Igrp_1-3          (naturally coded; _Igrp_1 omitted)
Logit estimates                           Number of obs   =        189
                                          LR chi2(3)      =       6.54
                                          Prob > chi2     =     0.0879
Log likelihood = -114.06375               Pseudo R2       =     0.0279

─────────────┬───────────────────────────────────────────────────────────
     outcome │ Odds Ratio  Std. Err.       z    P>|z|     [95% Conf. Interval]
─────────────┼───────────────────────────────────────────────────────────
     _Igrp_2 │  2.106974   .9932407     1.58   0.114     .8363679    5.307878
     _Igrp_3 │  1.767748   .6229325     1.62   0.106     .8860686    3.526738
         age │  .9612592   .0311206    -1.22   0.222     .9021588    1.024231
─────────────┴───────────────────────────────────────────────────────────
```

See [R] **xi**.

At other times, we will want to convert categorical variables to indicator variables permanently, so let's consider how to do that.

We should ask ourselves how this variable is stored. Is it a set of numbers, with different numbers reflecting the different categories, or is it a string? Things will be easier if it is numeric, so if it is not, we use encode to convert it; see [U] **23.2 Categorical string variables**. Making categorical variables numeric is not really necessary, but it is a good thing to do because numeric variables can be stored more compactly than string variables. More importantly, all of Stata's statistical commands know how to deal with numeric variables; some do not know what to make out of a string.

Let's suppose that you have a categorical variable that divides your data into four groups. To make matters concrete, we will assume that an observation in your data is a state and that the categorical variable denotes the geographical region for each state. Each state is in one of the four census regions known as the Northeast, North Central, South, and West.

Typing one command will create four new variables, the first indicating whether the state is in the North Central, the second whether the state is in the Northeast, and so on. Such variables are sometimes called *dummy* variables, and you can use them in regressions to control for the effects of, for instance, geographic region.

Here is the dataset before we type this miraculous command:

```
. use http://www.stata-press.com/data/r9/states3
(State data)
. describe
Contains data from http://www.stata-press.com/data/r9/states3.dta
  obs:            50                          State data
  vars:            6                          17 Mar 2005 03:19
  size:         1,700 (99.8% of memory free)  (_dta has notes)

              storage  display    value
variable name  type    format     label    variable label

state          str8    %9s
reg            int     %8.0g      reg      Census Region
median_age     float   %9.0g               Median Age
marriage_rate  long    %12.0g              Marriages per 100,000
divorce_rate   long    %12.0g              Divorces per 100,000
region         str8    %9s                 Census Region

Sorted by:  reg
. label list reg
reg:
           1 N. Centr
           2 N. East
           3 South
           4 West
```

reg is the categorical variable, and in our example, it is numeric, although that is not important for what we are about to do. The regions are numbered 1 to 4, and a value label, also named reg, maps those numbers into the words: N. Centr, N. East, South, and West.

We can make the four indicator variables from this categorical variable by typing

```
. tabulate reg, generate(reg)
   Census |
   Region |      Freq.      Percent        Cum.

 N. Centr |         12        24.00       24.00
  N. East |          9        18.00       42.00
    South |         16        32.00       74.00
     West |         13        26.00      100.00

    Total |         50       100.00
```

```
. describe

Contains data from http://www.stata-press.com/data/r9/states3.dta
  obs:           50                          State data
  vars:          10                          17 Sep 2002 11:06
  size:       1,900 (99.8% of memory free)   (_dta has notes)

                storage  display    value
variable name   type     format     label    variable label

state           str8     %9s                  
reg             int      %8.0g      reg       Census Region
median_age      float    %9.0g                Median Age
marriage_rate   long     %12.0g               Marriages per 100,000
divorce_rate    long     %12.0g               Divorces per 100,000
region          str8     %9s                  Census Region
reg1            byte     %8.0g                reg==N. Centr
reg2            byte     %8.0g                reg==N. East
reg3            byte     %8.0g                reg==South
reg4            byte     %8.0g                reg==West

Sorted by:  reg
     Note:  dataset has changed since last saved
```

Typing `tabulate reg, generate(reg)` produced a table of the number of states in each region (which is, after all, what `tabulate` does), and because we specified the `generate()` option, it silently created four new variables—one for each line of the table.

Describing the data, we see that there are four new variables called `reg1`, `reg2`, `reg3`, and `reg4`. They are called this because we said `generate(reg)`. If we had said `tabulate reg, gen(junk)`, they would have been called `junk1`, `junk2`, `junk3`, and `junk4`.

Each of the new variables is stored as a `byte`, and each has been automatically labeled for us. The variable `reg1` takes on the value 1 if the state is in the North Central and 0 otherwise. (It also takes on the value "missing" if `reg` is missing for the observation, which never occurs in our data.)

Just to be clear about the relationship of `reg1` to `reg`, here is a tabulation:

```
. tabulate reg reg1
    Census |     reg==N. Centr
    Region |      0          1  |    Total

   N. Centr |      0         12  |       12
   N. East  |      9          0  |        9
     South  |     16          0  |       16
      West  |     13          0  |       13

     Total  |     38         12  |       50
```

If `reg1` is 1, the region is North Central, and vice versa.

25.2   Using indicator variables in estimation

Indicator variables allow you to control for the effects of a variable in a regression. Using the indicator variables we generated in the previous example, we can control for region in the following regression:

$$y_j = \beta_0 + \beta_1 age_j + \beta_2 \delta_{2j} + \beta_3 \delta_{3j} + \beta_4 \delta_{4j} + \epsilon_j$$

where y_j represents the marriage rate in state j; age_j represents the median age of the state's population; and δ_{ij} is 1 if state j is in region i and 0 otherwise. We also eliminate the state of Nevada from our regression.

```
. regress marriage_rate median_age reg1 reg3 reg4 if state!="NEVADA"
```

Source	SS	df	MS		Number of obs =	49
					F(4, 44) =	8.71
Model	.000232847	4	.000058212		Prob > F =	0.0000
Residual	.000294193	44	6.6862e-06		R-squared =	0.4418
					Adj R-squared =	0.3910
Total	.000527039	48	.00001098		Root MSE =	.00259

marriage_r~e	Coef.	Std. Err.	t	P>|t|	[95% Conf. Interval]	
median_age	-.0008815	.0002723	-3.237	0.002	-.0014303	-.0003326
reg1	.0003526	.0012315	0.286	0.776	-.0021293	.0028345
reg3	.002699	.0011637	2.319	0.025	.0003537	.0050442
reg4	.0022186	.0014201	1.562	0.125	-.0006434	.0050807
_cons	.039585	.0085496	4.630	0.000	.0223544	.0568155

We see from the results above that the marriage rate, after controlling for age, is significantly higher in region 3, the South.

25.2.1   Testing the significance of indicator variables

After seeing these results, you might wonder if region, taken as a whole, significantly contributes to the explanatory power of the regression. We can find out using the test command:

```
. test reg1=0

 ( 1)  reg1 = 0.0

       F(  1,    44) =    0.08
            Prob > F =    0.7760
. test reg3=0, accumulate

 ( 1)  reg1 = 0.0
 ( 2)  reg3 = 0.0

       F(  2,    44) =    4.07
            Prob > F =    0.0239
. test reg4=0, accumulate

 ( 1)  reg1 = 0.0
 ( 2)  reg3 = 0.0
 ( 3)  reg4 = 0.0

       F(  3,    44) =    2.86
            Prob > F =    0.0478
```

We typed three commands. The first, test reg1=0, tested the coefficient on the variable reg1 against 0. The resulting F test showed the same significance level as the corresponding t test presented in the regression output.

We next typed test reg3=0, accumulate, which tests whether reg3 is zero and accumulates that test with any previous tests. Thus we are now testing the hypothesis that reg1 and reg3 are jointly zero. The F statistic is 4.07, and the result is significant at the 2.4% level; thus at any significance level above 2.4%, it appears that reg1 and reg3 are not both zero.

We finally typed `test reg4=0, accumulate`. As before, this command tests the newly introduced constraint that `reg4` is zero and accumulates that with the previous tests. We are now testing whether `reg1`, `reg3`, and `reg4` are jointly zero. The F statistic is 2.86, and its significance level is roughly 4.8%; thus at the 5% level, we can reject the hypothesis that, taken together, region has no effect on the marriage rate after controlling for age.

Stata's `test` command has a shorthand for tests that two or more variables are simultaneously equal to zero. Type `test` followed by the names of the variables:

```
. test reg1 reg3 reg4
 ( 1)   reg1 = 0.0
 ( 2)   reg3 = 0.0
 ( 3)   reg4 = 0.0
        F(  3,    44) =    2.86
             Prob > F =    0.0478
```

❏ Technical Note

Sometimes tests of this kind are embedded in an ANOVA or ANCOVA model. In the language of ANOVA, we are doing a one-way layout after controlling for the effect of age. Hence, we did not have to fit the model with `regress`: we could have used Stata's `anova` command. We would have typed `anova marriage_rate median_age reg, continuous(median_age)` to obtain the ANOVA table and the desired test directly. We could have seen the underlying regression after estimation of the model by then typing `regress` without arguments. Typing any estimation command without arguments is taken as a request to reshow the last estimation results. Since all of Stata's estimation commands are tightly coupled, after fitting an ANOVA or ANCOVA model, you can ask `regress` to show you the underlying regression coefficients.

❏

25.2.2   Importance of omitting one of the indicators

Some people prefer to fit models with dummy variables in the context of ANOVA, and others prefer the regression context. The choice is yours.

If you opt for the regression method, however, remember to leave one of the indicator variables out of the regression so that the coefficients have the interpretation of changes from a base group. If you fail to follow that advice, your regression will still be correct, but it is important that you understand what it is you are estimating and testing.

In the example above, for instance, we omitted `reg2`, the dummy for the Northeast. Let's rerun that regression and include the `reg2` dummy:

```
. regress marriage_rate median_age reg1-reg4 if state!="NEVADA"
```

Source	SS	df	MS		Number of obs =	49
					F(4, 44) =	8.71
Model	.000232847	4	.000058212		Prob > F =	0.0000
Residual	.000294193	44	6.6862e-06		R-squared =	0.3910
					Adj R-squared =	0.3910
Total	.000527039	48	.00001098		Root MSE =	.00259

marriage_r~e	Coef.	Std. Err.	t	P>|t|	[95% Conf. Interval]	
median_age	-.0008815	.0002723	-3.237	0.002	-.0014303	-.0003326
reg1	.0399376	.0080754	4.946	0.000	.0236628	.0562124
reg2	.039585	.0085496	4.630	0.000	.0223544	.0568155
reg3	.0422839	.0080922	5.225	0.000	.0259752	.0585926
reg4	.0418036	.0076958	5.432	0.000	.0262938	.0573135
_cons	(dropped)					

If you compare the top half of the regression output with that of the previous example, you will find that they are identical. You will also find that the estimated coefficient, standard error, and related statistics for the variable median_age are identical, yet the estimates for each of the reg variables are different.

You will also note that the models have been fitted on different variables. In the first case, dummies for regions 1, 3, and 4 were included, along with a constant. In the last case, dummies for regions 1, 2, 3, and 4 were included, and the constant was mysteriously dropped by regress. There are the same number of variables in each regression, but their identities have changed. In one case, there is a constant; in the other, there is a dummy for region 2.

Let's first explain why the constant was dropped; it was dropped because it was unnecessary. You can think of this model as making four different predictions for a given median age, one for each region. Each prediction is given by −.0008815 multiplied by the median age plus a constant for the region. The constant for the region is the estimated coefficient for the region from the table above. We can write this mathematically as

$$-.0008815age + c_i$$

For instance, for region 2, $c_2 = .039585$.

An overall constant is unnecessary in the sense that it is arbitrary, and zero is a good choice if you are choosing an arbitrary number. Let's choose another arbitrary number, say, 1. Then, each prediction would be given by −.0008815 multiplied by the median age plus a (different) constant for the region plus 1:

$$-.0008815age + c_i' + 1$$

The estimated constants for each of the regions would then have to change by exactly 1, so that

$$c_i' + 1 = c_i$$

The constant was dropped because including all four dummy variables in the regression made it unnecessary. In our first model, we turned the problem around. We included a constant but left out the dummy for region 2.

Let's spend a moment proving that the results are identical. The model we just fitted indicates that the marriage rate in region 2 is given by

$$-.0008815age + .039585$$

Now, turn back to the previous model. The marriage rate in region 2 is equal to the same thing! The constant in the original regression is identical to the coefficient on `reg2` in the second regression.

Let's look at region 1. The model we just fitted indicates that the marriage rate is given by

$$-.0008815 age + .0399376$$

In the first model, it is

$$-.0008815 age + .0003526 + .039585 = -.0008815 age + .0399376$$

which is again the same thing! If you perform the calculations for region 3 and region 4, you will discover two more equivalencies.

The models are identical in the sense that they make the same predictions. They differ, however, in other ways. In the first model, the coefficients on `reg1`, `reg3`, and `reg4` measure the difference between that region and region 2. In the second model, the coefficients measure the region's level directly.

Notice that the t statistic on the coefficient for `reg2` in the second model is equal to the t statistic on the constant in the first—namely, 4.630—as it should be. Both test the same hypothesis—that the marriage rate is zero in Northeastern states with a median age of zero. (Yes, there are no such states, and yes, the hypothesis sounds silly. Perhaps you prefer the hypothesis stated as "the constant for the Northeast is zero".)

The comparison of the t statistics for region 1, however, does not yield such equivalent results. In the first model, the t statistic is 0.286; in the second model, the statistic is 4.946. They are different because they test different hypotheses. In the first case, the statistic tests whether the constant for region 1 is the same as that for region 2; in the second case, it tests the hypothesis that the constant for region 1 is zero.

The differences in meaning of these statistical tests carry over to any `test` commands that you type. In the technical note above, we tested that `reg1`, `reg3`, and `reg4` were all simultaneously zero. We obtained an F statistic of 2.86. If we were to type the same statements after estimating our second model, we would obtain a whopping F statistic of 13.54! What we are testing, however, is different. If we wanted to perform the same test—namely, that all the regions have the same intercept—we would type `test reg1=reg2`; followed by `test reg3=reg2, accumulate`; followed by `test reg4=reg2, accumulate`. That test gives identical results.

26 Overview of Stata estimation commands

Contents

26.1   Introduction

Estimation commands fit models such as linear regression, probit, and the like. Stata has many such commands, so many that it is easy to overlook a few. Some of these commands differ greatly from each other, others are gentle variations on a theme, and still others are equivalent to each other.

Estimation commands share features that this chapter will not discuss; see [U] **20 Estimation and postestimation commands**. Especially see [U] **20.14 Obtaining robust variance estimates**, which discusses an alternative calculation for the estimated variance matrix (and hence standard errors) that many of Stata's estimation commands provide, and [U] **20.10 Performing hypothesis tests on the coefficients**.

Here, however, this chapter will put aside all of that—and all issues of syntax—and deal solely with matching commands to their statistical concepts. Nor will it cross-reference specific commands. To find the details on a particular command, look up its name in the index.

26.2   Linear regression with simple error structures

Considering models of the form

$$y_j = \mathbf{x}_j \boldsymbol{\beta} + \epsilon_j$$

for a continuous y variable. In this category, estimation is restricted to when σ_ϵ^2 is constant across observations j. The model is called the linear regression model, and the estimator is often called the (ordinary) least squares estimator.

regress is Stata's linear regression command. (regress produces the robust estimate of variance as well as the conventional estimate, and regress has a collection of commands that can be run after it to explore the nature of the fit.)

In addition, the following commands will do linear regressions, as does regress, but offer special features:

1. ivreg fits instrumental variables models.

2. areg fits models $y_j = \mathbf{x}_j \boldsymbol{\beta} + \mathbf{d}_j \boldsymbol{\gamma} + \epsilon_j$, where $\mathbf{d}_j$ is a mutually exclusive and exhaustive dummy variable set. areg obtains estimates of $\boldsymbol{\beta}$ (and associated statistics) without ever forming $\mathbf{d}_j$, meaning that it also does not report the estimated $\boldsymbol{\gamma}$. If your interest is in fitting fixed-effects models, Stata has a better command—xtreg—discussed in [U] **26.14.1 Linear regression with panel data** below. Most users who find areg appealing will probably want to use xtreg because it provides more useful summary and test statistics. areg literally duplicates the output that regress would produce if you were to generate all the dummy variables. This means, for instance, that the reported R^2 includes the effect of $\boldsymbol{\gamma}$.

3. boxcox obtains maximum likelihood estimates of the coefficients and the Box–Cox transform parameters in a model of the form

$$y_i^{(\theta)} = \beta_0 + \beta_1 x_{i1}^{(\lambda)} + \beta_2 x_{i2}^{(\lambda)} + \cdots + \beta_k x_{ik}^{(\lambda)} + \gamma_1 z_{i1} + \gamma_2 z_{i2} + \cdots + \gamma_l z_{il} + \epsilon_i$$

where $\epsilon \sim N(0, \sigma^2)$. Here the *depvar* y is subject to a Box–Cox transform with parameter θ. Each of the *indepvars* $x_1, x_2, \ldots, x_k$ is transformed by a Box–Cox transform with parameter λ. The $z_1, z_2, \ldots, z_l$ are independent variables that are not transformed. In addition to the general form specified above, boxcox can fit three other versions of this model defined by the restrictions $\lambda = \theta$, $\lambda = 1$, and $\theta = 1$.

4. tobit allows estimation of linear regression models when y_i has been subject to left-censoring, right-censoring, or both. For instance, say that y_i is not observed if $y_i < 1000$, but for those observations, it is known that $y_i < 1000$. tobit fits such models.

   ivtobit does the same but allows for endogenous regressors.

5. cnreg (censored-normal regression) is a generalization of tobit. The lower and upper censoring points, rather than being constants, are allowed to vary observation by observation. cnreg can fit any model tobit can fit.

6. intreg (interval regression) is a generalization of cnreg. In addition to allowing open-ended intervals, intreg allows closed intervals. Rather than observing y_j, it is assumed that y_{0j} and y_{1j} are observed, where $y_{0j} \leq y_j \leq y_{1j}$. Survey data might report that a subject's monthly income was in the range \$1,500 to \$2,500. intreg allows such data to be used to fit a regression model. intreg allows $y_{0j} = y_{1j}$ and so can reproduce results reported by regress. intreg allows y_{0j} to be $-\infty$ and y_{1j} to be $+\infty$ and so can reproduce results reported by cnreg and tobit.

7. `truncreg` fits the regression model when the sample is drawn from a restricted part of the population and so is similar to `tobit`, except that in this case, the independent variables are not observed. Under the normality assumption for the whole population, the error terms in the truncated regression model have a truncated-normal distribution.

8. `cnsreg` allows you to place linear constraints on the coefficients.

9. `eivreg` adjusts estimates for errors in variables.

10. `nl` provides the nonlinear least-squares estimator of $y_j = f(\mathbf{x}_j, \boldsymbol{\beta}) + \epsilon_j$.

11. `rreg` fits robust regression models, which are not to be confused with regression with robust standard errors. Robust standard errors are discussed in [U] **20.14 Obtaining robust variance estimates**. Robust regression concerns point estimates more than standard errors, and it implements a data-dependent method for downweighting outliers.

12. `qreg` produces quantile-regression estimates, a variation that is not linear regression at all but is an estimator of $y_j = \mathbf{x}_j\boldsymbol{\beta} + \epsilon_j$. In the basic form of this model, sometimes called median regression, $\mathbf{x}_j\boldsymbol{\beta}$ measures not the predicted mean of y_j conditional on $\mathbf{x}_j$, but its median. As such, `qreg` is of most interest when ϵ_j does not have constant variance. `qreg` allows you to specify the quantile, so you can produce linear estimates for the predicted 1st, 2nd, ..., 99th percentile.

Another command, `bsqreg`, is identical to `qreg` but presents bootstrapped standard errors.

The `sqreg` command estimates multiple quantiles simultaneously; standard errors are obtained via the bootstrap.

The `iqreg` command estimates the difference between two quantiles; standard errors are obtained via the bootstrap.

13. `vwls` (variance-weighted least squares) produces estimates of $y_j = \mathbf{x}_j\boldsymbol{\beta} + \epsilon_j$, where the variance of ϵ_j is calculated from group data or is known *a priori*. As such, `vwls` is of most interest to categorical-data analysts and physical scientists.

26.3 ANOVA, ANCOVA, MANOVA, and MANCOVA

ANOVA and ANCOVA are related to linear regression, but we classify them separately. The related Stata commands are `anova`, `oneway`, and `loneway`. The `manova` command provides MANOVA and MANCOVA (multivariate ANOVA and ANCOVA).

`anova` fits ANOVA and ANCOVA models, one-way and up—including two-way factorial, three-way factorial, etc.—and fits nested and mixed-design models and repeated-measures models.

`oneway` fits one-way ANOVA models. It is quicker at producing estimates than `anova`, although `anova` is so fast that this probably does not matter. The important difference is that `oneway` can report multiple-comparison tests.

`loneway` is an alternative to `oneway`. The results are numerically the same, but `loneway` can deal with more levels (limited only by dataset size; `oneway` is limited to 376 levels and `anova` to 798, but for `anova` to reach 798 requires a lot of memory), and `loneway` reports some additional statistics, such as the intraclass correlation.

`manova` fits MANOVA and MANCOVA models, one-way and up—including two-way factorial, three-way factorial, etc.—and it fits nested and mixed-design models.

26.4   Generalized linear models

The generalized linear model is

$$g\{E(y_j)\} = \mathbf{x}_j\boldsymbol{\beta}, \qquad y_j \sim F$$

where $g()$ is called the link function and F is a member of the exponential family, both of which you specify before estimation. `glm` fits this model.

The GLM framework encompasses a surprising array of models known by other names, including linear regression, Poisson regression, exponential regression, and others. Stata provides dedicated estimation commands for many of these. Stata has, for instance, `regress` for linear regression, `poisson` for Poisson regression, and `streg` for exponential regression, and that is not all of the overlap.

`glm` by default uses maximum likelihood estimation and alternatively estimates via iterated reweighted least squares (IRLS) when the `irls` option is specified. For each family F there is a corresponding link function $g()$, called the canonical link, for which IRLS estimation produces results identical to maximum likelihood estimation. You can, however, match families and link functions as you wish, and, when you match a family to a link function other than the canonical link, you obtain a different but valid estimator of the standard errors of the regression coefficients. The estimator you obtain is asymptotically equivalent to the maximum likelihood estimator, which, in small samples, produces slightly different results.

For example, the canonical link for the binomial family is logit. `glm, irls` with that combination produces results identical to the maximum-likelihood `logit` (and `logistic`) command. The binomial family with the probit link produces the probit model, but probit is not the canonical link in this case. Hence, `glm, irls` produces standard error estimates that differ slightly from those produced by Stata's maximum-likelihood `probit` command.

Many researchers feel that the maximum-likelihood standard errors are preferable to IRLS estimates (when they are not identical), but they would have a difficult time justifying that feeling. Maximum likelihood probit is an estimator with (solely) asymptotic properties; `glm, irls` with the binomial family and probit link is an estimator with (solely) asymptotic properties, and in finite samples, the standard errors differ a little.

Still, we recommend that you use Stata's dedicated estimators whenever possible. IRLS—the theory—and `glm, irls`—the command—are all-encompassing in their generality, meaning that they rarely use quite the right jargon or provide things in quite the way you wish they would. The narrower commands, such as `logit`, `probit`, `poisson`, etc., focus on the issue at hand and are invariably more convenient.

`glm` is useful when you want to match a family to a link function that is not provided elsewhere.

`glm` also offers a number of estimators of the variance–covariance matrix that are consistent, even when the errors are heteroskedastic or autocorrelated. Another advantage of a `glm` version of a model over a model-specific version is that many of these VCE estimators are available only for the `glm` implementation. In addition, you can also obtain the ML–based estimates of the VCE from `glm`.

26.5   Binary-outcome qualitative dependent variable models

There are lots of ways to write these models, such as

$$\Pr(y_j \neq 0) = F(\mathbf{x}_j\boldsymbol{\beta})$$

where F is some cumulative distribution. Two popular choices for $F()$ are the normal and logistic, and the models are called the probit and logit (or logistic regression) models. A third is the complementary log–log function; maximum likelihood estimates are obtained by Stata's `cloglog` command.

The two parent commands for the maximum likelihood estimator of probit and logit are `probit` and `logit`, although `logit` has a sibling, `logistic`, that provides the same estimates but displays results in a slightly different way.

Do not read anything into the names logit and logistic. Logit and logistic have two completely interchanged definitions in two scientific camps. In the medical sciences, logit means the minimum χ^2 estimator, and logistic means maximum likelihood. In the social sciences, it is the other way around. From our experience, it appears that neither reads the other's literature, since both talk (and write books) asserting that logit means one thing and logistic the other. Our solution is to provide both `logit` and `logistic`, which do the same thing, so that each camp can latch on to the maximum likelihood command under the name each expects.

There are two slight differences between `logit` and `logistic`. `logit` reports estimates in the coefficient metric, whereas `logistic` reports exponentiated coefficients—odds ratios. This is in accordance with the expectations of each camp and makes no substantial difference. The other difference is that `logistic` has a family of post-`logistic` commands that you can run to explore the nature of the fit. Actually, that is not exactly true because all the commands for use after `logistic` can also be used after `logit`.

If you have not already selected `logit` or `logistic` as your favorite, we recommend that you try `logistic`. Logistic regression (logit) models are more easily interpreted in the odds-ratio metric.

In addition to `logit` and `logistic`, Stata provides `glogit`, `blogit`, and `binreg` commands.

`blogit` is the maximum likelihood estimator (the same as `logit` or `logistic`) but applied on data organized in a different way. Rather than having individual observations, your data are organized so that each observation records the number of observed successes and failures.

`glogit` is the weighted-regression, grouped-data estimator.

`binreg` can be used to model either individual-level or grouped data in an application of the generalized linear model. The family is assumed to be binomial, and each link provides a distinct parameter interpretation. In addition, `binreg` offers several options for setting the link function according to the desired biostatistical interpretation. The available links and interpretation options are

Option	Implied link	Parameter
or	logit	Odds ratios $= \exp(\beta)$
rr	log	Risk ratios $= \exp(\beta)$
hr	log complement	Health ratios $= \exp(\beta)$
rd	identity	Risk differences $= \beta$

Related to logit, the skewed logit estimator scobit adds a power to the logit link function and is estimated by Stata's `scobit` command.

Turning to probit, you have three choices: `probit`, `dprobit`, and `ivprobit`. `probit` and `dprobit` are maximum likelihood estimators, and it makes no substantial difference which you use; they differ only in how they report results. `probit` reports coefficients and `dprobit` reports changes in probabilities. Many researchers find changes in probabilities easier to interpret.

`ivprobit` is for use with endogenous regressors.

As in the logit case, Stata also provides `bprobit` and `gprobit`. `bprobit` is a maximum likelihood estimator—equivalent to `probit` or `dprobit`—but works with data organized in the different way outlined above. `gprobit` is the weighted-regression, grouped-data estimator.

Continuing with probit, `hetprob` fits heteroskedastic probit models. In these models, the variance of the error term is parameterized.

In addition, Stata's `biprobit` command fits bivariate probit models, meaning two correlated outcomes. `biprobit` also fits partial-observability models in which only the outcomes $(0, 0)$ and $(1, 1)$ are observed.

26.6   Conditional logistic regression

`clogit` is Stata's conditional logistic regression estimator. In this model, observations are assumed to be partitioned into groups, and a predetermined number of events occur in each group. The model measures the risk of the event according to the observation's covariates, $\mathbf{x}_j$. The model is used in matched case–control studies (`clogit` allows $1 : 1$, $1 : k$, and $m : k$ matching) and is also used in natural experiments whenever observations can be grouped into pools in which a fixed number of events occur.

26.7   Multiple-outcome qualitative dependent variable models

For more than two outcomes, Stata provides ordered logit, ordered probit, rank-ordered logit, multinomial logistic regression, McFadden's choice model (conditional fixed-effects logistic regression), and nested logistic regression.

`oprobit` and `ologit` provide maximum-likelihood ordered probit and logit. These are generalizations of probit and logit models known as the proportional-odds model and are used when the outcomes have a natural ordering from low to high. The idea is that there is an unmeasured $z_j = \mathbf{x}_j\boldsymbol{\beta}$, and the probability that the kth outcome is observed is $\Pr(c_{k-1} < z_j < c_k)$, where $c_0 = -\infty$, $c_k = +\infty$, and $c_1, \ldots, c_{k-1}$ along with $\boldsymbol{\beta}$ are estimated from the data.

`rologit` fits the rank-ordered logit model for rankings. This model is also known as the Plackett–Luce model, the exploded logit model, and choice-based conjoint analysis.

`slogit` fits the stereotype logit model for data that is not truly ordered, as in the case of `ologit`, but for which you are not sure that it is unordered, in which case `mlogit` would be appropriate.

`mlogit` fits maximum-likelihood multinomial logistic models, also known as polytomous logistic regression. It is intended for use when the outcomes have no natural ordering and you know only the characteristics of the outcome chosen (and, perhaps, the chooser).

`clogit` fits McFadden's choice model, also known as conditional logistic regression. In the context denoted by the name *McFadden's choice model*, the model is used when the outcomes have no natural ordering, just as multinomial logistic regression, but the characteristics of the outcomes chosen and not chosen are known (along with, perhaps, the characteristics of the chooser).

In the context denoted by the name *conditional logistic regression*—mentioned above—subjects are members of pools, and one or more are chosen, typically to be infected by some disease or to have some other unfortunate event befall them. Thus the characteristics of the chosen and not chosen are known, and the issue of the characteristics of the chooser never arises. Said either way, it is the same model.

In their choice-model interpretations, `mlogit` and `clogit` assume that the odds ratios are independent of any other, unspecified, alternatives. Since this assumption is frequently rejected by the data, the nested logit model is a very useful generalization. `nlogit` fits a nested logit model using full maximum likelihood. The model may contain one or more levels.

asmprobit is for use with outcomes that have no natural ordering and with regressors that are alternative specific. It is weakly related to mlogit. Unlike mlogit, asmprobit does not assume the independence of the irrelevant alternatives (IIA).

mprobit is also for use with outcomes that have no natural ordering but with models that do not have alternative-specific regressors.

26.8   Simple count dependent variable models

These models concern dependent variables that count the number of occurrences of an event. In this category, we include Poisson and negative binomial regression. For the Poisson model,

$$E(\text{count}) = E_j \exp(\mathbf{x}_j \boldsymbol{\beta})$$

where E_j is the exposure time. poisson fits this model.

Negative-binomial regression refers to estimating with data that are a mixture of Poisson counts. One derivation of the negative binomial model is that individual units follow a Poisson regression model but there is an omitted variable that follows a gamma distribution with variance α. Negative-binomial regression estimates β and α. nbreg fits such models. A variation on this, unique to Stata, allows you to model α. gnbreg fits those models.

Zero inflation refers to count models in which the number of 0 counts is more than would be expected in the regular model, and that is due to there being a probit or logit process that must first generate a positive outcome before the counting process can begin.

Stata's zip command fits zero-inflated Poisson models.

Stata's zinb command fits zero-inflated negative binomial models.

ztp and ztnb fit zero-truncated Poission and negative binomial models. In zero-inflated models, you observe too many zeros, so you fit a separate model to them. In zero-truncated models, you do not observe the zeros.

26.9   Linear regression with heteroskedastic errors

We now consider the model $y_j = \mathbf{x}_j \boldsymbol{\beta} + \epsilon_j$, where the variance of ϵ_j is nonconstant.

First, regress can fit such models if you specify the robust option. What Stata calls robust is also known as the White correction for heteroskedasticity.

For scientists who have data where the variance of ϵ_j is known *a priori*, vwls is the command. vwls produces estimates for the model given each observation's variance, which is recorded in a variable in the data.

Finally, as mentioned above, qreg performs quantile regression, which in the presence of heteroskedasticity is most of interest. Median regression (one of qreg's capabilities) is an estimator of $y_j = \mathbf{x}_j \boldsymbol{\beta} + \epsilon_j$ when ϵ_j is heteroskedastic. Even more usefully, you can fit models of other quantiles and so model the heteroskedasticity. Also see the sqreg and iqreg commands; sqreg estimates multiple quantiles simultaneously. iqreg estimates differences in quantiles.

26.10   Stochastic frontier models

`frontier` fits stochastic production or cost frontier models on cross-sectional data. The model can be expressed as

$$y_i = \mathbf{x}_i \boldsymbol{\beta} + v_i - s u_i$$

where

$$s = \begin{cases} 1, & \text{for production functions} \\ -1, & \text{for cost functions} \end{cases}$$

u_i is a non-negative disturbance standing for technical inefficiency in the production function or cost inefficiency in the cost function. While the idiosyncratic error term v_i is assumed to have a normal distribution, the inefficiency term is assumed to be one of the three distributions: half-normal, exponential, or truncated-normal. In addition, when the non-negative component of the disturbance is assumed to be either half-normal or exponential, `frontier` can fit models in which the error components are heteroskedastic conditional on a set of covariates. When the non-negative component of the disturbance is assumed to be from a truncated-normal distribution, `frontier` can also fit a conditional mean model, where the mean of the truncated-normal distribution is modeled as a linear function of a set of covariates.

For panel-data stochastic frontier models, see [U] **26.14.1 Linear regression with panel data**.

26.11   Linear regression with systems of equations (correlated errors)

By correlated errors, we mean that observations are grouped, and that within group, the observations might be correlated but, across groups, they are uncorrelated. `regress` with the `robust` and `cluster()` options can produce "correct" estimates, which is to say, inefficient estimates with correct standard errors and lots of robustness; see [U] **20.14 Obtaining robust variance estimates**. Obviously, if you know the correlation structure (and are not mistaken), you can do better, so `xtreg` and `xtgls` are also of interest in this case; we discuss them in [U] **26.14.1 Linear regression with panel data** below.

Turning to simultaneous multiple-equation models, Stata can produce three-stage least-squares (3SLS) and two-stage least-squares (2SLS) estimates using the `reg3` and `ivreg` commands. Two-stage models can be estimated by either `reg3` or `ivreg`. Three-stage models require use of `reg3`. The `reg3` command can produce constrained and unconstrained estimates.

In the case where we have correlated errors across equations but no endogenous right-hand-side variables,

$$y_{1j} = \mathbf{x}_{1j} \boldsymbol{\beta} + \epsilon_{1j}$$
$$y_{2j} = \mathbf{x}_{2j} \boldsymbol{\beta} + \epsilon_{2j}$$
$$\vdots$$
$$y_{mj} = \mathbf{x}_{mj} \boldsymbol{\beta} + \epsilon_{mj}$$

where $\epsilon_{k.}$ and $\epsilon_{l.}$ are correlated with correlation ρ_{kl}, a quantity to be estimated from the data. This is called Zellner's seemingly unrelated regressions, and `sureg` fits such models. In the case where $\mathbf{x}_{1j} = \mathbf{x}_{2j} = \cdots = \mathbf{x}_{mj}$, the model is known as multivariate regression, and the corresponding command is `mvreg`.

Estimation in the presence of autocorrelated errors is discussed in [U] **26.13 Models with time-series data**.

26.12   Models with endogenous sample selection

What has become known as the Heckman model refers to linear regression in the presence of sample selection: $y_j = \mathbf{x}_j \boldsymbol{\beta} + \epsilon_j$ is not observed unless some event occurs that itself has probability $p_j = F(\mathbf{z}_j \boldsymbol{\gamma} + \nu_j)$, where ϵ and ν might be correlated and $\mathbf{z}_j$ and $\mathbf{x}_j$ may contain variables in common.

`heckman` fits such models by maximum likelihood or Heckman's original two-step procedure.

This model has recently been generalized to replace the linear regression equation with another probit equation, and that model is fitted by `heckprob`.

Another important case of endogenous sample selection is the treatment-effects model, which considers the effect of an endogenously chosen binary treatment on another endogenous, continuous variable, conditional on two sets of independent variables. `treatreg` fits a treatment-effects model using either a two-step consistent estimator or full maximum likelihood.

26.13   Models with time-series data

ARIMA refers to models with autoregressive integrated moving-average processes, and Stata's `arima` command fits models with ARIMA disturbances via the Kalman filter and maximum likelihood. These models may be fitted with or without confounding covariates.

Stata's `prais` command performs regression with AR(1) disturbances using the Prais–Winsten or Cochrane–Orcutt transformation. Both two-step and iterative solutions are available, as well as a version of the Hildreth–Lu search procedure. The Prais–Winsten estimates for the model are an improvement over the Cochrane–Orcutt estimates in that the first observation is preserved in the estimation. This is particularly important with trended data in small samples.

`prais` automatically produces the Durbin–Watson d-statistic, which can also be obtained after `regress` using `estat dwatson`.

`newey` produces linear regression estimates with the Newey–West variance estimates that are robust to heteroskedasticity and autocorrelation of specified order.

Stata provides estimators for regression models with autoregressive conditional heteroskedastic (ARCH) disturbances

$$y_t = \mathbf{x}_t \boldsymbol{\beta} + \mu_t$$

where μ_t is distributed $N(0, \sigma_t^2)$ and σ_t^2 is given by some function of the lagged disturbances.

Stata's `arch`, `aparch`, and `egarch` commands provide different parameterizations of the conditional heteroskedasticity. All three of these commands also allow ARMA disturbances and/or multiplicative heteroskedasticity.

Stata provides `var` and `svar` for fitting vector autoregression (VAR) and structural vector autoregression (SVAR) models. See [TS] **var** for information on Stata's suite of commands for forecasting, specification testing, and inference on VAR and SVAR models. Stata also provides `vec` for fitting vector error-correction models; see [TS] **vec**. See [TS] **irf** for information on Stata's suite of commands for estimating, analyzing, and presenting impulse–response functions and forecast error variance decompositions. There is also a set of commands for performing Granger causality tests, lag-order selection, and residual analysis.

26.14   Panel-data models

26.14.1   Linear regression with panel data

This section could just as well be called "linear regression with complex error structures". Commands in this class begin with the letters xt.

xtreg fits models of the form

$$y_{it} = \mathbf{x}_{it}\boldsymbol{\beta} + \nu_i + \epsilon_{it}$$

xtreg can produce the between-regression (random-effects) estimator, the within-regression (fixed-effects) estimator, or the GLS random-effects (matrix-weighted average of between and within results) estimator. In addition, it can produce the maximum-likelihood random-effects estimator.

xtregar can produce the within estimator and a GLS random-effects estimator when the ϵ_{it} are assumed to follow an AR(1) process.

xtivreg contains the between-2SLS estimator, the within-2SLS estimator, the first-differenced-2SLS estimator, and two GLS random-effects-2SLS estimators to handle cases in which some of the covariates are endogenous.

xtmixed is a generalization of xtreg that allows for multiple levels of panels, random coefficients, variance-component estimation in general. In the xtmixed framework, residuals (random effects) can occur anywhere and have any level of subscript.

xtabond is for use with dynamic panel-data models (models in which there are lagged dependent variables) and can produce the one-step, one-step robust, and two-step Arellano–Bond estimators. xtabond can handle predetermined covariates, and it reports both the Sargan and autocorrelation tests derived by Arellano and Bond.

xtgls produces generalized least-squares estimates for models of the form

$$y_{it} = \mathbf{x}_{it}\boldsymbol{\beta} + \epsilon_{it}$$

where you may specify the variance structure of ϵ_{it}. If you specify that ϵ_{it} is independent for all i and t, xtgls produces the same results as regress up to a small-sample degrees-of-freedom correction applied by regress but not by xtgls.

You may choose among three variance structures concerning i and three concerning t, producing a total of nine different models. Assumptions concerning i deal with heteroskedasticity and cross-sectional correlation. Assumptions concerning t deal with autocorrelation and, more specifically, AR(1) serial correlation.

Alternative methods report the OLS coefficients and a version of the GLS variance–covariance estimator. xtpcse produces panel-corrected standard error (PCSE) estimates for linear cross-sectional time-series models, where the parameters are estimated by OLS or Prais–Winsten regression. When you are computing the standard errors and the variance–covariance estimates, the disturbances are, by default, assumed to be heteroskedastic and contemporaneously correlated across panels.

In the jargon of GLS, the random-effects model fitted by xtreg has exchangeable correlation within i—xtgls does not model this particular correlation structure. xtgee, however, does.

xtgee fits population-averaged models, and it optionally provides robust estimates of variance. Moreover, xtgee allows other correlation structures. One that is of particular interest to those with lots of data goes by the name *unstructured*. The within-panel correlations are simply estimated in an unconstrained way. [U] **26.14.3 Generalized linear models with panel data** will discuss this estimator further since it is not restricted to linear regression models.

`xthtaylor` uses instrumental variables estimators to estimate the parameters of panel-data random-effects models of the form

$$y_{it} = \mathbf{X}_{1it}\boldsymbol{\beta}_1 + \mathbf{X}_{2it}\boldsymbol{\beta}_2 + \mathbf{Z}_{1i}\boldsymbol{\delta}_1 + \mathbf{Z}_{2i}\boldsymbol{\delta}_2 + u_i + e_{it}$$

The individual effects u_i are correlated with the explanatory variables $\mathbf{X}_{2it}$ and $\mathbf{Z}_{2i}$ but are uncorrelated with $\mathbf{X}_{1it}$ and $\mathbf{Z}_{1i}$, where $\mathbf{Z}_1$ and $\mathbf{Z}_2$ are constant within panel.

`xtfrontier` fits stochastic production or cost frontier models for panel data. You may choose from a time-invariant model or a time-varying decay model. In both models, the non-negative inefficiency term is assumed to have a truncated-normal distribution. In the time-invariant model, the inefficiency term is constant within panels. In the time-varying decay model, the inefficiency term is modeled as a truncated-normal random variable multiplied by a specific function of time. In both models, the idiosyncratic error term is assumed to have a normal distribution. The only panel-specific effect is the random inefficiency term.

26.14.2   Censored linear regression with panel data

`xttobit` fits random-effects tobit models and generalizes that to observation-specific censoring.

`xtintreg` performs random-effects interval regression and generalizes that to observation-specific censoring. Interval regression, in addition to allowing open-ended intervals, also allows closed intervals.

26.14.3   Generalized linear models with panel data

[U] **26.4 Generalized linear models** above discussed the model

$$g\{E(y_j)\} = \mathbf{x}_j\boldsymbol{\beta}, \qquad y_j \sim F$$

where $g()$ is the link function and F is a member of the exponential family, both of which you specify before estimation. This model can be further generalized to work with cross-sectional time-series data, so it can be rewritten as

$$g\{E(y_{it})\} = \mathbf{x}_{it}\boldsymbol{\beta}, \qquad y_{it} \sim F \text{ with parameters } \theta_{it}$$

This is referred to as the GEE method for panel-data models, where GEE stands for Generalized Estimating Equations. `xtgee` fits this model and allows you to specify the correlation structure of the errors.

If you specify that errors are independent within i, `xtgee` is equivalent to `glm`. Thus, since `glm` can reproduce (to name a few) the estimates produced by `regress`, `logit`, and `poisson`, so can `xtgee`.

If you specify errors are exchangeable within i, `xtgee` fits equal-correlation models. This means that with the identity link and Gaussian family, `xtgee` can reproduce the models fitted by `xtreg`. The only difference is that `xtgee` can provide standard errors that are robust to the correlations not being exchangeable.

`xtgee` provides other correlation structures, including multiplicative, AR(m), stationary(m), nonstationary(m), unstructured, and fixed (meaning user-specified). Unstructured should be of particular interest if you have large datasets even if you ultimately plan to impose a structure such as exchangeability (equal correlation). If relaxing the equal-correlation assumption in a large dataset causes your results to change importantly, there is an issue before you worthy of some thought.

`xtgee` provides 175 models from which to choose.

26.14.4 Qualitative dependent-variable models with panel data

xtprobit fits random-effects probit regression via maximum likelihood. It also fits population-averaged models via GEE. This last is nothing more than xtgee with the binomial family, probit link, and exchangeable error structure.

xtlogit fits random-effects logistic regression models via maximum likelihood. It also fits conditional fixed-effects models via maximum likelihood. Finally, as with xtprobit, it fits population-averaged models via GEE.

xtcloglog estimates random-effects complementary log-log regression via maximum likelihood. It also fits population-averaged models via GEE.

clogit is also of interest since it provides the conditional fixed-effects logistic estimator.

26.14.5 Count dependent-variable models with panel data

xtpoisson fits two different random-effects Poisson regression models via maximum likelihood. The two distributions for the random effect are gamma and normal. It also fits conditional fixed-effects models, and it fits population-averaged models via GEE. This last is nothing more than xtgee with the Poisson family, log link, and exchangeable error structure.

xtnbreg fits random-effects negative binomial regression models via maximum likelihood (the distribution of the random effects is assumed to be beta). It also fits conditional fixed-effects models, and it fits population-averaged models via GEE.

26.14.6 Random-coefficient models with panel data

xtrc fits Swamy's random-coefficients linear-regression model. In this model, rather than only the intercept varying across groups, all the coefficients are allowed to vary.

26.15 Survival-time (failure-time) models

Commands are provided to fit Cox proportional hazards models, as well as several parametric survival models including exponential, Weibull, Gompertz, log-normal, log-logistic, and generalized gamma (see [ST] **stcox** and [ST] **streg**). The commands all allow for right-censoring, left-truncation, gaps in histories, and time-varying regressors. The commands are appropriate for use with single- or multiple-failure per subject data. Conventional and robust standard errors are available with and without clustering.

Both the Cox model and the parametric models (as fitted using Stata) allow for two additional generalizations. First, the models may be modified to allow for latent random effects, or *frailties*. Second, the models may be stratified in the sense that the baseline hazard function may vary completely over a set of strata. The parametric models also allow the modeling of ancillary parameters.

stcox and streg require that the data be stset so that the proper response variables may be established. After you stset the data, the response is taken as understood, and you need only supply the regressors (and other options) to stcox and streg.

26.16   Survey data

Stata's `svy` command fits statistical models for complex survey data. `svy` is a prefix command, so to obtain linear regression, you type

      . svy: regress ...

or to obtain probit regression, you type

      . svy: probit ...

but first you must type a `svyset` command to define the survey design characteristics. Prefix `svy` works with many estimation commands, and everything is documented together in the *Survey Data Reference Manual*.

`svy` supports the following variance-estimation methods:

1. Taylor-series linearization

2. Balanced repeated replication (BRR)

3. Jackknife

See [SVY] **variance estimation** for details.

`svy` supports the following survey-design characteristics:

1. with and without replacement sampling

2. sampling weights

3. stratification

4. poststratification

5. clustering

6. multiple stages of clustering without replacement

7. BRR and jackknife replication weights

See [SVY] **svyset** for details.

Subpopulation estimation is available for all estimation commands.

Tabulations and summary statistics are also available, including means, proportions, ratios, and totals over multiple subpopulations, and direct standardization of means, proportions, and ratios.

26.17   Multivariate and cluster analysis

Most of Stata's multivariate capabilities are to be found in the *Multivariate Statistics Reference Manual*, although there are some exceptions.

1. `mvreg` fits multivariate regressions.

2. `manova` fits MANOVA and MANCOVA models, one-way and up—including two-way factorial, three-way factorial, etc.—and it fits nested and mixed-design models. Also see [U] **26.3 ANOVA, ANCOVA, MANOVA, and MANCOVA** above.

3. `canon` estimates canonical correlations and their corresponding loadings. Canonical correlation attempts to describe the relationship between two sets of variables.

4. `pca` extracts principal components and reports eigenvalues and loadings. Some people consider principal components a descriptive tool—in which case standard errors as well as coefficients are relevant—and others look at it as a dimension-reduction technique.

5. `factor` fits factor models and provides principal factors, principal-component factors, iterated principal-component factors, and maximum-likelihood solutions. Factor analysis is concerned with finding a small number of common factors $\hat{\mathbf{z}}_k$, $k = 1, \ldots, q$ that linearly reconstruct the original variables $\mathbf{y}_i$, $i = 1, \ldots, L$.

6. `tetrachoric`, in conjunction with `pca` or `factor`, allow you to perform PCA or factor analysis on binary data.

7. `rotate` provides a wide variety of orthogonal and oblique rotations after `factor` or `pca`. Rotations are often used to produce more interpretable results.

8. `procrustes` performs Procrustes analysis, one of the standard methods of multidimensional scaling. It can perform orthogonal or oblique rotations, as well as translation and dilation.

9. `mds` performs metric multidimensional scaling for dissimilarity between observations with respect to a set of variables. A wide variety of dissimilarity measures are available and, in fact, are the same as those for `cluster`.

10. `ca` performs correspondence analysis, an exploratory multivariate technique for analyzing cross-tabulations and the relationship between rows and columns.

11. `cluster` provides cluster analysis; both hierarchical and partition clustering methods are available. Strictly speaking, cluster analysis does not fall into the category of statistical estimation. Rather, it is a set of techniques for exploratory data analysis. Stata's cluster environment has many different similarity and dissimilarity measures for continuous and binary data.

26.18   Pharmacokinetic data

There are four estimation commands for analyzing pharmacokinetic data. See [R] **pk** for an overview of the pk system.

1. `pkexamine` calculates pharmacokinetic measures from time-and-concentration subject-level data. `pkexamine` computes and displays the maximum measured concentration, the time at the maximum measured concentration, the time of the last measurement, the elimination time, the half-life, and the area under the concentration-time curve (AUC).

2. `pksumm` obtains the first four moments from the empirical distribution of each pharmacokinetic measurement and tests the null hypothesis that the distribution of that measurement is normally distributed.

3. `pkcross` analyzes data from a crossover design experiment. When analyzing pharmaceutical trial data, if the treatment, carryover, and sequence variables are known, the omnibus test for separability of the treatment and carryover effects is calculated.

4. `pkequiv` performs bioequivalence testing for two treatments. By default, `pkequiv` calculates a standard confidence interval symmetric about the difference between the two treatment means. `pkequiv` also calculates confidence intervals symmetric about zero and intervals based on Fieller's theorem. Additionally, `pkequiv` can perform interval hypothesis tests for bioequivalence.

26.19   Specification search tools

There are three other commands that are not really estimation commands but are combined with estimation commands to assist in specification searches: `stepwise`, `fracpoly`, and `mfp`.

`stepwise:`, one of Stata's prefix commands, provides stepwise estimation. You can use the `stepwise` prefix with some, but not all, estimation commands. In [R] **stepwise** is a table of the estimation commands that are currently supported, but do not take it too literally. It was accurate as of the day Stata 9 was released, but, if you install the official updates, `stepwise` may now work with other commands, too. If you want to use `stepwise` with some estimation command, you should try it. Either it will work, or you will get the message that the estimation command is not supported by `stepwise`.

`fracpoly` and `mfp` are commands to assist you in performing fractional-polynomial functional specification searches.

26.20   Obtaining new estimation commands

This chapter has discussed all the estimation commands included in Stata 9 the day it was released; by now, there may be more. To obtain an up-to-date list, type `search estimation`.

And, of course, you can always write your own commands; see [R] **ml**.

26.21   Reference

Gould, W. W. 2000. sg124: Interpreting logistic regression in all its forms. *Stata Technical Bulletin* 53: 19–29. Reprinted in *Stata Technical Bulletin Reprints*, vol. 9, pp. 257–270.

27 Commands everyone should know

27.1   Forty-one commands

Putting aside the statistical commands that might particularly interest you, here are 41 commands everyone should know:

Getting online help
     help, net search, search   [U] **4 Stata's online help and search facilities**

Keeping Stata up to date
     ado, net, update   [U] **28 Using the Internet to keep up to date**

Operating system interface
     pwd, cd   [D] **cd**

Using and saving data from disk
     use, save   [D] **save**
     append, merge   [U] **22 Combining datasets**
     compress   [D] **compress**

Inputting data into Stata   [U] **21 Inputting data**
     input   [D] **input**
     edit   [D] **edit**
     infile   [D] **infile (free format)**; [D] **infile (fixed format)**
     infix   [D] **infix (fixed format)**
     insheet   [D] **insheet**

Basic data reporting
     describe   [D] **describe**
     codebook   [D] **codebook**
     list   [D] **list**
     browse   [D] **edit**
     count   [D] **count**
     inspect   [D] **inspect**
     table   [R] **table**
     tabulate   [R] **tabulate oneway** and [R] **tabulate twoway**

(*Continued on next page*)

Data manipulation	[U] **13 Functions and expressions**
generate, replace	[D] **generate**
egen	[D] **egen**
rename	[D] **rename**
drop, keep	[D] **drop**
sort	[D] **sort**
encode, decode	[D] **encode**
order	[D] **order**
by	[U] **11.5 by varlist: construct**
reshape	[D] **reshape**
Keeping track of your work	
log	[U] **15 Printing and preserving output**
notes	[D] **notes**
Convenience	
display	[R] **display**

27.2   The by construct

If you do not understand the by *varlist*: construct, _n, and _N, and their interaction, and if you process data where observations are related, you are missing out on something. See

[U] **13.7 Explicit subscripting**
[U] **11.5 by varlist: construct**

For instance, say you have a dataset with multiple observations per person, and you want the average value of each person's blood pressure (bp) for the day. You could

```
. egen avgbp = mean(bp), by(person)
```

but you should understand that you could also

```
. by person, sort: gen avgbp = sum(bp)/_N
. by person: replace avgbp = avgbp[_N]
```

Yes, typing two commands is more work than typing just one, but understanding the two-command construct is the key to generating more complicated things that no one ever thought about adding to egen.

For instance, say your dataset also contains time recording when each observation was made. If you want to add the total time the person is under observation (last time minus first time) to each observation, type

```
. by person (time), sort: gen ttl = time[_N]-time[1]
```

Or, suppose you want to add how long it has been since the person was last observed to each observation:

```
. by person (time), sort: gen howlong = time - time[_n-1]
```

If instead you wanted how long it would be until the next observation, type

```
. by person (time), sort: gen whennext = time[_n+1] - time
```

by *varlist*:, _n, and _N are often the solution to difficult calculations.

28 Using the Internet to keep up to date

Contents

28.1   Overview

Stata has the ability to read files over the Internet. Just to prove that to yourself, type the following:

```
. use http://www.stata.com/manual/chapter28, clear
```

You have just reached out and gotten a dataset from our web site. The dataset is not in HTML format, nor does this have anything to do with your browser. We just copied the Stata data file `chapter32.dta` onto our server, and now people all over the world can use it. If you have a web page, you can do the same thing. It is a very convenient way to share datasets with colleagues.

Now type the following:

```
. update from http://www.stata.com
```

We promise nothing bad will happen. `update` will read a short file from www.stata.com that will allow Stata to report whether your copy of Stata is up to date. Is your copy up to date? Now you know. If it is not, we will show you how to update it—it is no more difficult than typing `update`.

Now type the following:

```
. net from http://www.stata.com
```

That will go to www.stata.com and tell you what is available from our user-download site. The material there is not official, but it is useful. More usefully, type

```
. search kernel regression, net
```

or equivalently,

```
. net search kernel regression
```

That will search the entire web for additions to Stata having to do with kernel regression, whether it be from the *Stata Journal*, *Stata Technical Bulletin*, Statalist, archive sites, or user private sites.

To summarize, Stata has the ability to read files over the Internet:

1. You can share datasets, do-files, etc., with colleagues all over the world. This requires no special expertise, but you do need to have a web page.

2. You can update Stata; it is free, easy, and nearly instant.

3. You can find and add new features to Stata; it is also free, easy, and nearly instant.

Finally, you can create a site to distribute new features for Stata.

28.2   Sharing datasets (and other files)

There is just nothing to it: you copy the file as-is (in binary) onto the server and then let your colleagues know the file is there. This works for .dta files, .do files, .ado files, and, in fact, all files.

On the receiving end, you can use the file (if it is a .dta dataset) or you can copy it:

```
. use http://www.stata.com/manual/chapter32, clear
. copy http://www.stata.com/manual/chapter32.dta mycopy.dta
```

Stata includes a copy-file command and it works over the Internet just as use does; see [D] **copy**.

❏ Technical Note

If you are concerned about transmission errors, you can create a checksum file before you copy the file onto the server. In placing chapter32.dta on our site, we started with chapter32.dta in our working directory and typed

```
. checksum chapter32.dta, save
```

This created the new file chapter32.sum. We then placed both files on our server. We did not have to create this second file, but, since we did, when you use the data, Stata will be able to detect transmission errors and warn you if there are problems.

How would Stata know? chapter32.sum is a very short file containing the result of a mathematical calculation made on the contents of chapter32.dta. When your Stata receives chapter32.dta, it repeats the calculation and then compares that result with what is recorded in chapter32.sum. If the results are different, then there must have been a transmission error.

Whether you create a checksum file is optional.

See [D] **checksum**.

❏

28.3   Official updates

Although we follow no formal schedule for the release of updates, the fact is that we update Stata about once every two weeks. You do not have to update that often, although we recommend that you do. There are two ways to check whether your copy of Stata is up to date:

type
```
. update query
```

or

Select **Help > Official Updates**
Click on *http://www.stata.com*

After that, you will either

    type: or:

    . update ado   click on *update ado-files*

or

    type: or:

    . update executable   click on *update executable*

or

    type: or:

    . update all   click on *update ado-files and executable*

and which, if any, of those things need doing will be obvious.

    After you have updated your Stata, to find out what has changed

    type: or:

    . help whatsnew   Select **Help > What's New?**

28.3.1 Example

When you type update from http://www.stata.com or when you select **Help > Official Updates**, and click on *http://www.stata.com*, Stata presents a report:

```
. update from http://www.stata.com
(contacting http://www.stata.com)

Stata executable
    folder:                 C:\Program Files\Stata9\
    name of file:           wstata.exe
    currently installed:    27 Apr 2005
    latest available:       27 Apr 2005

Ado-file updates
    folder:                 C:\Program Files\Stata9\ado\updates\
    names of files:         (various)
    currently installed:    15 May 2006
    latest available:       15 May 2006

Recommendation
    Do nothing; all files up-to-date.
```

There are two components of official Stata: the binary Stata executable and the ado-files that we shipped with it. Ado-files are just programs written in Stata. For instance, when you use generate, you are using a command that was compiled into the Stata executable. When you use stcox, you are using a command that was implemented as an ado-file.

Both components of our Stata are up to date.

28.3.2   Updating ado-files

When you obtain the above report, you might see

```
. update from http://www.stata.com
(contacting http://www.stata.com)

Stata executable
    folder:               C:\Program Files\Stata9\
    name of file:         wstata.exe
    currently installed:  27 Apr 2005
    latest available:     27 Apr 2005

Ado-file updates
    folder:               C:\Program Files\Stata9\ado\updates\
    names of files:       (various)
    currently installed:  27 Apr 2005
    latest available:     15 May 2006

Recommendation
    Type -update ado-
```

If you go with the point-and-click alternative, at the bottom of the screen you will see

```
Recommendation
    update ado-files
```

where *update ado-files* is in blue and is therefore clickable.

Anyway, what you are to do next is type `update ado` or click on *update ado-files*. Either way, you will see something like the following:

```
. update ado
(contacting http://www.stata.com)

Ado-file update log
    1.   verifying C:\Program Files\Stata9\ado\updates\ is writeable
    2.   obtaining list of files to be updated
    3.   downloading relevant files to temporary area
         downloading filename.ado
         downloading filename.hlp
         ...
         downloading filename.ado
    4.   examining files
    5.   installing files
    6.   setting last date updated

Updates successfully installed.

Recommendation
    Type -help whatsnew- to learn about the new features
```

That is all there is to it, but do type `help whatsnew` to learn about the new features. (If you go the point-and-click path, click on *whatsnew*.)

Here is what happens if you type `update ado` and you are already up to date:

```
. update ado
(contacting http://www.stata.com)
ado-files already up to date
```

28.3.3 Frequently asked questions about updating the ado-files

1. Could something go wrong and make my Stata become unusable?

   No. The updates are copied to a temporary place on your computer, Stata examines them to make sure they are complete before copying them to the official place. Thus, either the updates are installed or they are not.

2. I do not believe you. Pretend that something you did not anticipate goes wrong, such as the power fails at the instant Stata is doing the local disk to local disk copy.

   If the improbable should happen, you can erase the update directory and then your Stata is back to being just as it was shipped. Updates go into a different directory than the originals and the originals are never erased.

   Stata tells you where it is installing your updates. You can also find out by typing sysdir. The directory you want is the one listed opposite UPDATES.

   (By the way, power failure should not cause a problem; the marker that the update is applied is set last, so you could also just type update ado again and Stata would refetch the partially installed update.)

3. How much is downloaded?

   A typical update is 100k to 300k. Ado-files are small; the biggest file that is copied is probably the database for search.

4. I am using Unix or a networked version of Stata. When I try to update ado, I am told that the directory is not writeable. Can I copy the updates into another directory and then copy them to the official directory myself?

   Yes, assuming you are a system administrator. Type 'update ado, into(*dirname*)'. Stata will download the updates just as it would ordinarily, but will place them in the directory you specify. We recommend that *dirname* be a new, empty directory, because later you will need to copy the entire contents of the directory to the official place. The official place is the directory listed next to UPDATES if you type sysdir. When you copy the files, copy over any existing files. Previously existing files in the official update directory are just previous updates. Also remember to make the files globally readable if necessary. See [R] **update**.

28.3.4 Updating the executable

Ado-file updates are released every other week; updates for the executable are rarer than that. If the executable needs updating, Stata will mention it when you type update:

```
. update from http://www.stata.com
(contacting http://www.stata.com)

Stata executable
    folder:                C:\Program Files\Stata9\
    name of file:          wstata.exe
    currently installed:   27 Apr 2005
    latest available:      15 May 2006

Ado-file updates
    folder:                C:\Program Files\Stata9\ado\updates\
    names of files:        (various)
    currently installed:   27 Apr 2005
    latest available:      27 Apr 2005

Recommendation
    Type -update executable-
```

Here is what happens when you type update executable:

```
. update executable
(contacting http://www.stata.com)

Executable update log
    1.  verifying C:\Program Files\Stata9\ is writeable
    2.  downloading new executable

New executable successfully downloaded

Instructions
    1.  Type -update swap-
```

Just follow the instructions, which will vary depending on your computer. In this case, update swap is a command that automatically copies the newly downloaded executable over the current one. It then briefly restarts Stata to begin using the new executable.

28.3.5   Frequently asked questions about updating the executable

1. If I understand this, update executable does not really install the update; it just copies one file onto my computer, and that one file happens to be the new executable, right?

   Probably. There can be more than one file such as a DLL. All the files are copied to the same place. In the case where no DLLs are downloaded, this statement is true. On most systems, using the update swap command after downloading a new executable will perform the copy and actually install the executable.

2. How big is the downloaded file?

   2 to 14 megabytes, depending on operating system.

3. What happens if I type update executable and my executable is already up to date?

   Nothing. You are told "executable already up to date".

4. I am using Unix or a networked version of Stata. When I try to update executable, I am told that the directory is not writeable. Can I download the updated executable to another directory and then copy it to the official directory myself?

   Yes, assuming you are a system administrator.

   Type 'update executable, into(*dirname*)'. We recommend that *dirname* be a new, empty directory, because there may be more than one file and later you will need to copy all of them to the official place. The official place is the directory listed next to STATA if you type sysdir. When you copy the files, copy over any existing files; we recommend that you make a backup of the originals first. See [R] **update**.

28.3.6   Updating both ado-files and the executable

When you type update, you may be told you need to update both ado-files and executable:

```
. update from http://www.stata.com
(contacting http://www.stata.com)

Stata executable
    folder:               C:\Program Files\Stata9\
    name of file:         wstata.exe
    currently installed:  27 Apr 2005
    latest available:     15 May 2006
```

```
Ado-file updates
    folder:             C:\Program Files\Stata9\ado\updates\
    names of files:     (various)
    currently installed: 27 Apr 2005
    latest available:   15 May 2006
Recommendation
    Type -update all-
```

Typing `update all` is the same as typing `update ado` and then typing `update executable`. You could type the separate commands if you preferred. The order does not matter.

Note, you could skip the `update from` step. You could just type `update all` and follow the instructions. If nothing needed updating, you would see

```
. update all
```

```
> update ado
(contacting http://www.stata.com)
ado-files already up to date
```

```
> update executable
(contacting http://www.stata.com)
executable already up to date

. _
```

28.4   Downloading and managing additions by users

Try the following:

type

    `. net from http://www.stata.com`

or

    Select **Help > SJ and User-written Programs**
    Click on one of the links

28.4.1   Downloading files

We are not the only ones developing additions to Stata. Stata is supported by a large and highly competent user community. An important part of this is the *Stata Journal* (SJ) and the *Stata Technical Bulletin* (STB). The *Stata Journal* is a refereed, quarterly journal containing articles of interest to Stata users. For more details and subscription information, visit the *Stata Journal* web site at *http://www.stata-journal.com/*.

The *Stata Journal* is a printed and electronic journal with corresponding software. If you want the journal, you must subscribe, but the software is available for free from our web site at *http://www.stata-journal.com*.

The predecessor to the *Stata Journal* was the *Stata Technical Bulletin* (STB). The STB was also a printed and electronic journal with corresponding software. Individual STB journals may still be purchased. The STB software is available for free from our web site at *http://www.stata.com*.

Below are instructions for installing the *Stata Journal* and the *Stata Technical Bulletin* software from our web site.

Installing the Stata Journal software

1. Select **Help > SJ and User-written Programs**.

2. Click on *Stata Journal*.

3. Click on *sj2-2*.

4. Click on *st0001_1*.

5. Click on *click here to install*.

or type

1. Type: . net from http://www.stata-journal.com/software

2. Type: . net cd sj2-2

3. Type: . net describe st0001_1

4. Type: . net install st0001_1

The above could be shortened to

```
. net from http://www.stata-journal.com/software/sj2-2
. net describe st0001_1
. net install st0001_1
```

Alternatively, you could type

```
. net sj 2-2
. net describe st0001_1
. net install st0001_1
```

Installing the STB software

1. Select **Help > SJ and User-written Programs**.

2. Click on *STB*.

3. Click on *stb58*.

4. Click on *sg84_3*.

5. Click on *click here to install*.

or type

1. Type: . net from http://www.stata.com

2. Type: . net cd stb

3. Type: . net cd stb58

4. Type: . net describe sg84_3

5. Type: . net install sg84_3

The above could be shortened to

```
. net from http://www.stata.com/stb/stb58
. net describe sg84_3
. net install sg84_3
```

28.4.2   Managing files

You now have the `concord` command, because we just downloaded and installed it. Convince yourself of this by typing

```
. help concord
```

and you might try it out, too. Let's now list the additions you have installed—that is probably just `concord`—and then get rid of `concord`.

In command mode, you can type

```
. ado dir
[1] package sg84_3 from http://www.stata.com/stb/stb58
      STB-58 sg84_3.  Concordance correlation coefficient: minor corrections
```

If you had more additions installed, they would be listed. Now knowing that you have *sg84_3* installed, you can obtain a more thorough description by typing

```
. ado describe sg84_3
```

```
[1] package sg84_3 from http://www.stata.com/stb/stb58
```

```
TITLE
      STB-58 sg84_3.  Concordance correlation coefficient:  minor corrections
DESCRIPTION/AUTHOR(S)
      STB insert by Thomas J. Steichen, RJRT
                    Nicholas J. Cox, University of Durham, UK
      Support:  steicht@rjrt.com, n.j.cox@durham.ac.uk
      After installation, see help concord
INSTALLATION FILES
      c/concord.ado
      c/concord.hlp
INSTALLED ON
      5 Oct 2002
```

You can erase *sg84_3* by typing

```
. ado uninstall sg84_3
package sg84_3 from http://www.stata.com/stb/stb58
      STB-58 sg84_3.  Concordance correlation coefficient:  minor corrections
(package uninstalled)
```

You can do all of this from the point-and-click interface, too. Pull down **Help** and select **SJ and User-written Programs** and then click on *List*. From there, you can click on *sg84_3* to see the detailed description of the package and from there you can click on *click here to uninstall* if you want to erase it.

For more information on the `ado` command and the corresponding menu, see [R] **net**.

28.4.3   Finding files to download

There are two ways to find useful files to download. One is simply to thumb through sites. That is inefficient but entertaining. If you want to do that,

1. Select **Help > SJ and User-written Programs**.

2. Click on *Other Locations*.

3. Click on *links*.

What you are doing is starting at our download site and then working out from there. We maintain a list of other sites and those sites will have more links. You can do this from command mode, too:

```
. net from http://www.stata.com
. net cd links
```

The efficient way to find files—at least if you know what you are looking for—is to search. There are two ways to do that. If you suspect what you are looking for might already be in Stata (or published in the SJ), use Stata's `search` command:

```
. search concordance correlation
```

Equivalently, you could select **Help > Search**. Either way, you will learn about *sg84_3* and you can even click to install it.

If you want to search for additions over the net, which is to say, the SJ and archive sites and user sites, type

```
. net search concordance correlation
```

or select **Help > Search**, and this time click *Search net resources*, rather than the default "*Search documentation and FAQs*".

28.5   Making your own download site

There are two reasons you may wish to create your own download site:

1. You have datasets and the like, you want to share them with colleagues, and you want to make it easier for colleagues to download the files.

2. You have written Stata programs, etc., that you wish to share with the Stata user community.

Making a download site is easy; the full instructions are found in [R] **net**.

At the beginning of this chapter, we pretended that you had a dataset you wanted to share with colleagues. We said you just had to copy the dataset onto your server and then let your colleagues know the dataset is there.

Let's now pretend that you had two datasets, `ds1.dta` and `ds2.dta`, and you wanted your colleagues to be able to learn about and fetch the datasets using the `net` command or by pulling down **Help** and selecting **SJ and User-written Programs**.

First, you would copy the datasets to your homepage just as before. Then you would create three more files, one to describe your site named `stata.toc` and two more to describe each "package" you want to provide:

── top of stata.toc ───────────

```
v 3
d My name and affiliation (or whatever other title I choose)
d Datasets for the PAR study
p ds1 The base dataset
p ds2 The detail dataset
```

── end of stata.toc ───────────

── top of ds1.pkg ───────────

```
v 3
d ds1.  The base dataset
d My name or whatever else I wanted to put
d This dataset contains the baseline values for ...
p ds1.dta
```

── end of ds1.pkg ───────────

——————————————————————————————————— top of ds2.pkg ———————

```
v 3
d ds1.  The detail dataset
d My name or whatever else I wanted to put
d This dataset contains the follow-up information ...
p ds2.dta
```

——————————————————————————————————— end of ds2.pkg ———————

Here is what users would see when they went to your site:

```
. net from http://www.myuni.edu/hande/~aparker
```
———
```
http://www.myuni.edu/hande/~aparker
My name and whatever else I wanted to put
```
———
```
Datasets for the PAR study

PACKAGES you could -net describe-:
    ds1                The base dataset
    ds2                The detail dataset
```
———
```
. net describe ds1
```
———
```
package ds1 from http://www.myuni.edu/hande/~aparker
```
———
```
TITLE
    ds1.  The base dataset
DESCRIPTION/AUTHOR(S)
    My name and whatever else I wanted to put
    This dataset contains the baseline values for ...
ANCILLARY FILES                              (type net get ds1)
    ds1.dta
```
———
```
. net get ds1
checking ds1 consistency and verifying not already installed...

copying into current directory...
    copying  ds1.dta
ancillary files successfully copied.

. _
```

See [R] **net**.

Subject and author index

This is the subject and author index for the *Stata User's Guide.* You may also want to consult the combined subject index in the *Stata Quick Reference and Index,* which indexes the *Getting Started with Stata for Macintosh Manual,* the *Getting Started with Stata for Unix Manual,* the *Getting Started with Stata for Windows Manual,* the *Stata Base Reference Manual,* the *Stata Data Management Reference Manual,* the *Stata Graphics Reference Manual,* the *Stata Programming Reference Manual,* the *Stata Longitudinal/Panel Data Reference Manual,* the *Stata Multivariate Statistics Reference Manual,* the *Stata Survey Data Reference Manual,* the *Stata Survival Analysis & Epidemiological Tables Reference Manual,* the *Stata Time-Series Reference Manual,* and this manual.
Readers interested in Mata topics should see the index at the end of the *Mata Reference Manual.*

Semicolons set off the most important entries from the rest. Sometimes no entry will be set off with semicolons; this means all entries are equally important.

& (and), *see* logical operators
| (or), *see* logical operators
~ (not), *see* logical operators
! (not), *see* logical operators
== (equality), *see* relational operators
!= (not equal), *see* relational operators
~= (not equal), *see* relational operators
< (less than), *see* relational operators
<= (less than or equal), *see* relational operators
> (greater than), *see* relational operators
>= (greater than or equal), *see* relational operators
* abbreviation character, *see* abbreviations
~ abbreviation character, *see* abbreviations
? abbreviation character, *see* abbreviations
- abbreviation character, *see* abbreviations

A

.a, .b, ..., .z, *see* missing values
abbreviations, [U] **11.2 Abbreviation rules**;
      [U] **11.1.1 varlist**, [U] **11.4 varlists**
aborting command execution, [U] **9 The Break key**,
      [U] **10 Keyboard use**
Abramowitz, M., [U] **13 Functions and expressions**
Access, Microsoft, reading data from, [U] **21.4 Transfer programs**
addition operator, *see* arithmetic operators
.ado filename suffix, [U] **11.6 File-naming conventions**
ado-files, [U] **3.5 The Stata Journal and the Stata Technical Bulletin**, [U] **17 Ado-files**,
      [U] **18.11 Ado-files**,
   downloading, [U] **28 Using the Internet to keep up to date**

ado-files, *continued*
   installing, [U] **17.6 How do I install an addition?**
   location, [U] **17.5 Where does Stata look for ado-files?**
   long lines, [U] **18.11.2 Comments and long lines in ado-files**
   official, [U] **28 Using the Internet to keep up to date**
adopath command, [U] **17.5 Where does Stata look for ado-files?**
adosize, [U] **18.11 Ado-files**
algebraic expressions, functions, and operators,
      [U] **13 Functions and expressions**,
      [U] **13.3 Functions**
_all, [U] **11.1.1 varlist**
alphanumeric variables, *see* string variables
analytic weights, [U] **11.1.6 weight**,
      [U] **20.16.2 Analytic weights**
and operator, [U] **13.2.4 Logical operators**
append command, [U] **22 Combining datasets**
appending data, [U] **22 Combining datasets**
arithmetic operators, [U] **13.2.1 Arithmetic operators**
auto.dta, [U] **1.2.1 Sample datasets**
autocode() function, [U] **25.1.2 Converting continuous to categorical variables**
[aweight=*exp*] modifier, [U] **11.1.6 weight**,
      [U] **20.16.2 Analytic weights**

B

_b[], [U] **13.5 Accessing coefficients and standard errors**
b() function, [U] **24.3.2 Specifying particular dates (date literals)**
BASE directory, [U] **17.5 Where does Stata look for ado-files?**
Belsley, D. A., [U] **18 Programming Stata**
Binder, D. A., [U] **20 Estimation and postestimation commands**
bitesti command, [U] **19 Immediate commands**
biyear() function, [U] **24.3.5 Extracting components of time**
biyearly() function, [U] **25 Dealing with categorical variables**
bofd() function, [U] **24.3.4 Translating between time units**
Break key, [U] **9 The Break key**, [U] **16.1.4 Error handling in do-files**
built-in variables, [U] **11.3 Naming conventions**,
      [U] **13.4 System variables (_variables)**
by *varlist*: prefix, [U] **11.5 by varlist: construct**;
      [U] **13.7 Explicit subscripting**, [U] **27.2 The by construct**
by-groups, [U] **11.5 by varlist: construct**
byte, [U] **12.2.2 Numeric storage types**

C

char command, [U] **12.8 Characteristics**
character data, *see* string variables
characteristics, [U] **12.8 Characteristics**,
    [U] **18.3.6 Extended macro functions**,
    [U] **18.3.13 Referencing characteristics**
cii command, [U] **19 Immediate commands**
clear option, [U] **11.2 Abbreviation rules**
cmdlog command, [U] **15 Printing and preserving output**
_coef[], [U] **13.5 Accessing coefficients and standard errors**
coefficients (from estimation),
    accessing, [U] **13.5 Accessing coefficients and standard errors**
    estimated linear combinations, *see* linear combinations of estimators
Cook, I., [U] **1 Read this—it will help**
columns of matrix, names, [U] **14.2 Row and column names**
combining datasets, [U] **22 Combining datasets**
command arguments, [U] **18.4 Program arguments**
command parsing, [U] **18.4 Program arguments**
command timings, [U] **8 Error messages and return codes**
commands,
    abbreviating, [U] **11.2 Abbreviation rules**
    aborting, [U] **9 The Break key**, [U] **10 Keyboard use**
    editing and repeating, [U] **10 Keyboard use**
    immediate, [U] **19 Immediate commands**
comments in programs, do-files, etc.,
    [U] **16.1.2 Comments and blank lines in do-files**, [U] **18.11.2 Comments and long lines in ado-files**
concatenating strings, [U] **13.2.2 String operators**
confidence intervals, [U] **20.6 Specifying the width of confidence intervals**
courses in Stata, [U] **3.7 NetCourses**
Cox, N. J., [U] **1 Read this—it will help**,
    [U] **23 Dealing with strings**

D

d() function, [U] **24.3.2 Specifying particular dates (date literals)**
daily() function, [U] **24.3.6 Creating time variables**
data, [U] **12 Data**
    appending, *see* appending data
    characteristics of, *see* characteristics
    combining, *see* combining datasets
    exporting, *see* exporting data
    importing, *see* importing data
    inputting, *see* importing data
    labeling, *see* labeling data
    large, dealing with, *see* memory
    missing values, *see* missing values

data, *continued*
    reading, *see* reading data from disk
    strings, *see* string variables
    survey, *see* survey data
data entry, *see* reading data from disk
database, reading data from other software,
    [U] **21.4 Transfer programs**
datasets, sample, [U] **1.2.1 Sample datasets**
date,
    displaying, [U] **24.2.3 Displaying dates**;
        [U] **12.5.3 Date formats**
    elapsed, [U] **24.2.2 Conversion into elapsed dates**
    formats, [U] **24.2.3 Displaying dates**;
        [U] **12.5.3 Date formats**
    functions, [U] **24.2.2 Conversion into elapsed dates**,
        [U] **24.2.4 Other date functions**
    inputting, [U] **24.2.1 Inputting dates**
    variables, [U] **24 Dealing with dates**
date() function, [U] **24.2.2.2 The date() function**
datelist, [U] **11.1.9 datelist**
day() function, [U] **24.2.4 Other date functions**,
    [U] **24.3.5 Extracting components of time**
dBASE, reading data from, [U] **21.4 Transfer programs**
.dct filename suffix, [U] **11.6 File-naming conventions**
Deaton, A., [U] **20 Estimation and postestimation commands**
describe command, [U] **12.6 Dataset, variable, and value labels**
difference of estimated coefficients, *see* linear combinations of estimators
difference operator, [U] **11.4.3 Time-series varlists**
digits, controlling the number displayed,
    [U] **12.5 Formats: controlling how data are displayed**
directories, [U] **11.6 File-naming conventions**,
    [U] **18.3.11 Constructing Windows filenames using macros**
    location of ado-files, [U] **17.5 Where does Stata look for ado-files?**
discard command, [U] **18.11.3 Debugging ado-files**
display command, [U] **19 Immediate commands**
display formats, [U] **12.5 Formats: controlling how data are displayed**; [U] **24.2.3 Displaying dates**
division operator, *see* arithmetic operators
.do filename suffix, [U] **11.6 File-naming conventions**
do command, [U] **16 Do-files**
do-files, [U] **16 Do-files**, [U] **18.2 Relationship between a program and a do-file**
    long lines, [U] **18.11.2 Comments and long lines in ado-files**
documentation, [U] **1 Read this—it will help**
documentation, keyword search on, [U] **4 Stata's online help and search facilities**
dofb(), dofd(), dofm(), dofq(), dofw(), and
    dofy() functions, [U] **24.3.4 Translating between time units**
double, [U] **12.2.2 Numeric storage types**

H

Hadi, A. S., [U] **18 Programming Stata**

halfyear() function, [U] **24.3.5 Extracting components of time**

haver command, [U] **21 Inputting data**

Haynam, G. E., [U] **13 Functions and expressions**

help command, [U] **4 Stata's online help and search facilities**, [U] **7 —more— conditions**
  writing your own, [U] **18.11.6 Writing online help**

help—I don't know what to do, [U] **3 Resources for learning and using Stata**

Heyde, C. C., [U] **1 Read this—it will help**

.hlp files, [U] **4 Stata's online help and search facilities**, [U] **18.11.6 Writing online help**

http://www.stata.com, [U] **3.2 The http://www.stata.com web site**

Huber, P. J., [U] **20 Estimation and postestimation commands**

hypertext help, [U] **4 Stata's online help and search facilities**, [U] **18.11.6 Writing online help**

I

if *exp*, [U] **11 Language syntax**

immediate commands, [U] **19 Immediate commands**; [U] **18.4.5 Parsing immediate commands**

importance weights, [U] **11.1.6 weight**, [U] **20.16.4 Importance weights**

importing data, [U] **21.4 Transfer programs**

in *range* modifier, [U] **11 Language syntax**

index search, [U] **4 Stata's online help and search facilities**

indicator variables, [U] **25.1.3 Converting categorical to indicator variables**, [U] **25.2 Using indicator variables in estimation**

Informix, reading data from, [U] **21.4 Transfer programs**

inputting data from a file, *see* reading data from disk

installation,
  of official updates, [U] **28 Using the Internet to keep up to date**
  of SJ and STB, [U] **3.6 Updating and adding features from the web**, [U] **17.6 How do I install an addition?**

int, [U] **12.2.2 Numeric storage types**

Intercooled Stata, [U] **5 Flavors of Stata**

Internet, [U] **3.2 The http://www.stata.com web site**
  installation of updates from, [U] **28 Using the Internet to keep up to date**

interrupting command execution, [U] **10 Keyboard use**

[iweight=*exp*] modifier, [U] **11.1.6 weight**, [U] **20.16.4 Importance weights**

J

Johnson, N. L., [U] **1 Read this—it will help**

joinby command, [U] **22 Combining datasets**

joining datasets, *see* combining datasets

K

Kent, J. T., [U] **20 Estimation and postestimation commands**

keyboard
  entry, [U] **10 Keyboard use**
  search, [U] **4 Stata's online help and search facilities**

Kish, L., [U] **20 Estimation and postestimation commands**

Kleiner, B., [U] **1.2.1 Sample datasets**

Kotz, S., [U] **1 Read this—it will help**

Kuh, E., [U] **18 Programming Stata**

L

label command, [U] **12.6 Dataset, variable, and value labels**

label values, [U] **12.6 Dataset, variable, and value labels**; [U] **13.9 Label values**

labeling data, [U] **12.6 Dataset, variable, and value labels**

labeling data in other languages, [U] **12.6.4 Labels in other languages**

lag operator, [U] **11.4.3 Time-series varlists**

lagged values, [U] **13.7 Explicit subscripting**, [U] **13.7.1 Generating lags and leads**, [U] **13.8.1 Generating lags and leads**

language syntax, [U] **11 Language syntax**

lead operator, [U] **11.4.3 Time-series varlists**

lead values, *see* lagged values

Leone, F. C., [U] **13 Functions and expressions**

less than (or equal) operator, [U] **13.2.3 Relational operators**

limits, [U] **6 Setting the size of memory**

Lin, D. Y., [U] **20 Estimation and postestimation commands**

linear combinations of estimators, [U] **20.11 Obtaining linear combinations of coefficients**

lines, long, in do-files and ado-files, [U] **18.11.2 Comments and long lines in ado-files**

listserver, [U] **3.4 The Stata listserver**

loading data, *see* reading data from disk

local command, [U] **18.3.1 Local macros**, [U] **18.3.9 Advanced local macro manipulation**

log command, [U] **15 Printing and preserving output**; [U] **15.2 Placing comments in logs**, [U] **16.1.2 Comments and blank lines in do-files**

.log filename suffix, [U] **11.6 File-naming conventions**

log files, *see* log command

logical operators, [U] **13.2.4 Logical operators**

long, [U] **12.2.2 Numeric storage types**

Long, J. S., [U] **20 Estimation and postestimation commands**